Carbon Nanotube-Based Sensors

Carbon Nanotube-Based Sensors: Fabrication, Characterization, and Implementation highlights the latest research and developments on carbon nanotubes (CNTs) and their applications in sensors and sensing systems. It offers an overview of CNTs, including their synthesis, functionalization, characterization, and toxicology. It then delves into the fabrication and various applications of CNT-based sensors.

FEATURES

- Defines the significance of different forms of CNT-based sensors synthesized for diverse engineering applications and compares the feasibility of their generation
- Helps readers evaluate different types of fabrication techniques to generate CNTs and their subsequent sensing
- Discusses fabrication of low-cost, efficient CNTs-based sensors that can be used for diverse applications and sheds light on synthesis methods for a range of printing techniques
- Highlights challenges and advances in security-related issues using CNTs-based sensors

This book is aimed at researchers in the fields of materials and electrical engineering who are interested in the development of sensor technology for industrial, biomedical, and related applications.

Anindya Nag is an assistant professor affiliated with the Faculty of Electrical and Computer Engineering, Technische Universität Dresden. He earned a PhD in electrical and electronics engineering at Macquarie University.

Alivia Mukherjee is a postdoctoral fellow at the University of Alberta. She earned a PhD in chemical engineering at the University of Saskatchewan.

Carbon Nanotube-Based Sensors

Fabrication, Characterization, and Implementation

Edited by
Anindya Nag and Alivia Mukherjee

CRC Press
Taylor & Francis Group
Boca Raton London New York

CRC Press is an imprint of the
Taylor & Francis Group, an **informa** business

Designed cover image: © Shutterstock

First edition published 2024
by CRC Press
2385 NW Executive Center Drive, Suite 320, Boca Raton FL 33431

and by CRC Press
4 Park Square, Milton Park, Abingdon, Oxon, OX14 4RN

CRC Press is an imprint of Taylor & Francis Group, LLC

ISBN: 978-1-032-45231-9 (hbk)
ISBN: 978-1-032-45232-6 (pbk)
ISBN: 978-1-003-37607-1 (ebk)

DOI: 10.1201/9781003376071

Typeset in Times
by Apex CoVantage, LLC

Contents

Contributors

Md Eshrat E. Alahi
Walailak University
Tha Ngio, Nakhon Si Thammarat,
 Thailand

Mehmet Ercan Altinsoy
Technische Universität Dresden
Dresden, Germany

Pratidhwani Biswal
Thermal Joining Engineering
Fraunhofer Institute for Large
 Structures in Production
 Engineering IGP
Rostock, Germany

Aniket Chakraborthy
Technische Universität Dresden
Dresden, Germany

Merlin R. Charlotte
Vellore Institute of Technology
Tamil Nadu, India

H. Harija
Technische Universität Dresden
Dresden, Germany

Kavithanjali M
Vellore Institute of Technology
Tamil Nadu, India

Mahalakshmi
Vellore Institute of Technology
Tamil Nadu, India

Alisha Mary Manoj
Vellore Institute of Technology
Tamil Nadu, India

Alivia Mukherjee
University of Alberta
Alberta, Canada

Anindya Nag
Technische Universität Dresden
Dresden, Germany

Suresh Nuthalapati
Technische Universität Dresden
Dresden, Germany

Andreas Richter
Technische Universität Dresden
Dresden, Germany

Somaye Seraj
University of Saskatchewan
Saskatchewan, Canada

Fahmida Wazed Tina
Nakhon Si Thammarat Rajabhat
 University
Tha Ngio, Nakhon Si Thammarat,
 Thailand

Leema Rose Viannie
Vellore Institute of Technology
Tamil Nadu, India

Arash Yahyazadeh
University of Saskatchewan
Saskatchewan, Canada

Preface

In the era of nanotechnology, using nanomaterials-assisted devices to improve the quality of life has become a cornerstone in recent years. Scientists are constantly trying to optimize the quality of the existing sensing prototypes in terms of their robustness, sensitivity, and longevity. Some advantages of using these sensors have been their low production cost, roll-to-roll production, and multiple applications. They have been formed using different printing techniques, each of which has been able to produce lightweight, environmentally friendly devices. Among the flexible prototypes synthesized and used for various industrial, environmental, and domestic applications, the Carbon Nanotubes (CNTs)-based flexible sensors have been one of the most efficient due to their excellent electromechanical characteristics. The CNTs have been used for the last three decades to fabricate effective, thin films for sensing applications. Some of the techniques used to synthesize CNTs are Chemical Vapour Deposition (CVD), laser ablation, and electric arc discharge. Each of these techniques has generated CNTs with high quality and high yield due to their rapid process, easy customization due to their facile nature, ability to grow the nanotubes in variant surfaces, and high suitability for mass production. With time, the CNTs have been produced with well-defined orientation suitable for the resultant flexible electronics. Due to the high aspect-ratio and outstanding electrical and mechanical conductivity, they have formed pure and composite sensors via the integration with polymers.

CNTs have been primarily considered in two physiochemical types—Single-Walled Carbon Nanotubes (SWCNTs) and Multi-Walled Carbon Nanotubes (MWCNTs)—each considered significantly due to their distinct advantages. While the SWCNTs have thermal conductivity compared to MWCNTs, the electrical behaviour of the former type is conducted by their chirality. Due to a higher proportion of metallic nature over semiconducting nature in MWCNTs, the latter has better signal integrity, lower power consumption, and lower resistance, as compared to SWCNTs. Apart from these two types, Double-Walled Carbon Nanotubes (DWCNTs) and Few-Walled Carbon Nanotubes (FWCNTs) have also been considered due to their superior thermal and chemical stability and mechanical flexibility. Before forming the flexible sensors, the CNTs sometimes have been processed using chemical routes to functionalize them with certain organic groups. These additional groups assert additional functionalities like reactivity, solubility, and avenues for further chemical modification. The functionalized CNTs-based sensors have been considered primarily for chemical and biological sensing due to their altered cellular interaction pathways, thus increasing their selectivity and specificity towards targeted molecules.

The conjugation of these functionalized or non-functionalized CNTs has been carried out with polymers that can have proper adhesion with the CNTs' electrodes and form uniform nanocomposites as the polymer matrix. Some polymers, like polydimethylsiloxane (PDMS) and polyimide (PI), have been considerably used alongside CNTs to form the resultant flexible sensors. A number of printing techniques, including screen printing, inkjet printing, and 3D printing, have been used as the fabrication processes to develop the thin-film sensors. Each of these processes has been chosen based on the required physical dimension of the prototypes and their target application. Among the applications tested with CNTs, the biocompatible nature of the CNTs has allowed them to be used to develop wearable sensors. These sensors have been predominantly used to monitor gross and fine motor skills and electrochemical ions released from the body. The periodic and ubiquitous monitoring with the CNTs-based flexible sensors in controlled and real-time situations has allowed the production and popularization of point-of-care (POC) devices. Other interesting applications include energy harvesting with these sensors. With the exponential rise in urbanization and the use of electronic devices, the need to harvest energy has become paramount. Due to the ability of the flexible sensors to operate as nanogenerators and energy harvesters, the CNTs-based sensors have been deployed as piezoelectric and triboelectric sensors to harvest energy. These sensors have been able to perform with a wide operating range, high linearity, low hysteresis, and low recovery time. This book highlights the work done on fabricating and utilizing CNTs-based sensors for diverse applications. With the initial chapters highlighting the work done on the synthesis and characterization of CNTs, the succeeding chapters elucidate the employment of CNTs-based sensors for different applications.

1 Introduction

Pratidhwani Biswal, Anindya Nag, and Mehmet Ercan Altinsoy

1.1 INTRODUCTION

With the exponential advancement in the electronic world, sensors play a vital role in improving the quality of human life in the modern world. The entire range of sensors can be roughly classified into two types; namely, rigid (1, 2) and flexible (3, 4) sensors. The former category includes microelectromechanical system (MEMS)-based sensors (5) that are formed on single-crystal silicon substrates (6). The processing of the raw material is done in cleanrooms using a conventional photolithography process (7). Although these MEMS-based sensors have been used thoroughly for industrial (8, 9) and environmental (10, 11) applications, there are some associated limitations that have led the researchers to opt for alternatives. The flexible sensors (12, 13) have been devised with a wide range of processed material. In the current world, the flexible electronic devices have been attached to surfaces with various mechanical flexibility to detect their respective movement (14, 15). This is in contrast to the MEMS-based sensors, which are only created for stationary surfaces (16). Moreover, the use of conformal contact sensors to detect strain, stress, temperature, light, moisture, chemical species and biological species has made it possible to continuously monitor the conditions underlying human activity (17). The field of nanotechnology (18, 19) has been exploited in order to process different kinds of nanomaterial (20–22) to form the conductive parts of the sensing prototypes. The wide range of nanomaterial can be categorized into two classes – carbon-based allotropes (23–25) and metallic nanomaterial (26–28). The first part includes some of the efficient allotropes of carbon elements like Carbon Nanotubes (CNTs) (29–31), graphene (32–34) and graphite (35–37). Apart from the excellent electromechanical characteristics, each of these allotropes is highly preferred to form wearable sensors (38, 39) due to their biocompatible and biodegradable nature. The second category includes the metallic nanomaterials which are formulated as nanowires (40, 41), nanobeads (42, 43), nanoparticles (44, 45) and others. Among all these nanomaterials, CNTs have been quite significant for the last two decades (46) in developing a range of sensing prototypes. The CNTs-based flexible sensors have been formed using certain printing techniques (47), some of which include screen printing (48, 49), inkjet printing (50, 51), 3D printing (52, 53) and gravure printing (54, 55). A high demand of Internet of Things (IoT) has caused a shift in interest towards big data collection and analysis. The IoT-based flexible sensing systems (56) can be developed by integrating flexible and stretchable devices with various communication protocols.

Flexible sensors provide a wide range of applications, such as artificial neurons, electronic skin, human-machine interaction systems (57, 58), robotics (59, 60), artificial intelligence (61, 62) and in-vitro diagnostics (63, 64). Carbon-based material has

DOI: 10.1201/9781003376071-1

been proved to be one of the most preferable materials in the field of wearable flexible devices. Examples of carbon-based material include CNTs, diamond, graphene, graphite and fullerene. Among these, carbon nanotubes are considered as the promising material in the field of flexible electronics (65).

Carbon is the element that has six electrons in the atomic orbitals as $1s^2$, $2s^2$ and $2p^2$. It enables a hybridization of sp, sp^2 or sp^3 forms (66). The properties of CNTs are derived from graphene. The sp^2 hybridization state of carbon atoms in graphene, which are hexagonally arranged, forms the basic structure of carbon nanotubes (67–69). It is formed by fabrication of a rolled up graphene sheet. Nanotubes are also classified into Single-Walled Carbon Nanotubes (SWCNTs) and Multi-Walled Carbon Nanotubes (MWCNTs). The fabrication of these nanomaterials is comprised of various processes such as chemical deposition of vapour, discharge using an electric arc and a laser ablation mechanism. Each of these techniques produces CNTs with different specifications. These mechanisms involve steps such as functionalization, chemical addition, doping and filing (70).

CNTs have strong mechanical, thermal and electrical properties due to their hollow structure (71). These physical properties, along with the characteristic optical features of CNTs, make them highly suitable to be used in sensors (72, 73). Figure 1.1

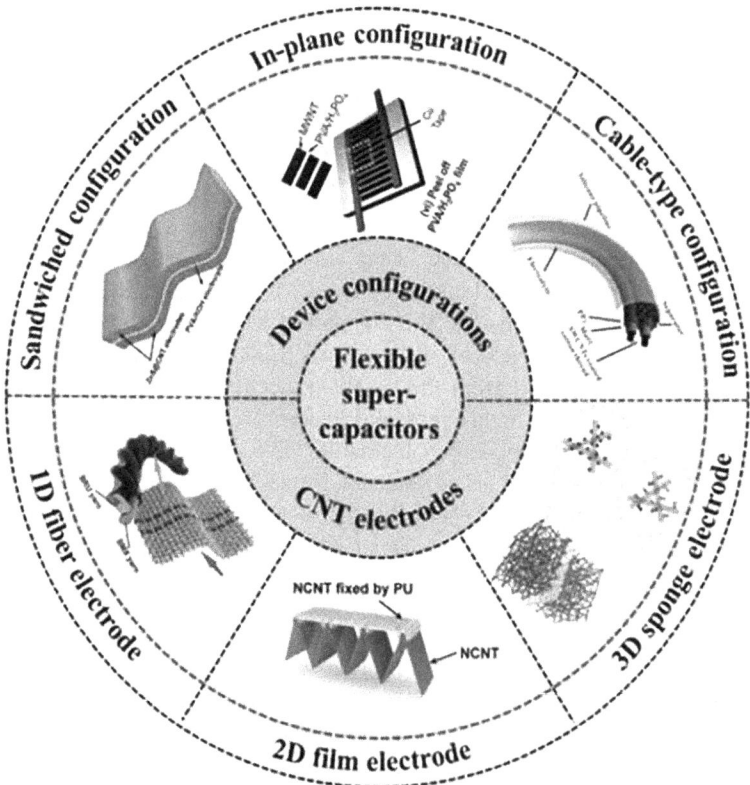

FIGURE 1.1 Application spectrum of the CNTs-based flexible sensors based on different electrode configurations (74).

(74) represents the application spectrum of the CNTs-based flexible sensors on the basis of different electrode configurations. Each of these configurations operate with respect to certain working mechanisms that are chosen with respect to the specific application. Most importantly, using CNTs in sensors leads to a change in electrical resistivity, inductance and capacitance, and electrical obstruction due to piezo-resistivity, thermo-resistivity and magneto-resistivity respectively (75). These characteristics make it a suitable candidate for usage in high-performance, flexible, integrated circuits (ICs) and complementary metal-oxide-semiconductor (CMOS) circuits (76). Its applications in biosensors include the building of an artificial arm and usage as reinforcement phase in composites (77).

One common method to fabricate flexible sensors using CNTs is their dispersion within the matrix. This enables CNTs to produce stacking with other conjugated material, allowing for an effortless and compact combination (78). The quality of the matrix material affects the performance of sensors. CNTs often operate as conductive and reinforced components in the matrix. These nanotubes, when assembled into aligned structures, offer applications in the field of flexible sensors, energy conversion devices, storage devices and implantable electronics (79).

1.2   CARBON NANOTUBES

Although both SWCNTs and MWCNTs have been thoroughly used to develop flexible sensors, they differ in their electromechanical characteristics as a result of corresponding differences in the fabrication parameters. These two types of CNTs, along with their sub-categories, have been explained in the following section.

1.2.1   TYPES OF CARBON NANOTUBES

CNTs can be classified based on their physical structures. For example, in simple terms, CNTs can be categorized as long or short, based on their length. In technical terms, they can be classified as single-walled (80), double-walled (81) or multi-walled (82) on the basis of the number of concentric, cylindrical layers in their nanostructure. For example, different physical configurations of SWCNTs are used for sensorial applications (83). Theoretically, in accordance with the temperature at which SWCNTs are synthesized, they can have a diameter between 0.2 nm to 0.4 nm. Based on the difference in the rolling up process of carbon nanotubes, there exist three types of CNTs – armchair Carbon Nanotubes, zigzag Carbon Nanotubes and chiral Carbon Nanotubes. The chiral vector describing the rolled-up graphene sheet and affecting its electrical properties is represented by a pair of indices (n, m). The indices, n and m, represent the number of unit vectors arranged along two directions in the graphene honeycomb crystal lattice. These structural characteristics represent the amount of twist in a nanotube. If m = 0, the nanotubes are referred to as zigzag; for example, (1, 0), (2, 0) or (5, 0). If n = m, the nanotubes are termed as armchairs; for example, (1, 1), (2, 2) or (3, 3). And in other cases, the nanotubes are referred to as chiral nanotubes; for example, (4, 2), (3, 2) or (4, 1). The armchair and zigzag nanotubes have the maximum degree of symmetry (84, 85). The diameter of SWCNTs is directly proportional to the temperature. In general, SWCNTs consist of only 10 atoms around the circumference and has a tube thickness of one-atom only (86).

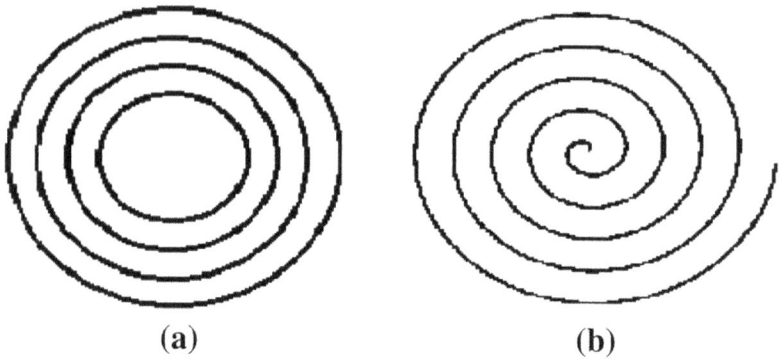

(a)   **(b)**

FIGURE 1.2   Two different configurations of MWCNTs: a) Russian doll model and b) Parchment model (70).

MWCNTs are referred to a group of concentric SWCNTs cylinders with varying diameters. While the outer diameter of MWCNTs ranges between 2 nm to 100 nm, the inner diameter ranges from 1 nm to 3 nm (87). Due to the insufficient space, it is usually not possible for the tubes of all diameters to fit into one another.

But it is crucial to have a gap of 3.44 mm between the surfaces of any two concentric tubes. Due to the sp^2 hybridization in MWCNTs, a delocalized electron cloud is generated along the wall. The interaction between the adjacent cylindrical layers in MWCNTs is due to the presence of this delocalized electron cloud. This results in less flexibility and more structural defects (88). MWCNTs can be produced using two different models – the Russian Doll model and Parchment structural model (89) – as shown in Figure 1.2 (70).

The Russian Doll model refers to CNTs that have concentric nanotubes of variable diameter inside of it. On the other hand, with the Parchment model, a single graphene sheet is wrapped repeatedly around itself in the shape of a spiral. As the outer wall of MWCNTs has many layers, it shields the inner CNTs from chemical reactions and increases tensile strength. As SWCNTs has only one layer, it can be easily depleted (90, 91). MWCNTs is often used in most nanotube applications nowadays because it can be produced more easily and cost-effectively on a wide scale than SWCNTs. DWCNTs are comprised of two SWCNTs aligned coaxially. Due to the presence of a second wall, they have better durability but a complex electronic structure (92, 93). DWCNTs are great candidates for nanotube-based sensor systems due to their concentric structure, which makes it possible to simultaneously use the superior conductivity of the non-functionalized inner wall and the chemical reactivity of the outer wall (94). A visual representation of the differences between SWCNTs, DWCNTs and MWCNTs in shown in Figure 1.3 (95). Table 1.1 (96) highlights the use of CVD technique to form various kinds of CNTs.

1.2.2   SYNTHESIS OF CARBON NANOTUBES

The synthesis of CNTs have been primarily carried out using four distinct techniques; namely, Chemical Vapour Deposition (CVD) (97, 98), Laser Ablation (99, 100)

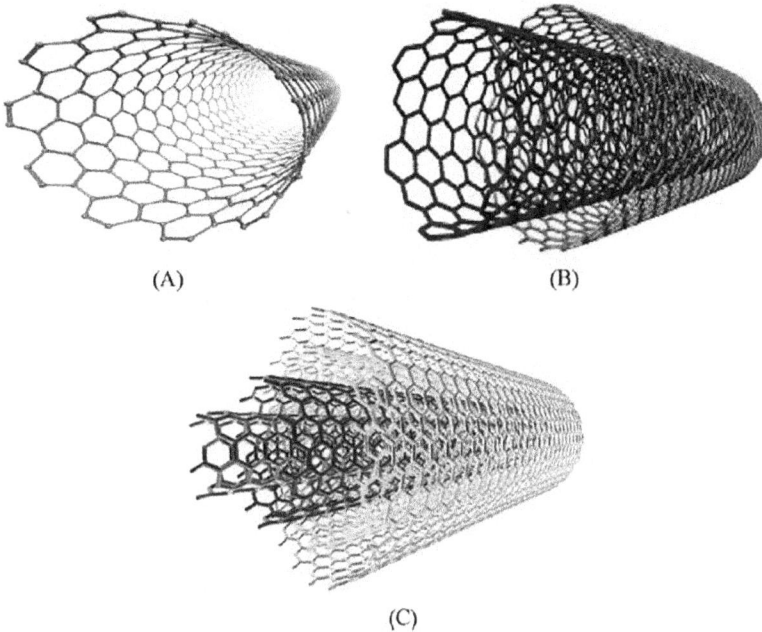

(A)　　　　(B)

(C)

FIGURE 1.3 Three different types of Carbon Nanotubes: a) SWCNT, b) DWCNTs and c) MWCNTs (95).

TABLE 1.1
Comparative Study of Different Types of CNTs Formed Using CVD Method (96)

	SWCNTs	DWCNTs	MWCNTs
Purity	95%	90%	95%
Impurities	MWCNTs, Ash, Metal	MWCNTs, Ash, Metal	Ash, Metal
Length	~30 µm	~50 µm	~50 µm
Diameters	1-2 nm	2-4nm	8-15 nm
Aspect ratio	~20k	~17k	~2k

and Electric Arc Discharge (101, 102). These processes have been extensively used and optimized in terms of experimental conditions such as processed material, temperature of the chamber and gaseous conditions.

1.2.2.1 Chemical Vapour Deposition

CVD process is used to create CNTs, where hydrocarbon gas such as methane, ethylene or acetylene is used as the source of carbon (103). In the process, vapourized reactants undergo chemical reactions to produce a nanomaterial that is deposited on a substrate (104). This approach involves placing a quartz tube within a furnace that is kept at a high temperature of about 500–900°C. This furnace is heated by an RF heater. The quartz tube filled with an inert gas is used to put the crucible inside the furnace. The crucible contains the substrate coated with catalyst nanoparticles. The pumping of hydrocarbon gas is done into the quartz tube, which further leads to pyrolysis of the gas and production of vapourized carbon atoms.

These carbon atoms adhere to the substrate, connect with one another by the Vander Waal force of attraction and develop on the substrate as multi-walled carbon nanotubes (MWCNTs) (105). For synthesis of SWCNTs, nanoparticles of Fe, Co and Ni are required as catalysts (106). The merits of CVD method are high purity of the material obtained and the ease with which the course of reaction may be controlled on a large scale (107).

1.2.2.2 Laser Ablation

Laser ablation can be defined as a Physical Vapour Deposition technique, where a material is vapourized into a gas, and subsequently, deposited on the surface of a substrate (108). Laser ablation uses a laser source to evaporate a graphite target. This technique involves the placement of a graphite target in the middle of a quartz chamber that is filled with argon gas and kept at a temperature of 1200°C. The vapourization of the graphite target is performed by using a continuous laser source or a pulsed laser source. The flow of argon gas sweeps the vapourized target atoms in the direction of a cooled copper collector, causing the deposition of carbon atoms and growth as CNTs. Figure 1.4 (109) depicts the schematic diagram of the laser ablation process of synthesis of the CNTs. In case of a continuous laser beam, the carbon atoms are constantly vapourized, whereas, in the case of a pulsed laser beam, the vapourization of carbon atoms is directly proportional to the shots of pulsed laser beam, which helps to monitor the amount of CNTs synthesized.

This technique can be used to synthesize both SWCNTs and MWCNTs. Similar to the electric discharge method, the production of single-walled carbon nanotubes requires the use of catalyst nanoparticles such as Fe, Co and Ni. This process has the ability to produce SWCNTs with a high degree of purity and quality. On the other side, the drawback is that the nanotubes made using this method contain some

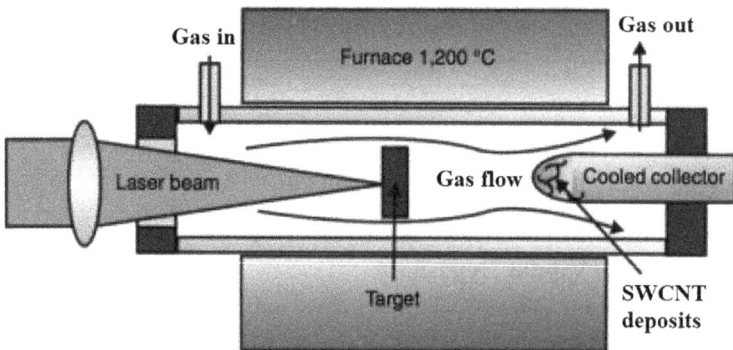

FIGURE 1.4   Laser ablation method for synthesis of carbon nanotubes (109).

branching. Also, the laser ablation approach is not economically suitable, since it requires high-purity graphite rods, strong laser strengths and produces fewer nanotubes per day compared to the arc-discharge method (110).

1.2.2.3 Electric Arc Discharge

In the third approach, a voltage is applied between pure graphite electrodes in a quartz chamber. Figure 1.5 (111) showcases the methodology used to synthesize CNTs using electric arc discharge method. The quartz chamber is attached to a vacuum pump and a diffusion pump that supplies inert gas. The chamber is first completely evacuated by the vacuum pump, followed by the diffusion pump filling the chamber with helium gas (112). The electrodes are spaced 1 millimetre apart and placed inside the quartz chamber at 500 torr of circulating helium gas. An electric arc is created when the electrodes come in contact to one another. As the arc's energy is transferred to the anode, the carbon atoms from the pure graphite anode are ionized, creating C+ ions and plasma. These positively charged carbon ions migrate in the direction of the cathode, where they are reduced and deposited, which then grow as CNTs. The length of the anode shortens as the CNTs increase, but the electrodes are regulated to always maintain a 1-mm space between them. If the electrodes are properly cooled, homogeneous CNTs deposition forms on the cathode. To achieve this, it is important that the inert gas is maintained at the correct pressure (113). There are two basic methods for the arc discharge deposition and synthesis of CNTs – synthesis using catalyst precursors and without using catalyst precursors (114). Usually, production of MWCNTs is without any usage of catalysts. However, in the production of SWCNTs, there is incorporation of nanoparticle catalyst such as Fe, Co and Ni in

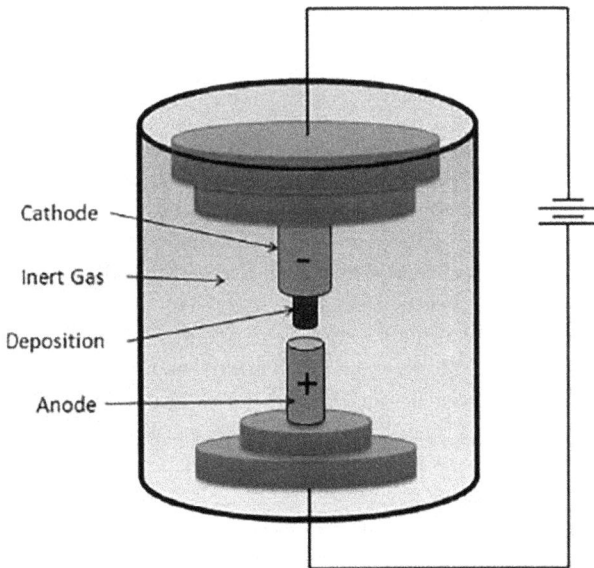

FIGURE 1.5 Electric arc discharge method for synthesis of carbon nanotubes (111).

the centre of the positive electrode. The possibility of producing a significant mass of nanotubes is the main benefit of the arc-discharge method. On the other hand, this method has a drawback that it gives limited control over the alignment of nanotubes, which is crucial for both their characterization and function (115).

1.2.3 CARBON NANOTUBES-BASED FLEXIBLE DEVICES

The CNTs-based flexible sensors have been developed and utilized extensively for a wide spectrum of applications. The nature of the CNTs used for each of these applications has been optimized as per the requirement. This section shows some of the primary areas in which the CNTs have been used over the years.

1.2.3.1 Electrochemical Sensors

Electrochemical sensors have been one of the primary CNTs-based sensors that have been developed and popularized. These sensors have been able to detect ions that are present in liquid samples (116, 117). Some of the sensing parameters like operating range, linear range, response time and sensitivity are of crucial importance for this sensing application. The chemical components present in the human body provide thorough and accurate biochemical information that acts as an asset for monitoring human health, environmental circumstances and the motion of human body (118). Due to its ability to offer practical insight into people's health, electrochemical sensors are considered as one of the most crucial tools for detection of biochemical signals. By measuring the chemical constituents, electrochemical sensors may be able to give real-time, quantified and diverse health information crucial for detecting the presence of any chronic disease or abnormal condition (119). CNTs-based electrochemical sensors offer advantages such as huge surface area, flexibility, better chemical stability and improved electrical conductivity. Hence, they are considered great candidates for usage in wearable or implantable biomedical devices.

1.2.3.2 Gas Sensors

CNTs-based gas sensors are currently pivotal to different commercial industries in the world. Due to the high aspect ratio of these elements, the sensors formed using CNTs are capable of detecting gases with very low concentrations (120, 121). In general, a device that reacts to changes in the gaseous concentration in the local environment is referred to as a chemical sensor. The sensitivity of the gas sensor is vital to focus on the sensing of concerned gas molecules. CNTs have been addressed as the promising choice because of their extraordinarily large surface area and nanoscale size (122). Moreover, the sensitivity of CNTs is more compared to bulk carbon materials. It has good sensitivity towards certain gases like NH_3 (123), NO_2 (124), H_2 (125), C_2H_4 (126), CO (127), SO_2 (128), H_2S (129) and O_2 (130). Properties such as excellent low-temperature sensitivity, rapid response and excellent selectivity are all characteristics of CNTs array gas sensors (131).

As an example, a certain type of CNTs-based gas sensor is the gas ionization sensor, which ionizes the gas which is being analyzed by using CNTs as field emission electrodes. This type of sensor shows high gas-selectivity and sensitivity (132). Another type relies on the electrical conductance of CNTs, which changes as a result of adsorption of the gas molecules. The functioning of CNTs-based gas sensors can be described as a relationship between electrical resistance of CNTs and the ambient gas concentration (133). When oxidizing molecules are adsorbed onto the p-type semiconducting CNT, the absorption rate of gas molecules increases, thus leading to decrease in the electrical resistance of the CNTs (134). The electron transfer between the CNTs and the oxidizing or reducing gas molecules deposited on the CNTs surfaces affects the electrical conductance of p-type semiconducting CNT. These gas sensors can be used to create integrated circuits due to their high sensitivity and quick response time even at room temperature.

1.2.3.3 Biosensors

CNTs are employed in the manufacturing of biosensors due to their chemical and dimensional compatibility with biomolecules (135). These characteristics speed up the flow of electrons between the electrode surface and biomolecules (136–138) and catalyze the process (139). Biosensors are used to describe biological processes or to identify the presence of biomolecules. In contrast to other sensors, biosensors contain a sensing element made up of living materials such proteins, oligo- or polynucle-otides, or microbes (140). The electrochemical biosensors involving carbon material are the most common type of biosensing prototypes, as these biosensors have sim-ple calibration requirements and are often used to detect the presence of biomole-cules in solutions (141). To detect the presence of glucose and other biomolecules, enzyme-containing CNTs-based biosensors have been developed (142). These sen-sors typically work on the enzymatic catalysis process involving either production or consumption of electrons (143).

The nano-size of CNTs sensors and the corresponding requirement of low amount of substrate for a detectable response are its key advantages. For the identification of biological species, CNTs configured as field effect transistors (FET) are popular. Figure 1.6 (144) represents a type of FET-equipped CNTs-based gas sensor. They have the advantages of potential biocompatibility, size compatibility and sensitivity to even the slightest electrical disturbance (145). These biosensors serve a wide range

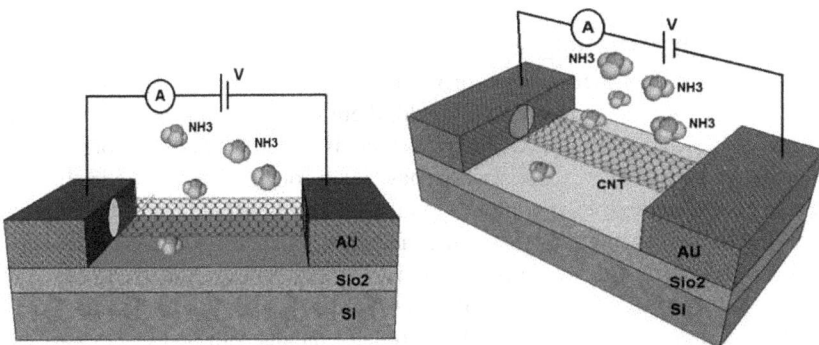

FIGURE 1.6 Schematic diagram of the FET-equipped, CNTs-based gas sensor (144).

of applications in the field of biology, including the detection of substances such as protein, glucose, enzyme, antigen and antibody, DNA, bacterium and hormone.

1.2.3.4 Strain and Pressure Sensors

Flexible CNTs-based sensors exhibit many advantages such as high mechanical flexibility, high stretchability, high bendability, lightweight body, better aspect ratio, excellent permeability and suitability for integration into smart textiles. Strain and pressure sensors are utilized in flexible, wearable devices for a wide range of applications such as detection of movement on human skin (146). A number of factors, such as sensitivity, stretchability, linearity, strain or pressure range, stability and hysteresis, are studied to determine the performance of flexible strain sensors (147). A high stretchability that can withstand better strain is used to detect the joint movements of hands, arms and legs (148). On the other hand, high sensitivity towards even a small strain is vital to detect subtle movements in the chest and neck during breathing, swallowing and speaking (149). A common strain sensor works by emitting an electrical signal from a mechanically deformed strain. Strain sensors have three categories: piezoelectric (150, 151), capacitive (152, 153) and resistive (154, 155). Usually, piezoelectric type of strain sensors is not popular due to poor flexibility and stretchability. An external stimulus can be converted into an electrical signal using a resistive strain sensor (156). The main components of the resistive strain sensor include a flexible substrate and conductive sensitive membrane. When an external strain is applied, the electrical resistance varies because of structural deformation of the conductive material, thus enabling detection of the corresponding altered strains.

On the other hand, a capacitive sensor works on the principle of detection of a change in capacitance. It consists of a flexible dielectric central layer and two conductive materials acting as electrodes on either side. The presence of an external stress leads to a deformation, and consequently, a variation between both electrodes, thus leading to a change in the values of capacitance of the sensors. Figure 1.7 (157) shows an example of CNTs-based strain sensors, where flexible strain sensors were formed using cellulose fibers coated with MWCNTs. It is seen that, after a compressive strain is applied in a perpendicular direction in these sensors, the prototypes changed their position and were capable of retracting to the original position after the strain is released.

1.2.3.5 Energy Harvesting

One of the most imminent usages of CNTs is in the sector of harvesting energy from different sources. In the present world, there's a deficit between the amount of energy required and that fulfilled. Today, as there is a need for finding more and more renewable sources of energy, researchers have been constantly working in the sector of energy harvesting (158). The micro- and nano-sized particles have been utilized to develop sensing prototypes that can convert the mechanical and thermal energy into electrical energy and store them. Some of the advantages of these nanogenerators include simple structure, lightweight build and high-power density. Among the different nanomaterials used to develop these nanogenerators, CNTs have been highly efficient to develop nanogenerators due to their high surface charge density and surface potential. As an example, work done by Badatya et al. (159) showcased

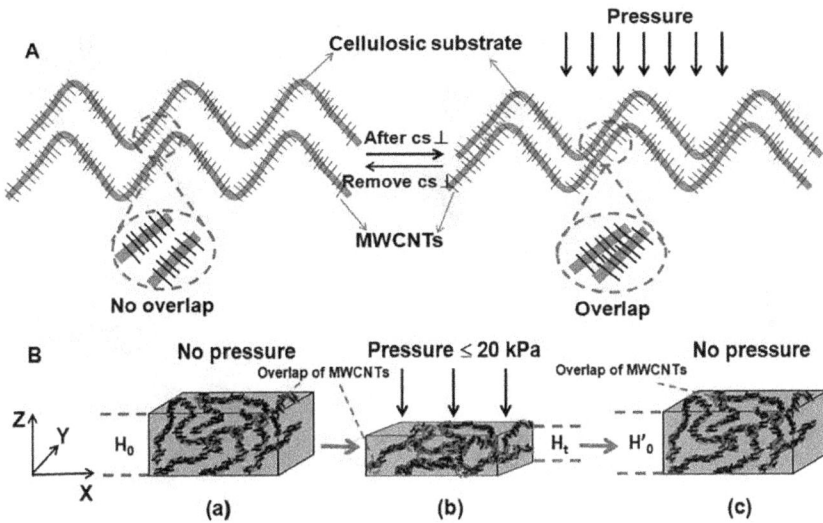

FIGURE 1.7 Schematic diagram of the (A) elucidation of the working mechanism of the CNTs-based strain sensors and (B) macroscopic response of these sensors under loading and unloading tests (157).

the development of humidity sustainable, hydrophobic piezoelectric nanogenerators (PENG) formed using Poly(vinylidene fluoride) (PVDF) and CNTs. The solution route process used here led to the formation of sensors that had a density and piezo-electric charge coefficient of 0.15 g/cm^3 and 9.4 pC/N respectively. Figure 1.8 (159) shows the overall representation of the physical structure of these PVDF/CNTs-based flexible sensors and their ability to generate voltage under mechanical deformations. The prototypes were able to operate successfully under different humidity conditions and to obtain an output voltage of 8 V with a condition of 60% RH. The CNTs-based energy harvesting devices have also been used in other applications, like supercapacitors and lithium-ion batteries. The energy harvesting devices have been prominently used as wearable sensors (160). The pressure exerted by specific joints of the body allows a certain degree of voltage to be induced with these nanogenerators.

1.3 EXISTING CHALLENGES OF THE CURRENT DEVICES

Although a lot of work has been done on the development of CNTs-based flexible sensors, there are still some issues that need to be addressed. Since the popularization of CNTs in the 90s, a lot of research has been carried out on other types of carbon allotropes. Since then, other efficient allotropes such as graphene (161, 162) have been produced in laboratories as well as in industries on a large scale and have been able to replicate the work done by CNTs with a higher efficiency. Thus, it is advantageous to amalgamate CNTs with other nanomaterials to increase the overall efficiency. A mixture of graphene, CNTs and other metallic nanomaterials would also broaden

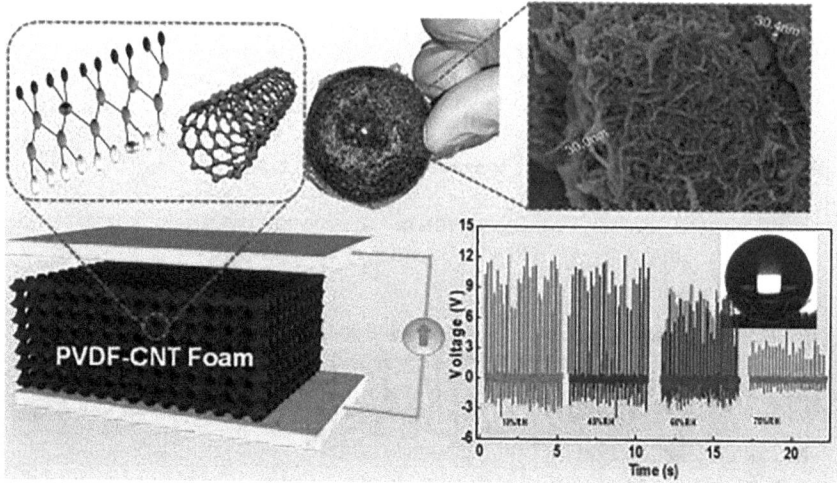

FIGURE 1.8 Representation of the physical structure and electron microscopic image of CNTs/PDVDF-based PENG (159).

the application spectrum of the resultant prototypes. Even though the CNTs perform effectively as nanocomposites (163, 164), certain attributes like electrical conductivity, non-homogeneous distribution of the nanofillers and irregular alignment of the nanotubes in the composites are some of the issues that need to be addressed. The incapability of CNTs to dissolve properly in different aqueous media is another problem that has existed for a while. Certain surfactants have been successful to solve this to a certain degree, although the presence of surfactants breaks the mechanical integrity of the CNTs. In spite of the fact that CNTs-based flexible sensors have been in use for quite a while now, their application has not yet been popularized on a commercial scale. Most of the work has only been carried out in laboratories under controlled temperature and humidity conditions. Similar to MEMS and other printed sensors, large-scale development of CNTs-based sensors is also necessary to exploit their electromechanical attributes in the academic and industrial sectors. The large-scale roll-to-roll production not only validates the reproducibility of the prototypes but also decreases the cost of prototype per sensing unit. The functionalization of CNTs with other groups like –OH, -COOH, -CHO should be worked on to increase their dispersion capacity in water and other media and to decrease cytotoxicity and agglomeration in the polymer matrix (165). Moreover, since a lot of work has been done on SWCNTs and MWCNTs, researchers should try to investigate the efficiency of DWCNTs and FWCNTs (few-walled carbon nanotubes) as electrochemical and strain sensors (166, 167).

1.4   CONCLUSION AND FUTURE PERSPECTIVES

The chapter provides an overview of the CNTs-based flexible sensors in terms of their fabrication and application. The CNTs-based sensors have been developed extensively using different techniques like CVD, laser ablation and electric arc discharge, and subsequently, used for different applications. CNTs have been primarily formed in three different forms; namely, SWCNTs, DWCNTs and MWCNTs. Each of these types are different with respect to each other in terms of electromechanical characteristics. Some of the advantages of these types of CNTs are high electrical conductivity, high aspect ratio, high elasticity, excellent tensile strength, good electron field emitters and excellent ability to form interfacial bonds with the nanocomposites with different polymer matrices. A lot of research is being carried out in order to optimize the quality of these CNTs-based sensors and extend the bandwidth of their application spectrum. These CNTs-based sensors are being integrated with different wireless-communication prototypes to form efficient sensing systems that can be deployed in harsh and partially accessible environment. As these CNTs-based flexible sensors are being fused with other sensors to form heterogeneous systems, they may assist to enhance the quality of human life on a completely different level.

ACKNOWLEDGEMENT

This work was supported by the Free State of Saxony and by the European Union (ESF Plus) by funding the research group 'MultiMOD.' This study was funded by the German Research Foundation (DFG, Deutsche Forschungsgemeinschaft) as part of Germany's Excellence Strategy – EXC 2050/1 – Project ID 390696704 – Cluster of Excellence 'Centre for Tactile Internet with Human-in-the-Loop' (CeTI) of Technische Universität Dresden.

REFERENCES

1. Tilli M, Paulasto-Krockel M, Petzold M, Theuss H, Motooka T, Lindroos V. Handbook of silicon based MEMS materials and technologies: Elsevier; 2020.
2. Ejeian F, Azadi S, Razmjou A, Orooji Y, Kottapalli A, Warkiani ME, et al. Design and applications of MEMS flow sensors: A review. Sensors and Actuators A: Physical. 2019;295:483–502.
3. Han T, Nag A, Afsarimanesh N, Mukhopadhyay SC, Kundu S, Xu Y. Laser-assisted printed flexible sensors: A review. Sensors. 2019;19(6):1462.
4. Nag A, Mukhopadhyay SC, Kosel J. Wearable flexible sensors: A review. IEEE Sensors Journal. 2017;17(13):3949–60.
5. Nag A, Zia AI, Mukhopadhyay S, Kosel J, editors. Performance enhancement of electronic sensor through mask-less lithography. 2015 9th International Conference on Sensing Technology (ICST): IEEE; 2015.
6. Nag A, Zia AI, Li X, Mukhopadhyay SC, Kosel J. Novel sensing approach for LPG leakage detection: Part I – Operating mechanism and preliminary results. IEEE Sensors Journal. 2015;16(4):996–1003.
7. Khumpuang S, Maekawa H, Hara S. Photolithography for minimal fab system. IEEJ Transactions on Sensors and Micromachines. 2013;133(9):272–7.
8. Nag A, Zia AI, Li X, Mukhopadhyay SC, Kosel J. Novel sensing approach for LPG leakage detection – part II: Effects of particle size, composition, and coating layer thickness. IEEE Sensors Journal. 2015;16(4):1088–94.
9. Alahi MEE, Xie L, Mukhopadhyay S, Burkitt L. A temperature compensated smart nitrate-sensor for agricultural industry. IEEE Transactions on Industrial Electronics. 2017;64(9):7333–41.

10. Barhoumi L, Baraket A, Nooredeen NM, Ali MB, Abbas MN, Bausells J, et al. Silicon nitride capacitive chemical sensor for phosphate ion detection based on copper phthalo-cyanine–acrylate-polymer. Electroanalysis. 2017;29(6):1586–95.

11. Tong J, Bian C, Li Y, Bai Y, Xia S, editors. Design of a MEMS-based total phosphorus sensor with a microdigestion system. 2010 4th International Conference on Bioinformatics and Biomedical Engineering: IEEE; 2010.

12. Nag A, Mukhopadhyay SC. Printed flexible sensors for academic research. Printed and flexible sensor technology: Fabrication and applications: IOP Publishing; 2021. p. 2-1-2-16.

13. Han T, Nag A, Simorangkir RB, Afsarimanesh N, Liu H, Mukhopadhyay SC, et al. Multifunctional flexible sensor based on laser-induced graphene. Sensors. 2019;19(16):3477.

14. Nag A, Mukhopadhyay SC, Kosel J. Flexible carbon nanotube nanocomposite sensor for multiple physiological parameter monitoring. Sensors and Actuators A: Physical. 2016;251:148–55.

15. Nag A, Simorangkir RB, Valentin E, Björninen T, Ukkonen L, Hashmi RM, et al. A transparent strain sensor based on PDMS-embedded conductive fabric for wearable sensing applications. IEEE Access. 2018;6:71020–7.

16. Malik A, Kandasubramanian B. Flexible polymeric substrates for electronic applications. Polymer Reviews. 2018;58(4):630–67.

17. Kim Y, Kweon OY, Won Y, Oh JH. Deformable and stretchable electrodes for soft electronic devices. Macromolecular Research. 2019;27(7):625–39.

18. Li X, Wang R, Wang L, Li A, Tang X, Choi J, et al. Scalable fabrication of carbon materials based silicon rubber for highly stretchable e-textile sensor. Nanotechnology Reviews. 2020;9(1):1183–91.

19. Jeevanandam J, Barhoum A, Chan YS, Dufresne A, Danquah MK. Review on nanoparticles and nanostructured materials: History, sources, toxicity and regulations. Beilstein journal of nanotechnology. 2018;9(1):1050–74.

20. He S, Zhang Y, Gao J, Nag A, Rahaman A. Integration of different graphene nanostructures with PDMS to form wearable sensors. Nanomaterials. 2022;12(6):950.

21. Nag A, Nuthalapati S, Mukhopadhyay SC. Carbon fiber/polymer-based composites for wearable sensors: A review. IEEE Sensors Journal. 2022;22:10235–45.

22. Nag A, Afsarimanesh N, Nuthalapati S, Altinsoy ME. Novel surfactant-induced MWCNTs/PDMS-based nanocomposites for tactile sensing applications. Materials. 2022;15(13):4504.

23. Gao J, He S, Nag A, Wong JWC. A review of the use of carbon nanotubes and graphene-based sensors for the detection of aflatoxin M1 compounds in milk. Sensors. 2021;21(11):3602.

24. Han T, Nag A, Mukhopadhyay SC, Xu Y. Carbon nanotubes and its gas-sensing applications: A review. Sensors and Actuators A: Physical. 2019;291:107–43.

25. Nag A, Alahi MEE, Mukhopadhyay SC. Recent progress in the fabrication of graphene fibers and their composites for applications of monitoring human activities. Applied Materials Today. 2021;22:100953.

26. Vijayaraghavan K, Ashokkumar T. Plant-mediated biosynthesis of metallic nanoparticles: A review of literature, factors affecting synthesis, characterization techniques and applications. Journal of Environmental Chemical Engineering. 2017;5(5):4866–83.

27. Shankar PD, Shobana S, Karuppusamy I, Pugazhendhi A, Ramkumar VS, Arvindnarayan S, et al. A review on the biosynthesis of metallic nanoparticles (gold and silver) using bio-components of microalgae: Formation mechanism and applications. Enzyme and Microbial Technology. 2016;95:28–44.

28. Schröfel A, Kratošová G, Šafařík I, Šafaříková M, Raška I, Shor LM. Applications of biosynthesized metallic nanoparticles–a review. Acta biomaterialia. 2014;10(10):4023–42.

29. Şenocak A, Tümay SO, Ömeroğlu İ, Şanko V. Crosslinker polycarbazole supported magnetite MOF@CNT hybrid material for synergetic and selective voltammetric determination of adenine and guanine. Journal of Electroanalytical Chemistry. 2022;905:115963.
30. Nurazzi N, Sabaruddin F, Harussani M, Kamarudin S, Rayung M, Asyraf M, et al. Mechanical performance and applications of CNTs reinforced polymer composites – A review. Nanomaterials. 2021;11(9):2186.
31. Li S, Li R, González OG, Chen T, Xiao X. Highly sensitive and flexible piezoresistive sensor based on c-MWCNTs decorated TPU electrospun fibrous network for human motion detection. Composites Science and Technology. 2021;203:108617.
32. Nag A, Mukhopadhyay SC, Kosel J. Sensing system for salinity testing using laser-induced graphene sensors. Sensors and Actuators A: Physical. 2017;264:107–16.
33. Nag A, Simorangkir RB, Gawade DR, Nuthalapati S, Buckley JL, O'Flynn B, et al. Graphene-based wearable temperature sensors: A review. Materials & Design. 2022:110971.
34. Nag A, Mukhopadhyay SC. Fabrication and implementation of printed sensors for taste sensing applications. Sensors and Actuators A: Physical. 2018;269:53–61.
35. Nag A, Alahi MEE, Feng S, Mukhopadhyay SC. IoT-based sensing system for phosphate detection using Graphite/PDMS sensors. Sensors and Actuators A: Physical. 2019;286:43–50.
36. Nag A, Feng S, Mukhopadhyay S, Kosel J, Inglis D. 3D printed mould-based graphite/PDMS sensor for low-force applications. Sensors and Actuators A: Physical. 2018;280:525–34.
37. Nag A, Afasrimanesh N, Feng S, Mukhopadhyay SC. Strain induced graphite/PDMS sensors for biomedical applications. Sensors and Actuators A: Physical. 2018;271:257–69.
38. Zhang H, He R, Niu Y, Han F, Li J, Zhang X, et al. Graphene-enabled wearable sensors for healthcare monitoring. Biosensors and Bioelectronics. 2022;197:113777.
39. Qiao Y, Li X, Hirtz T, Deng G, Wei Y, Li M, et al. Graphene-based wearable sensors. Nanoscale. 2019;11(41):18923–45.
40. Zhou Y, Wang Y, Guo Y. Cuprous oxide nanowires/nanoparticles decorated on reduced graphene oxide nanosheets: Sensitive and selective H2S detection at low temperature. Materials Letters. 2019;254:336–9.
41. Wang Y, Gong S, Wang SJ, Yang X, Ling Y, Yap LW, et al. Standing enokitake-like nanowire films for highly stretchable elastronics. ACS Nano. 2018;12(10):9742–9.
42. Kabe Y, Sakamoto S, Hatakeyama M, Yamaguchi Y, Suematsu M, Itonaga M, et al. Application of high-performance magnetic nanobeads to biological sensing devices. Analytical and Bioanalytical Chemistry. 2019;411(9):1825–37.
43. Li Y, Zhao X, Li P, Huang Y, Wang J, Zhang J. Highly sensitive Fe3O4 nanobeads/graphene-based molecularly imprinted electrochemical sensor for 17β-estradiol in water. Analytica Chimica Acta. 2015;884:106–13.
44. Khalifa Z, Zahran M, Zahran MA, Azzem MA. Mucilage-capped silver nanoparticles for glucose electrochemical sensing and fuel cell applications. RSC Advances. 2020;10(62):37675–82.
45. Zhao R, Liu C, Zhang X, Zhu X, Wei P, Ji L, et al. An ultrasmall Ru 2 P nanoparticles–reduced graphene oxide hybrid: An efficient electrocatalyst for NH 3 synthesis under ambient conditions. Journal of Materials Chemistry A. 2020;8(1):77–81.
46. Norizan MN, Moklis MH, Demon SZN, Halim NA, Samsuri A, Mohamad IS, et al. Carbon nanotubes: Functionalisation and their application in chemical sensors. RSC Advances. 2020;10(71):43704–32.
47. Khan S, Lorenzelli L, Dahiya RS. Technologies for printing sensors and electronics over large flexible substrates: A review. IEEE Sensors Journal. 2014;15(6):3164–85.
48. Xiao Y, Jiang S, Li Y, Zhang W. Screen-printed flexible negative temperature coefficient temperature sensor based on polyvinyl chloride/carbon black composites. Smart Materials and Structures. 2021;30(2):025035.

49. Marra F, Minutillo S, Tamburrano A, Sarto MS. Production and characterization of Graphene Nanoplatelet-based ink for smart textile strain sensors via screen printing technique. Materials & Design. 2021;198:109306.
50. Rosati G, Cunego A, Fracchetti F, Del Casale A, Scaramuzza M, De Toni A, et al. Inkjet printed interdigitated biosensor for easy and rapid detection of bacteriophage contamination: A preliminary study for milk processing control applications. Chemosensors. 2019;7(1):8.
51. Nayak L, Mohanty S, Nayak SK, Ramadoss A. A review on inkjet printing of nanoparticle inks for flexible electronics. Journal of Materials Chemistry C. 2019;7(29):8771–95.
52. He S, Feng S, Nag A, Afsarimanesh N, Han T, Mukhopadhyay SC. Recent progress in 3D printed mold-based sensors. Sensors. 2020;20(3):703.
53. Han T, Kundu S, Nag A, Xu Y. 3D printed sensors for biomedical applications: A review. Sensors. 2019;19(7):1706.
54. Wang H, Song Y, Miao L, Wan J, Chen X, Cheng X, et al., editors. Stamp-assisted gravure printing of micro-supercapacitors with general flexible substrates. 2019 IEEE 32nd International Conference on Micro Electro Mechanical Systems (MEMS): IEEE; 2019.
55. Park J, Nam D, Park S, Lee D. Fabrication of flexible strain sensors via roll-to-roll gravure printing of silver ink. Smart Materials and Structures. 2018;27(8):085014.
56. Ngo H-D, Hoang TH, Baeuscher M, Wang B, Mackowiak P, Grabbert N, et al., editors. A novel low cost wireless incontinence sensor system (screen-printed flexible sensor system) for wireless urine detection in incontinence materials. Multidisciplinary Digital Publishing Institute Proceedings; 2018.
57. Yin R, Wang D, Zhao S, Lou Z, Shen G. Wearable sensors-enabled human–machine interaction systems: From design to application. Advanced Functional Materials. 2021;31(11):2008936.
58. Tee BC, Wang C, Allen R, Bao Z. An electrically and mechanically self-healing composite with pressure-and flexion-sensitive properties for electronic skin applications. Nature Nanotechnology. 2012;7(12):825–32.
59. Nag A, Menzies B, Mukhopadhyay SC. Performance analysis of flexible printed sensors for robotic arm applications. Sensors and Actuators A: Physical. 2018;276:226–36.
60. Xie M, Hisano K, Zhu M, Toyoshi T, Pan M, Okada S, et al. Flexible multifunctional sensors for wearable and robotic applications. Advanced Materials Technologies. 2019;4(3):1800626.
61. Zang Y, Zhang F, Di C-a, Zhu D. Advances of flexible pressure sensors toward artificial intelligence and health care applications. Materials Horizons. 2015;2(2):140–56.
62. Pang C, Lee G-Y, Kim T-i, Kim SM, Kim HN, Ahn S-H, et al. A flexible and highly sensitive strain-gauge sensor using reversible interlocking of nanofibres. Nature Materials. 2012;11(9):795–801.
63. Vilela D, Romeo A, Sánchez S. Flexible sensors for biomedical technology. Lab on a Chip. 2016;16(3):402–8.
64. Kang I, Schulz MJ, Kim JH, Shanov V, Shi D. A carbon nanotube strain sensor for structural health monitoring. Smart Materials and Structures. 2006;15(3):737.
65. Chae SH, Lee YH. Carbon nanotubes and graphene towards soft electronics. Nano Convergence. 2014;1(1):1–26.
66. Eatemadi A, Daraee H, Karimkhanloo H, Kouhi M, Zarghami N, Akbarzadeh A, et al. Carbon nanotubes: Properties, synthesis, purification, and medical applications. Nanoscale Research Letters. 2014;9(1):1–13.
67. Ni W, Wang B, Wang H, Zhang Y. Fabrication and properties of carbon nanotube and poly (vinyl alcohol) composites. Journal of Macromolecular Science, Part B. 2006;45(4):659–64.
68. Prasek J, Drbohlavova J, Chomoucka J, Hubalek J, Jasek O, Adam V, et al. Methods for carbon nanotubes synthesis. Journal of Materials Chemistry. 2011;21(40):15872–84.

69. Esumi K, Ishigami M, Nakajima A, Sawada K, Honda H. Chemical treatment of carbon nanotubes. Carbon (New York, NY). 1996;34(2):279–81.
70. Gupta N, Gupta SM, Sharma S. Carbon nanotubes: Synthesis, properties and engineering applications. Carbon Letters. 2019;29(5):419–47.
71. Lan Y, Wang Y, Ren Z. Physics and applications of aligned carbon nanotubes. Advances in Physics. 2011;60(4):553–678.
72. Hu N, Fukunaga H, Atobe S, Liu Y, Li J. Piezoresistive strain sensors made from carbon nanotubes based polymer nanocomposites. Sensors. 2011;11(11):10691–723.
73. Behabtu N, Young CC, Tsentalovich DE, Kleinerman O, Wang X, Ma AW, et al. Strong, light, multifunctional fibers of carbon nanotubes with ultrahigh conductivity. Science. 2013;339(6116):182–6.
74. Zhu S, Ni J, Li Y. Carbon nanotube-based electrodes for flexible supercapacitors. Nano Research. 2020;13(7):1825–41.
75. Dervishi E, Li Z, Xu Y, Saini V, Biris AR, Lupu D, et al. Carbon nanotubes: Synthesis, properties, and applications. Particulate Science and Technology. 2009;27(2):107–25.
76. Tang J. Carbon nanotube-based flexible electronics. Flexible, Wearable, and Stretchable Electronics. 2020:137–56.
77. Mansoor M, Shahid M. Carbon nanotube-reinforced aluminum composite produced by induction melting. Journal of Applied Research and Technology. 2016;14(4):215–24.
78. Benda R, Zucchi G, Cancès E, Lebental B. Insights into the π–π interaction driven non-covalent functionalization of carbon nanotubes of various diameters by conjugated fluorene and carbazole copolymers. The Journal of Chemical Physics. 2020;152(6):064708.
79. He S, Hong Y, Liao M, Li Y, Qiu L, Peng H. Flexible sensors based on assembled carbon nanotubes. Aggregate. 2021;2(6):e143.
80. Yang X, Zhou Z, Zheng F, Zhang M, Zhang J, Yao Y, editors. A high sensitivity single-walled carbon-nanotube-array-based strain sensor for weighing. TRANSDUCERS 2009–2009 International Solid-State Sensors, Actuators and Microsystems Conference: IEEE; 2009.
81. Zhao J, Su Y, Yang Z, Wei L, Wang Y, Zhang Y. Arc synthesis of double-walled carbon nanotubes in low pressure air and their superior field emission properties. Carbon. 2013;58:92–8.
82. Nag A, Alahi M, Eshrat E, Mukhopadhyay SC, Liu Z. Multi-walled carbon nanotubes-based sensors for strain sensing applications. Sensors. 2021;21(4):1261.
83. Terrones M. Science and technology of the twenty-first century: Synthesis, properties, and applications of carbon nanotubes. Annual Review of Materials Research. 2003;33(1):419–501.
84. Nessim GD. Properties, synthesis, and growth mechanisms of carbon nanotubes with special focus on thermal chemical vapor deposition. Nanoscale. 2010;2(8):1306–23.
85. Schnorr JM, Swager TM. Emerging applications of carbon nanotubes. Chemistry of Materials. 2011;23(3):646–57.
86. Meyyappan M, Delzeit L, Cassell A, Hash D. Carbon nanotube growth by PECVD: A review. Plasma Sources Science and Technology. 2003;12(2):205.
87. Rastogi V, Yadav P, Bhattacharya SS, Mishra AK, Verma N, Verma A, et al. Carbon nanotubes: An emerging drug carrier for targeting cancer cells. Journal of Drug Delivery. 2014;2014.
88. Xia Z, Guduru P, Curtin W. Enhancing mechanical properties of multiwall carbon nanotubes via s p 3 interwall bridging. Physical Review Letters. 2007;98(24):245501.
89. Madani SY, Naderi N, Dissanayake O, Tan A, Seifalian AM. A new era of cancer treatment: Carbon nanotubes as drug delivery tools. International Journal of Nanomedicine. 2011;6:2963.
90. Aqel A, Abou El-Nour KM, Ammar RA, Al-Warthan A. Carbon nanotubes, science and technology part (I) structure, synthesis and characterisation. Arabian Journal of Chemistry. 2012;5(1):1–23.

91. Fan Z, Advani SG. Rheology of multiwall carbon nanotube suspensions. Journal of Rheology. 2007;51(4):585–604.

92. Shimada T, Sugai T, Ohno Y, Kishimoto S, Mizutani T, Yoshida H, et al. Double-wall carbon nanotube field-effect transistors: Ambipolar transport characteristics. Applied Physics Letters. 2004;84(13):2412–4.

93. Sun C-Y, Liu S-X, Liang D-D, Shao K-Z, Ren Y-H, Su Z-M. Highly stable crystalline catalysts based on a microporous metal– organic framework and polyoxometalates. Journal of the American Chemical Society. 2009;131(5):1883–8.

94. Moore KE, Tune DD, Flavel BS. Double-walled carbon nanotube processing. Advanced Materials. 2015;27(20):3105–37.

95. Rafique I, Kausar A, Anwar Z, Muhammad B. Exploration of epoxy resins, hardening systems, and epoxy/carbon nanotube composite designed for high performance materials: A review. Polymer-Plastics Technology and Engineering. 2016;55(3):312–33.

96. Liu L, Kong L, Yin W, Chen Y, Matitsine S. Microwave dielectric properties of carbon nanotube composites. Carbon nanotubes: IntechOpen; 2010.

97. Kumar M, Ando Y. Chemical vapor deposition of carbon nanotubes: A review on growth mechanism and mass production. Journal of Nanoscience and Nanotechnology. 2010;10(6):3739–58.

98. Fleming E, Du F, Ou E, Dai L, Shi L. Thermal conductivity of carbon nanotubes grown by catalyst-free chemical vapor deposition in nanopores. Carbon. 2019;145:195–200.

99. Chrzanowska J, Hoffman J, Małolepszy A, Mazurkiewicz M, Kowalewski TA, Szymanski Z, et al. Synthesis of carbon nanotubes by the laser ablation method: Effect of laser wavelength. physica status solidi (b). 2015;252(8):1860–7.

100. Ismail RA, Mohsin MH, Ali AK, Hassoon KI, Erten-Ela S. Preparation and characterization of carbon nanotubes by pulsed laser ablation in water for optoelectronic application. Physica E: Low-dimensional Systems and Nanostructures. 2020;119:113997.

101. Arora N, Sharma N. Arc discharge synthesis of carbon nanotubes: Comprehensive review. Diamond and Related Materials. 2014;50:135–50.

102. Jagannatham M, Sankaran S, Prathap H. Electroless nickel plating of arc discharge synthesized carbon nanotubes for metal matrix composites. Applied Surface Science. 2015;324:475–81.

103. Abdullah HB, Ramli I, Ismail I, Yusof NA. Hydrocarbon Sources for the carbon nanotubes production by chemical vapour deposition: A review. Pertanika Journal of Science & Technology. 2017;25(2).

104. Muhsin AE. Chemical vapor deposition of aluminum oxide (Al2O3) and beta iron disilicide (β-FeSi2) thin films. Doctor engineer approved dissertation, Faculty of engineering, Department of Mechanical Engineering, University of Duisburg-Essenger, German. 2007.

105. Pham VP, Jang H-S, Whang D, Choi J-Y. Direct growth of graphene on rigid and flexible substrates: Progress, applications, and challenges. Chemical Society Reviews. 2017;46(20):6276–300.

106. Dündar-Tekkaya E, Karatepe N. Effect of reaction time, weight ratio, and type of catalyst on the yield of single-wall carbon nanotubes synthesized by chemical vapor deposition of acetylene. Fullerenes, Nanotubes and Carbon Nanostructures. 2015;23(6):535–41.

107. Patole S, Alegaonkar P, Lee H-C, Yoo J-B. Optimization of water assisted chemical vapor deposition parameters for super growth of carbon nanotubes. Carbon. 2008;46(14):1987–93.

108. Schneider CW, Lippert T. Laser ablation and thin film deposition. Laser processing of materials: Springer; 2010. p. 89–112.

109. Venkataraman A, Amadi EV, Chen Y, Papadopoulos C. Carbon nanotube assembly and integration for applications. Nanoscale Research Letters. 2019;14(1):1–47.

110. Das R, Shahnavaz Z, Ali ME, Islam MM, Abd Hamid SB. Can we optimize arc discharge and laser ablation for well-controlled carbon nanotube synthesis? Nanoscale Research Letters. 2016;11(1):510.

111. Zhao W, Basnet B, Kim IJ. Carbon nanotube formation using zeolite template and applications. Journal of Advanced Ceramics. 2012;1(3):179–93.

112. Pham PV. Hexagon flower quantum dot-like Cu pattern formation during low-pressure chemical vapor deposited graphene growth on a liquid Cu/W substrate. ACS Omega. 2018;3(7):8036–41.

113. Martín N, Guldi DM. Carbon nanotubes and related structures: Synthesis, characterization, functionalization, and applications: John Wiley & Sons; 2010.

114. Szabó A, Perri C, Csató A, Giordano G, Vuono D, Nagy JB. Synthesis methods of carbon nanotubes and related materials. Materials. 2010;3(5):3092–140.

115. Purohit R, Purohit K, Rana S, Rana R, Patel V. Carbon nanotubes and their growth methods. Procedia Materials Science. 2014;6:716–28.

116. Cho G, Azzouzi S, Zucchi G, Lebental B. Electrical and electrochemical sensors based on carbon nanotubes for the monitoring of chemicals in water – A review. Sensors. 2022;22(1):218.

117. Liu R, Li B, Li F, Dubovyk V, Chang Y, Li D, et al. A novel electrochemical sensor based on β-cyclodextrin functionalized carbon nanosheets@ carbon nanotubes for sensitive detection of bactericide carbendazim in apple juice. Food Chemistry. 2022;384:132573.

118. Zhang P, Chen Y, Li Y, Zhang Y, Zhang J, Huang L. A flexible strain sensor based on the porous structure of a carbon black/carbon nanotube conducting network for human motion detection. Sensors. 2020;20(4):1154.

119. Kotru S, Klimuntowski M, Ridha H, Uddin Z, Askhar AA, Singh G, et al. Electrochemical sensing: A prognostic tool in the fight against COVID-19. TrAC Trends in Analytical Chemistry. 2021;136:116198.

120. Verma G, Gupta A. Recent development in carbon nanotubes based gas sensors. Journal of Materials NanoScience. 2022;9(1):3–12.

121. Rana MM, Ibrahim DS, Asyraf MM, Jarin S, Tomal A. A review on recent advances of CNTs as gas sensors. Sensor Review. 2017;37:127–36.

122. Anzar N, Hasan R, Tyagi M, Yadav N, Narang J. Carbon nanotube-A review on Synthesis, Properties and plethora of applications in the field of biomedical science. Sensors International. 2020;1:100003.

123. Norizan M, Zulaikha NS, Norhana A, Syakir MI, Norli A. Carbon nanotubes-based sensor for ammonia gas detection–an overview. Polimery. 2021;66(3):175–86.

124. Sedelnikova OV, Sysoev VI, Gurova OA, Ivanov YP, Koroteev VO, Arenal R, et al. Role of interface interactions in the sensitivity of sulfur-modified single-walled carbon nanotubes for nitrogen dioxide gas sensing. Carbon. 2022;186:539–49.

125. Dhall S, Jaggi N, Nathawat R. Functionalized multiwalled carbon nanotubes based hydrogen gas sensor. Sensors and Actuators A: Physical. 2013;201:321–7.

126. Kathirvelan J, Vijayaraghavan R. Development of prototype laboratory setup for selective detection of ethylene based on multiwalled carbon nanotubes. Journal of Sensors. 2014;2014.

127. Zhao W, Fam DWH, Yin Z, Sun T, Tan HT, Liu W, et al. A carbon monoxide gas sensor using oxygen plasma modified carbon nanotubes. Nanotechnology. 2012;23(42):425502.

128. Septiani NLW, Kaneti YV, Yuliarto B, Dipojono HK, Takei T, You J, et al. Hybrid nanoarchitecturing of hierarchical zinc oxide wool-ball-like nanostructures with multi-walled carbon nanotubes for achieving sensitive and selective detection of sulfur dioxide. Sensors and Actuators B: Chemical. 2018;261:241–51.

129. Suhail MH, Abdullah OG, Kadhim GA. Hydrogen sulfide sensors based on PANI/f-SWCNT polymer nanocomposite thin films prepared by electrochemical polymerization. Journal of Science: Advanced Materials and Devices. 2019;4(1):143–9.

130. Cava C, Salvatierra R, Alves D, Ferlauto A, Zarbin A, Roman L. Self-assembled films of multi-wall carbon nanotubes used in gas sensors to increase the sensitivity limit for oxygen detection. Carbon. 2012;50(5):1953–8.

131. Cantalini C, Valentini L, Armentano I, Kenny J, Lozzi L, Santucci S. Carbon nanotubes as new materials for gas sensing applications. Journal of the European Ceramic Society. 2004;24(6):1405–8.

132. Modi A, Koratkar N, Lass E, Wei B, Ajayan PM. Miniaturized gas ionization sensors using carbon nanotubes. Nature. 2003;424(6945):171–4.

133. Zhang W-D, Zhang W-H. Carbon nanotubes as active components for gas sensors. Journal of Sensors. 2009;2009.

134. Mousavi H. Gas adsorption effects on the electrical conductivity of semiconducting carbon nanotubes. Physica E: Low-dimensional Systems and Nanostructures. 2011;44(2):454–9.

135. Beitollahi H, Movahedifar F, Tajik S, Jahani S. A review on the effects of introducing CNTs in the modification process of electrochemical sensors. Electroanalysis. 2019;31(7):1195–203.

136. De Volder MF, Tawfick SH, Baughman RH, Hart AJ. Carbon nanotubes: Present and future commercial applications. Science. 2013;339(6119):535–9.

137. Kum MC, Joshi KA, Chen W, Myung NV, Mulchandani A. Biomolecules-carbon nanotubes doped conducting polymer nanocomposites and their sensor application. Talanta. 2007;74(3):370–5.

138. Chen A, Chatterjee S. Nanomaterials based electrochemical sensors for biomedical applications. Chemical Society Reviews. 2013;42(12):5425–38.

139. Rivas GA, Rubianes MD, Rodríguez MC, Ferreyra NF, Luque GL, Pedano ML, et al. Carbon nanotubes for electrochemical biosensing. Talanta. 2007;74(3):291–307.

140. Pandey P, Dahiya M. Carbon nanotubes: Types, methods of preparation and applications. Carbon. 2016;1(4):15–21.

141. Naresh V, Lee N. A review on biosensors and recent development of nanostructured materials-enabled biosensors. Sensors. 2021;21(4):1109.

142. Gao M, Dai L, Wallace GG. Biosensors based on aligned carbon nanotubes coated with inherently conducting polymers. Electroanalysis: An International Journal Devoted to Fundamental and Practical Aspects of Electroanalysis. 2003;15(13):1089–94.

143. Bollella P, Katz E. Enzyme-based biosensors: Tackling electron transfer issues. Sensors. 2020;20(12):3517.

144. Akbari E, Buntat Z, Ahmad MH, Enzevaee A, Yousof R, Iqbal SMZ, et al. Analytical calculation of sensing parameters on carbon nanotube based gas sensors. Sensors. 2014;14(3):5502–15.

145. Allen BL, Kichambare PD, Star A. Carbon nanotube field-effect-transistor-based biosensors. Advanced Materials. 2007;19(11):1439–51.

146. Huang K, Ning H, Hu N, Liu F, Wu X, Wang S, et al. Ultrasensitive MWCNT/PDMS composite strain sensor fabricated by laser ablation process. Composites Science and Technology. 2020:108105.

147. Nankali M, Nouri NM, Navidbakhsh M, Malek NG, Amindehghan MA, Shahtoori AM, et al. Highly stretchable and sensitive strain sensors based on carbon nanotube–elastomer nanocomposites: The effect of environmental factors on strain sensing performance. Journal of Materials Chemistry C. 2020;8(18):6185–95.

148. Yamada T, Hayamizu Y, Yamamoto Y, Yomogida Y, Izadi-Najafabadi A, Futaba DN, et al. A stretchable carbon nanotube strain sensor for human-motion detection. Nature Nanotechnology. 2011;6(5):296.

149. Liang B, Lin Z, Chen W, He Z, Zhong J, Zhu H, et al. Ultra-stretchable and highly sensitive strain sensor based on gradient structure carbon nanotubes. Nanoscale. 2018;10(28):13599–606.

150. Chen S, Luo J, Wang X, Li Q, Zhou L, Liu C, et al. Fabrication and piezoresistive/piezoelectric sensing characteristics of carbon nanotube/PVA/nano-ZnO flexible composite. Scientific Reports. 2020;10(1):1–12.

151. Li H, Zhang W, Ding Q, Jin X, Ke Q, Li Z, et al. Facile strategy for fabrication of flexible, breathable, and washable piezoelectric sensors via welding of nanofibers with multiwalled carbon nanotubes (MWCNTs). ACS Applied Materials & Interfaces. 2019;11(41):38023–30.

152. Hu X, Yang F, Wu M, Sui Y, Guo D, Li M, et al. A super-stretchable and highly sensitive carbon nanotube capacitive strain sensor for wearable applications and soft robotics. Advanced Materials Technologies. 2022;7(3):2100769.

153. Gu J, Kwon D, Ahn J, Park I. Wearable strain sensors using light transmittance change of carbon nanotube-embedded elastomers with microcracks. ACS Applied Materials & Interfaces. 2019;12(9):10908–17.

154. Yang H, Yuan L, Yao X, Zheng Z, Fang D. Monotonic strain sensing behavior of self-assembled carbon nanotubes/graphene silicone rubber composites under cyclic loading. Composites Science and Technology. 2020;200:108474.

155. Yan T, Wu Y, Yi W, Pan Z. Recent progress on fabrication of carbon nanotube-based flexible conductive networks for resistive-type strain sensors. Sensors and Actuators A: Physical. 2021;327:112755.

156. Wang C, Xia K, Wang H, Liang X, Yin Z, Zhang Y. Advanced carbon for flexible and wearable electronics. Advanced Materials. 2019;31(9):1801072.

157. Zhang H, Sun X, Hubbe M, Pal L. Flexible and pressure-responsive sensors from cellulose fibers coated with multiwalled carbon nanotubes. ACS Applied Electronic Materials. 2019;1(7):1179–88.

158. Riaz A, Sarker MR, Saad MHM, Mohamed R. Review on comparison of different energy storage technologies used in micro-energy harvesting, WSNs, low-cost microelectronic devices: Challenges and recommendations. Sensors. 2021;21(15):5041.

159. Badatya S, Bharti DK, Sathish N, Srivastava AK, Gupta MK. Humidity sustainable hydrophobic poly (vinylidene fluoride)-carbon nanotubes foam based piezoelectric nanogenerator. ACS Applied Materials & Interfaces. 2021;13(23):27245–54.

160. Hasan MN, Ahmad Asri MI, Saleh T, Muthalif AG, Mohamed Ali MS. Wearable thermoelectric generator with vertically aligned PEDOT: PSS and carbon nanotubes thermoelements for energy harvesting. International Journal of Energy Research. 2022;46(11):15824–36.

161. Nag A, Simorangkir RB, Sapra S, Buckley JL, O'Flynn B, Liu Z, et al. Reduced graphene oxide for the development of wearable mechanical energy-harvesters: A review. IEEE Sensors Journal. 2021;21:26415–25.

162. Alahi MEE, Nag A, Mukhopadhyay SC, Burkitt L. A temperature-compensated graphene sensor for nitrate monitoring in real-time application. Sensors and Actuators A: Physical. 2018;269:79–90.

163. Yuan C, Tony A, Yin R, Wang K, Zhang W. Tactile and thermal sensors built from carbon–polymer nanocomposites – A critical review. Sensors. 2021;21(4):1234.

164. Bozeya A, Makableh YF, Abu-Zurayk R, Khalaf A, Al Bawab A. Thermal and structural properties of high density polyethylene/carbon nanotube nanocomposites: A comparison study. Chemosensors. 2021;9(6):136.

165. Vardharajula S, Ali SZ, Tiwari PM, Eroğlu E, Vig K, Dennis VA, et al. Functionalized carbon nanotubes: Biomedical applications. International Journal of Nanomedicine. 2012;7:5361.

166. Duoc PND, Binh NH, Van Hau T, Thanh CT, Van Trinh P, Tuyen NV, et al. A novel electrochemical sensor based on double-walled carbon nanotubes and graphene hybrid thin film for arsenic (V) detection. Journal of Hazardous Materials. 2020;400:123185.

167. Thanh CT, Binh NH, Duoc PND, Thu VT, Van Trinh P, Anh NN, et al. Electrochemical sensor based on reduced graphene oxide/double-walled carbon nanotubes/octahedral Fe3O4/chitosan composite for glyphosate detection. Bulletin of Environmental Contamination and Toxicology. 2021;106(6):1017–23.

2 Synthesis of Carbon Nanotubes
Properties and Application

*Arash Yahyazadeh, Alivia Mukherjee,
and Somaye Seraj*

2.1 INTRODUCTION

Carbon Nanotubes (CNTs) belong to the relatively novel group of nanomaterial and were first described in 1991, along with the multi-walled carbon nanotubes (MWCNTs) [1]. In more recent history, the process to develop single-walled carbon nanotubes (SWCNTs), using a new arc evaporation method was reported by Iijima et al [2]. Structurally, the principle of fabrication of SWCNTs from one-atom-thick sheets (graphene) is shown in Figure 2.1, and the elastic moduli can be measured by the chiral vector. The diameter and the length of the CNTs range from a few nanometers to a few micrometers, respectively, which leads to a combination of outstanding properties.

The CNTs, owing to their promising electrical and physical properties, can be widely used in conductive films, supercapacitors, separation membranes, sensors, etc. As shown in Figure 2.2, a variety of methods have been developed to produce CNTs. Gupta et al. provided a comprehensive review of the process of synthesis and growth mechanism of CNTs. It was found that the production of high-quality nanotubes depends on multiple factors such as the source of hydrocarbon, catalyst, reaction temperature, reactor geometry and others [4].

The focus of this chapter is to provide an overview of the methods employed to synthesize CNTs and their realistic applications, including defect monitoring, displaying, sensing, along with their applications in solar cells, Lithium-Ion Batteries and electrochemical capacitors.

2.2 PRODUCTION OF CNTS

Methods of synthesis that involve high temperature such as arc growth and laser ablation are widely employed in producing high purity CNTs. However, processes involving low temperature such as the Chemical Vapor Deposition (CVD) method is now considered as one of the most common techniques for large scale CNTs production. Most of the techniques to synthesize CNTs require gases as catalyst and produce a number of impurities that may range from trace levels to significant amounts. Therefore, one of the most fundamental steps employed after the growth of CNTs is purification, which is mostly based on acid treatment [5].

DOI: 10.1201/9781003376071-2

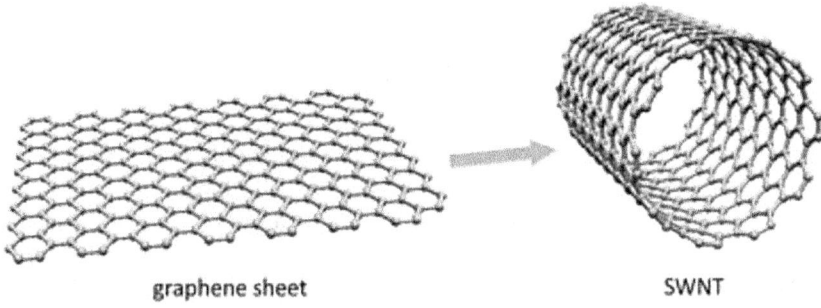

graphene sheet   SWNT

FIGURE 2.1   The fabrication of SWCNTs using graphene sheet. Reproduced with permission [3].

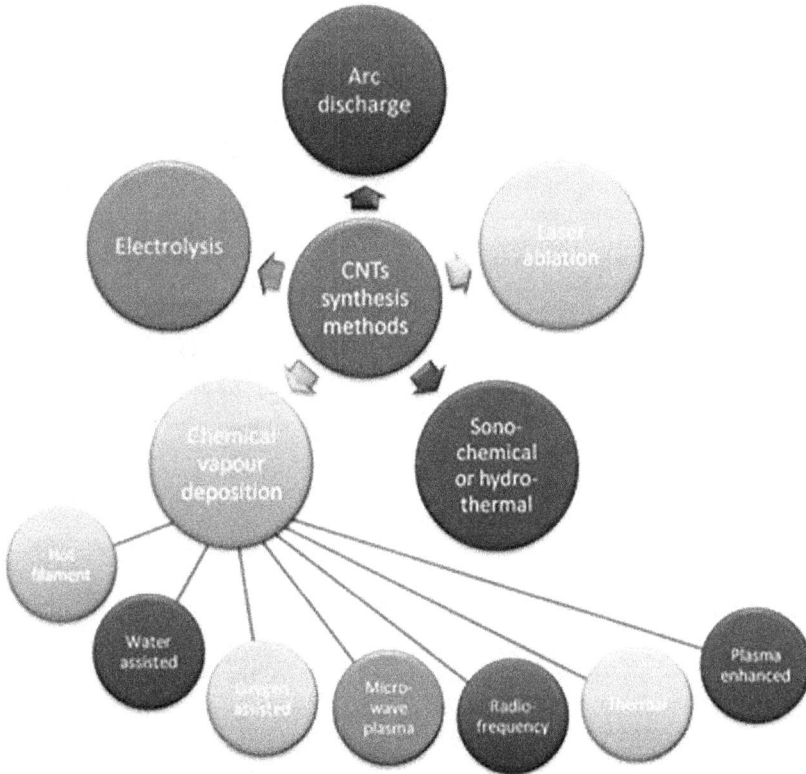

FIGURE 2.2   Production of CNTs by different routes. Reproduced with permission [3].

2.2.1   Catalytic Chemical Vapor Deposition Process

Catalytic Chemical Vapor Deposition (CCVD) is a process that is usually chosen to synthesize CNTs on a large scale. Recently, Hou et al. [6] discussed Floating Catalyst Chemical Vapor Deposition (FCCVD) method for the controlled synthesis of CNTs. The authors reported that the FCCVD technique operates in a way to control the topography, structure, quality and yield of CNTs. Additionally, it was found that the addition of sulfur-containing growth promoter can reduce the growth temperature of FCCVD, as it reduces the melting point of the catalyst particles. It was, however, concluded that, due to weak performance of the floating catalyst and the high growth temperature, the FCCVD technique cannot be commercialized. In another study, Double-Walled Carbon Nanotubes (DWCNTs) were synthesized using a water-assisted FCCVD technique to enhance electrical applications of CNT nanocomposites [7]. It was concluded that the addition of water resulted in more DWCNTs and the higher I_G/I_D ratio (better quality) from Raman spectroscopy. A list of some catalysts employed in CVD process to synthesize CNTs are summarized in Table 2.1.

2.2.2   Arc-Discharge Method

The arc-discharge synthesis of MWCNTs occurs at higher temperature (above 1700 °C) under sub-atmospheric helium pressure (between 50 and 700 mbar). Arc-discharge technique involves two usually water-cooled graphite electrodes to

---

TABLE 2.1

A Summary of Catalysts Used to Develop CNTs by the CVD Technique

Catalyst	Reaction Conditions*	Carbon Source	Formation of Carbon Structure	CNTs Yield/ Purity	References
NiMo/MgO	C_w = 0.5 g; T = 700 °C; t = 30 min; F = 1g	Waste plastics (polypropylene)	MWCNTs	77 %-87 %/-	[8]
Ferrocene	C_w = 0.2–4.6 mol m^{-3}; T = 800–1000 °C; t = 30 min; F = 2–37 kPa	Methane	SWCNTs	89 %- 92 %/ 80 %- 84 %	[9]
Cobalt nanoparticles	T = 750 °C; t = 15 min; F = 2 ml/min	Ethanol	MWCNTs	-	[10]
Co-Ni/Co-Fe	T = 1050 °C; t = 10 s; F = 200–800 ppm	Ethylene	SWCNTs	-	[11]
Co-Mo/MgO	C_w = 0.5 g; T = 700 °C; t = 60 min; F = 200 mL/min	Acetylene	MWCNTs	70 %/-	[12]
Fe$_3$O$_4$/SiO$_2$	C_w = 0.2 g; T = 800 °C; t = 60 min; F = 300 mL/min	Acetylene	MWCNTs	1 g/-	[13]

* C_w = Catalyst weight; T = Growth temperature; t = Reaction time; F = Feed rate

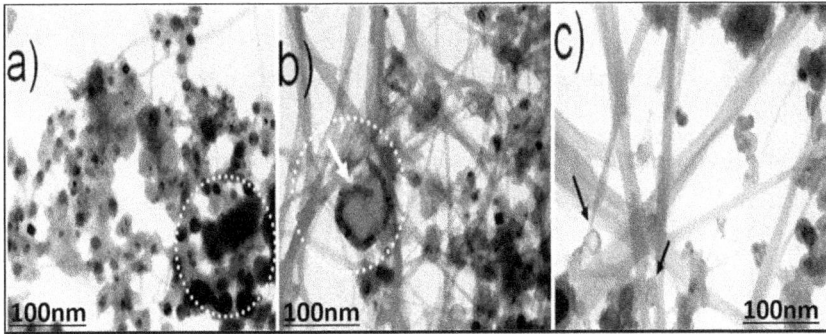

FIGURE 2.3 TEM micrographs for: (a) as-grown, (b) thermal oxidation and reflux purification and (c) reflux purification. Reproduced with permission [14].

generate plasma to synthesize CNTs. One of the features of arc discharge that is more significant in comparison to other methods is that it produces much less amorphous carbon. It was found that addition of catalyst (iron, cobalt, etc.) to target electrodes results in optimum synthesis of CNTs [4]. Ribeiro et al. [14] studied CNTs produced by electric arc-discharge method, which shows the minimum amount of amorphous structures formation. The morphology of the walls of CNTs confirmed that a large quantity of impurities had been removed actively and the tube bundles could be easily observed (Figure 2.3). It shows the thermal oxidation and reflux purification treatments can remover agglomeration of nanoparticles.

In another study, SWCNTs were obtained from bipolar, pulsed arc-discharge method without any cathode deposit [15]. The advantage of using a bipolar pulsed arc discharge is highly efficient production of SWCNTs. The authors concluded that the bipolar, pulsed-arc discharge is stable, and the pulse frequency increases the aspect ratio of SWCNTs. Furthermore, CNTs fabricated by arc-discharge and CVD methods were made to compare the physicochemical properties [16]. It was found that arc-synthesized CNTs had a high purity, and their growth rate increased at higher voltages and currents. On the other hand, MWCNTs synthesized by CVD method have diameters in the range of 10–50 nm and show narrow structures, as compared to arc-synthesized CNTs. Raniszewski et al. [17] showed that the addition of catalysts varied the arc-channel conductance in growth of CNTs in the arc-discharge method and the purity of the final product. It was reported that the synthesis of CNTs using plasma treatment leads to high stability, and there is a strong influence of the temperature on the size and structure of the deposit. They concluded that, by changing the plasma jet temperature trend, high performance CNTs should be achievable. Table 2.2 reviews the published research literature on carbonaceous nanomaterials synthesized by arc discharge technique.

2.2.3 LASER-ABLATION METHOD

Laser ablation was the first method adopted to synthesize SWCNTs in high yield in non-equilibrium conditions. Pulsed-laser vaporization (PLV) is suitable to produce

TABLE 2.2
A Summary of Carbonaceous Nanomaterials Production by Arc-Discharge Technique

Method	Product	Comments	Conditions	References
Electric arc discharge	CNTs, carbon nanofibers and carbon nano onions	Morphological analysis showed the diameter and length of CNTs ranges from 25 to 30 nm and 100 to 300 nm respectively.	Fe and Ni were used as catalysts and voltage was changed from 14 to 20 V under atmospheric pressure at 10 °C.	[18]
DC arc-discharge	SWCNTs	From the HRTEM images it implies that the synthesis of SWCNTs is based on the base growth mechanism.	Selenium promoted Y-Ni alloy catalyst was used under the current of 80–120 A for about 10 min resulting in 1.8 SWCNTs.	[19]
Electric arc discharge	SWCNTs and single-walled carbon nanohorn (SWCNH)	FE-SEM micrograph indicates well separated SWCNTs and nearly-spherical aggregate SWCNH in both air and inert atmospheres.	Graphite rods of 99% purity were used as both anode and cathode under a potential of 36 V with a direct current of 100 A.	[20]
Arc discharge	Aligned MWCNTs	The aligned CNTs in the soot treated with HCl revealed higher purity when compared to CNTs fabricated using CVD method.	A graphite rod of 1.72 g/cc density and nickel were employed as the anode and catalyst respectively using a current density of 157 A/cm^2 and 36 V.	[21]

a wide variety of carbon allotropes, including CNTs, carbon nanowalls, nanohorns and graphene [22]. Ismail et al. [23] synthesized MWCNTs by laser ablation in H$_2$O without a catalyst and employed these for photodetection applications. The average diameter and length of the colloidal CNTs prepared by 532 nm laser pulses were 20 nm and a few micrometers, respectively. As shown in Figure 2.4, the laser beam was focused on the graphite target to prohibit the molten metal and vapor to deposit the

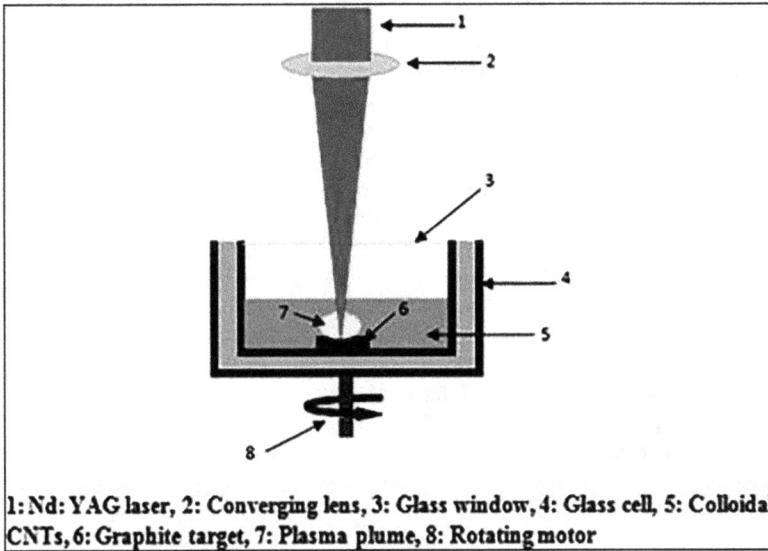

1: Nd: YAG laser, 2: Converging lens, 3: Glass window, 4: Glass cell, 5: Colloidal CNTs, 6: Graphite target, 7: Plasma plume, 8: Rotating motor

FIGURE 2.4 Schematic diagram of laser ablation in liquid technique by 523 nm and 1064 nm laser wavelengths. Reproduced with permission [24]. Table 2.3 provides production of CNTs using laser ablation technique under different operating conditions. Nanocomposite structures play a crucial role in purity of synthesized CNTs.

TABLE 2.3

A Summary of Carbonaceous Nanomaterials Production by Laser-Ablation Method Using Different Reaction Conditions

Method	Product	Comments	Conditions	References
Pulsed laser ablation	MWCNTs/Ag	MWCNTs/Ag nanocomposite was used for removal of hazardous 4-nitrophenol, methyl orange and methylene blue.	The ultra-fast pulsed laser was employed for the ablation of Ag plate submerged in CNTs solution. (1064 nm, 10 Hz and 7 ns).	[25]
Pulsed laser ablation	NiO-MWCNTs	NiO-MWCNTs nanocomposites were used for the catalytic degradation of methylene blue dye, owing to their high catalytic reduction capability and economic viability.	The ablation process was conducted at laser energy of 70 mJ with a pulse duration of 7 ns.	[26]

(Continued)

TABLE 2.3 *(Continued)*

A Summary of Carbonaceous Nanomaterials Production by Laser-Ablation Method Using Different Reaction Conditions

Method	Product	Comments	Conditions	References
Pulsed laser ablation	ZnO-MWCNTs	ZnO-MWCNTs nanocomposite was used for degradation of methylene blue dyes, and results confirmed that CNTs play a vital role for their high catalytic efficiency.	Solid-state Nd:YAG laser with a repetition rate of 10 Hz and the wavelength of 1064 nm was employed for the process of pulsed laser ablation.	[27]
Pulsed laser ablation	CdO-MWCNTs	The optical properties of CdO-MWCNTs nanocomposites were improved due to uniform embedding of CdO nanoparticles on the surface of CNTs.	Pulsed laser ablation of cleaned Cd sheets was adjusted to operate at a repetition rate of 10 Hz using a lens of focal length of 70 mm.	[28]

lens. It was concluded that CNTs synthesized with 1064 nm laser pulses were well graphitized and showed lower electrical resistivity.

2.2.4 OTHER METHODS

Another method to produce CNTs is electrolysis, which is based on electrowinning of alkali metals on a graphite cathode. Wang et al. [29] studied synthesis of CNTs through the electrolysis of CO_2 in molten carbonates. It was found that the number of walls in the CNTs was influenced by electrolysis time, and that the product showed an average diameter of 500 nm and length of 200 µm. During the electrolysis, the increasing concentration of the secondary binary components facilitates the formation of CNTs. Scanning electron microscopy (SEM) characterization confirms that increasing the concentration of Li_2CO_3 results in more agglomerated CNTs. It was concluded that electrolyses in lithium carbonates lead to formation of CNTs with a high aspect ratio. Furthermore, Novoselova et al. [30] used the novel electrolytic method to produce CNTs under an excess pressure of 15 bar at a temperature of 550 °C. It was found that, instead of CNTs, the cathode product shows nanographite and disordered graphite. Typical morphology of the electrolytic nanotubes shows that the outer diameter of CNTs changes from 5 to 250 nm, while the internal diameter changes from 2 to 140 nm. In another study, Schwandt et al. [31] developed molten salt electrolytic method of converting graphite into CNTs. As shown in Figure 2.5, a

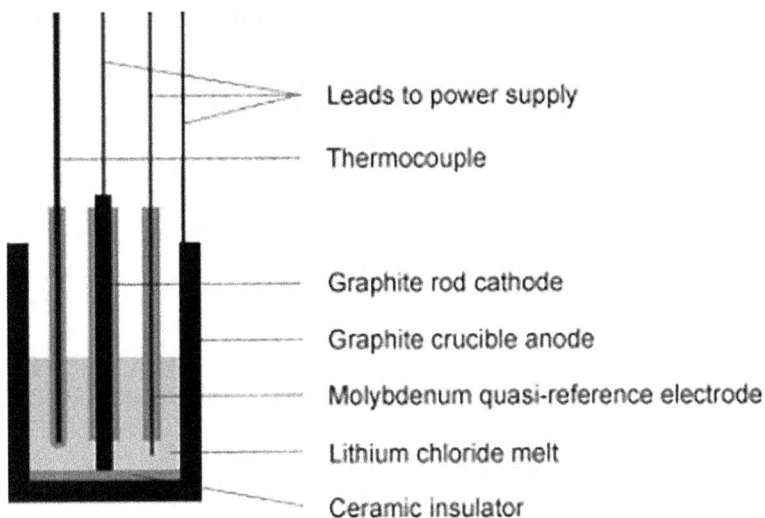

FIGURE 2.5 Schematic of the electrolytic cell used at 775 °C under Ar. Reproduced with permission [32].

graphite rod was shielded with an alumina sleeve, and they all were contacted with stainless steel rods to produce MWCNTs in the electrolytic fuel cell system.

Sonochemical/hydrothermal method can be used for the synthesis of nano-onions, nanorods, and CNTs at low temperature (150–180 °C). Veksha et al. [33] employed Fe_2O_3-Al_2O_3 catalysts to produce CNTs and H_2 from polyolefin plastics through hydrothermal synthesis using precipitating agents. The results showed formation of 4.28 g MWCNTs with an efficiency of 3.2 % plastic conversion and a stable hydrogen flow for 155 minutes. The presence of hydrothermal synthesis conditions leads to variations in the yield and size of the produced MWCNTs. Manafi et al. [34] reported low temperature synthesis of MWCNTs using a hydrothermal method. According to the experimental results, highly pure MWCNTs were prepared at 160 °C with length up to 2–5 μm and a diameter of 60 nm. For the subsequent hydrothermal growth of MWCNTs, ultrasonic pre-treatment step plays an important role. The Raman spectrum of the prepared CNTs indicates a rather defective structure, which may be attributed to the low temperature synthesis route.

In 2019, Hidalgo et al. [35] used a novel technique for the synthesis of CNTs using microwave radiation. They employed biochar as biological precursor due to its high carbon content, together with ferrocene for CNTs production. The results indicate that high concentration of biochar produced from hazelnut is more reactive for CNTs production due to higher aromaticity. In another study, Super-Long CNTs (SL-CNTs) were synthesized from cellulosic biomass using microwave radiation [36]. According to the experimental results, it was determined that the presence of inherent metallic species in biomass may have enhanced the production of CNTs due to the thermal effect of microwave treatment at high temperatures. Based on the microwave heating-mechanism, the produced volatile matter is transported from the hot center, leading to the increased growth of SL-CNTs.

2.3   STRUCTURE

Carbon Nanotubes are made up of sp^2-bonded carbon atoms and classified in the following two types: SWCNTs and MWCNTs. Table 2.4 provides a comparison between SWCNTs bundles and MWCNTs.

2.3.1   PROPERTIES

CNTs are unique nanostructures and can be employed in nano-electronic and nano-mechanical equipment. Due to C-C bonds in the honeycomb lattice, CNTs are estimated to have high stiffness and axial strength. It has been reported that SWCNTs and MWCNTs have high Young's modulus of 1 TPa and 1.2 TPa, respectively [37]. Generally, MWCNTs show higher Young's modulus owing to different diameters of the nanotube and van der Waals forces between the tubes. Table 2.5 provides a summary of the most important properties of SWCNTs and MWCNTs.

After highlighting the properties of CNTs, characterization techniques can provide information about microstructure, impurities content and morphology. Table 2.6 summarizes some of the widely preferred techniques to investigate the CNTs.

TABLE 2.4
Characterization of Commercially Available SWCNTs and MWCNTs

SWCNTs	MWCNTs
The diameter is less than 3 nm.	Diameter ranges between 5–100 nm.
SWCNT can only be produced on a laboratory scale.	It is easier to produce MWCNTs at a large scale and more economically.
The length of SWCNTs ranges from 100 to 1000 nm.	The length of MWCNTs ranges from 10 to 30 µm.
The interaction and dispersive nature of SWCNTs are excellent.	MWCNTs materials exhibit a strong dispersive nature.
SWCNTs have morphology in the visible and near infrared spectrum.	The optical spectrum in the visible and near infrared spectrum does not show length on the CNTs.
A low-frequency peak (<200 cm^{-1}) is the main signature of a SWCNTs.	The frequency of G-band leads to a broad asymmetric feature.
For SWCNTs, the reported surface areas range from 240–1250 m^2/g.	Surface areas for as-produced MWCNTs changes from 10 to 500 m^2/g.
SWCNTs have lower oxidation temperatures (350–500 °C)	Highly crystalline MWCNTs are more resistant to oxidation.
SWCNTs are found to be semiconductive and is suitable for modified electrodes.	Sole metallic layer in MWCNTs results in metallic features.

TABLE 2.5
The Physical Properties of CNTs

Properties	Unit	SWCNTs	MWCNTs	References
Density	Kg/m^3	2600	1600	[4]
Tensile strength	GPa	126	> 63	[38]
Young's modulus	Tpa	0.65–5.5	0.2–1.0	[38]
Thermal conductivity at 300 K	W/m K	3000–6000	2000–3000	[4]
Electrical conductivity	S/m	1×10^{6}–1×10^7	1×10^{6}–1×10^7	[39]
BET specific surface area	m^2/g	662	146	[40]
Thermal stability in air	°C	680	500	[41]

TABLE 2.6
Characterization Techniques of CNTs

Technique	Applications	References
XPS	To quantify functional groups and provide information on the chemical environment of surface species.	[42]
FTIR	To shed light on the formation of a chemical covalent bond between functional groups and CNTs.	[43]
Raman spectroscopy	To determine the size of crystal, clustering of the sp^2 phase and the introduction of crystal disorder.	[44]
AFM	To provide information about the elastic and electronic properties of CNTs at resolution lower than 1 nm.	[45]
TGA	To further investigate the number of functional groups adsorbed on the surface of CNTs.	[46]
TEM & HRTEM	To find defects along the walls and the morphology of CNTs.	[47]
XRD	Supplies information about the average structure and chirality of layers.	[45]

2.4 APPLICATIONS

CNTs, including SWCNTs and MWCNTs, show encouraging catalytic and electro-chemical properties due to their high specific surface area, super-high mechanical hardness and super-strong chemical inertness [48]. As shown in Figure 2.6, seven major applications of CNTs have been suggested in a variety of fields.

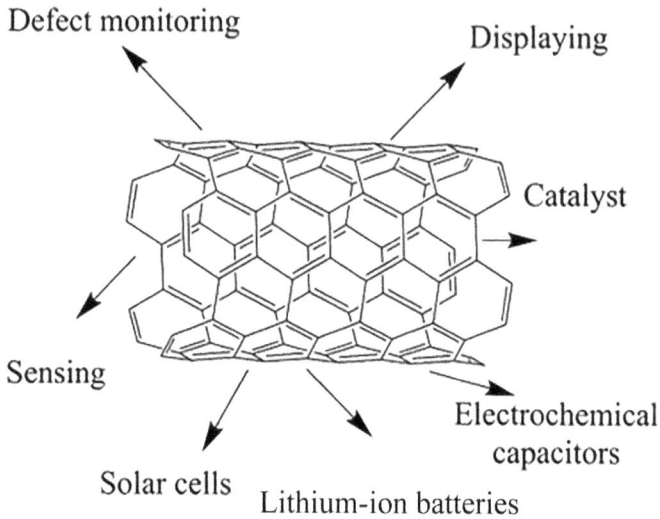

FIGURE 2.6 Potential applications of CNTs.

2.4.1 DEFECT MONITORING

CNTs, owing to their tube-like structure with high aspect ratio and excellent thermal conductivity, have been used to monitor defect in fiber-reinforced polymers (FRPs) [49]. Spinelli et al. [50] developed a structural resin reinforced with MWCNTs to monitor damage under axial and flexural stresses. It is worth noting that, for real-time structural health monitoring purposes, irreversible resistance changes because of the re-arrangement of the CNTs embedded in the polymer resin. In another study, Bregar et al. [51] employed CNTs embedded polymers to monitor real-time adhesion failure in various mechanical loads. It was determined that a reliable adhesion-failure measurement can be obtained by continuously monitoring the linearity of piezo-resistivity induced by the CNTs. Moreover, piezo-resistive strain sensors prepared with CNTs lead to optimize the overall strain sensing performance.

2.4.2 DISPLAYING

The application of CNTs in display equipment, such as liquid crystal displays (LCDs), thermochromic displays and field emission displays (FEDs), is an important part of modern IT [52]. Ma et al. [53] studied the electro-optical properties of LCD cells using modified MWCNTs. It was found that the chemically modified MWCNTs increased the intensity of light through the cell and drive voltage. In another study, Pagidi et al. [54] reported superior electro-optics by using functionalized CNTs (f-CNTs) in bire-fringent liquid crystal droplets. The results showed that polymer-wrapped f-CNTs exhibited enhanced field strength and elastic constant. It was concluded that variation in the concentration of f-CNTs also changed the scattering of feeble light in electro-optics and photonic devices.

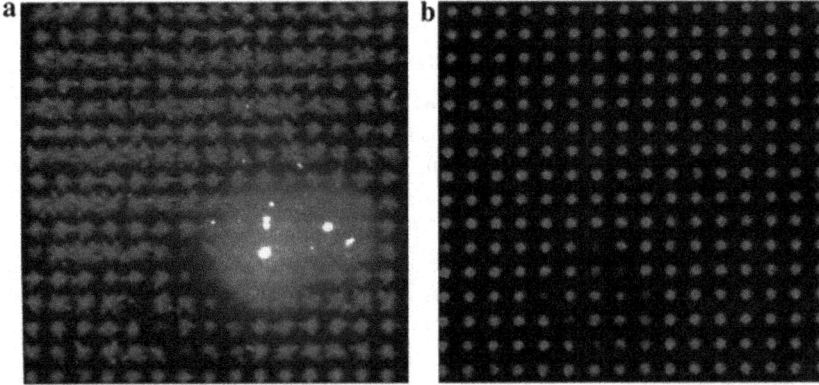

FIGURE 2.7   The SEE at different work modes. (a) Anode: DC voltage of 9900 V (b) Anode: pulse voltage of 9900 V. Reproduced with permission [55].

Liu et al. [55] used a triode CNTs FED to study the Secondary Electron Emission (SEE) under two different systems of work modes. As shown in Figure 2.7, the anode image shows bright spots (Figure 2.7a), and after a few minutes of exposure to the emission and transport of electrons, a breakdown can be seen.

2.4.3 SENSING

CNTs can find applications in nanoprobes and sensors due to the high sensitivity of their electronic properties towards target analytes [56]. Table 2.7 summarizes some of the significant works done on the gas-sensing applications of CNTs.

2.4.4 SOLAR CELLS

Solar cell devices absorb light and generate excitons. Subsequently, the conductive layers allow the dissociation of excitons in a strong electric field [68]. Cui et al. [69] employed SWCNTs for solar cells application, and it showed a power conversion efficiency (PCE) of 6%. The SWCNTs-Si solar cells reach the open-circuit voltage and short-circuit current of 0.55 V and 25.01 mA/cm^2, respectively, due to charge transfer from the SWCNTs enhanced by HNO_3. In another study, Wang et al. [70] used flexible solar cell based on Ti foil/TiO_2 nanotubes (TNTs) as transparent electrode for light illumination (Figure 2.8). The TNT/CNT perovskite solar cells show a higher charge separation efficiency after 100 bending cycles.

Application in hybrid CNTs/Si solar cells for photovoltaic devices has been investigated by Li et al. [71]. Furthermore, CNTs normally show p-type behavior in air and a high sheet resistance result in a high series resistance in solar cells. It was found that, due to an increase in the n-type doping of the SWCNTs thin film, the photovoltaic performance of CNTs/Si solar cell could be ascribed to the dopants. It was concluded that n-type and p-type SWCNTs thin films are more efficient than former photovoltaic equipment of this type.

TABLE 2.7

Comparative Study of Different Types of Gas Sensors Using CNTs

Sensor Material	Target Gas	Limit of Detection (ppm)	Concentration Range (ppm)	Response Time (seconds)	References
CNTs/alumina	NO_2	0.1	0.1–0.9	900	[57]
Co_3O_4-SWCNT	H_2S	10	5–150	300	[58]
Metal-phthalocyanine functionalized CNTs	H_2O_2	50	-	10	[59]
Pd/SWCNTs	H_2	10	10–1000	324	[60]
Polyallylamine-amino-CNTs	CO_2	75	1000–4000	3000	[61]
V_2O_5/MWCNTs	CH_4	40	40–100	138	[62]
MWCNTs/Si	C_2H_5OH	50	50–800	-	[63]
MWCNTs/Co	NH_3	14	100–800	30–50	[64]
ZnO/MWCNTs	CO	40	40–200	8–23	[65]
Phthalocyanine/MWCNTs	Chlorine	2	0.1–20	14	[66]
Functionalized SWCNTs	Trinitrotoluene	0.01	0.1–100	-	[67]

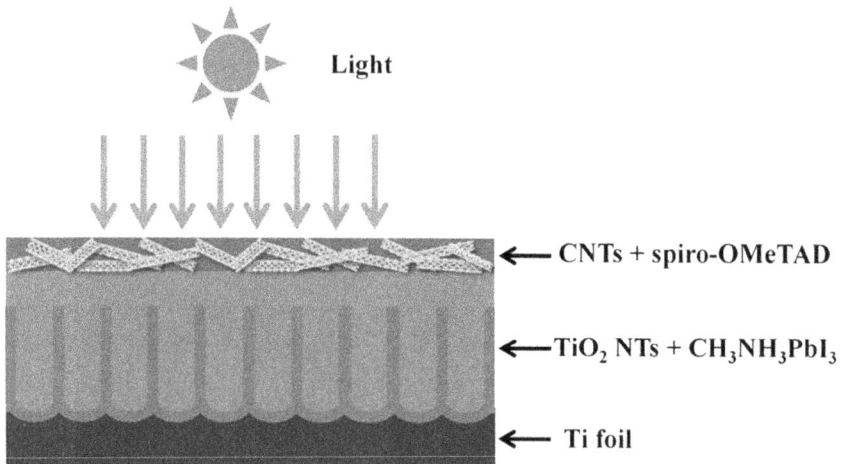

FIGURE 2.8 The cell configuration and CNT application as transparent electrode. Reproduced with permission [72].

2.4.5 LITHIUM-ION BATTERIES

Lithium-ion batteries can be employed for hybrid-electric-vehicles and remote-sensing applications because of their high energy density. CNTs are a promising conductive additive in CNT-based lithium-ion batteries to develop state-of-the-art electrodes [73]. In a recent study, Wang et al. [74] synthesized an all-carbon polymeric segment of CNTs as new material to be employed in applications for lithium-ion batteries. It was speculated that the anode material has significant structural and functional features that lead to a maximum reversible capacity of 557 mAhg^{-1}. Furthermore, the conjugated segments of CNTs correspond to a high current-density of 2000 mAg^{-1} after 1750 cycles. In another study, Zou et al. [75] used graphene and CNTs to enhance the performance of a lithium-ion power battery. Results showed that the addition of graphene and MWCNT can create 3D heat-transfer paths, which can decrease the thermal boundary resistance. Moreover, the addition of graphene and MWCNTs is substantial to the heat removal of the battery to avoid the overheating. Zhu et al. [76] reported the electrochemical performances of SnO_2/CNT composites prepared by Atomic Layer Deposition (ALD) in lithium-ion battery anodes. Moreover, the SnO_2/CNT shows excellent rate capability of ethanol electro-oxidation as well as high specific capacity of 1346.6 mAh g^{-1}. Chen et al. [77] showed the use of composite MWCNT/V_2O_5 nanostructures to explore energy, power and cycling stability for high performance Li-ion cathodes. As shown in Figure 2.9, V_2O_5-coated MWCNT sponge formed a coin cell battery and confirmed excellent cycling stability after 100 cycles.

2.4.6 ELECTROCHEMICAL CAPACITORS

The range of commercial applications of electrochemical capacitor technology leads to commercial production of various pure carbon electrodes with high surface area. CNTs can be a promising candidate for supercapacitors [78], owing to their broad range of capacitance values and high-rate performance. Lee et al. [79] studied MWCNTs/MnO_2 electrodes as an electrochemical capacitor. It was found that voltametric charge of MnO_2 within the MWCNTs electrode shows high cycling stability up to 1000 cycles. It was concluded that the ability to generate high electrode-capacitance

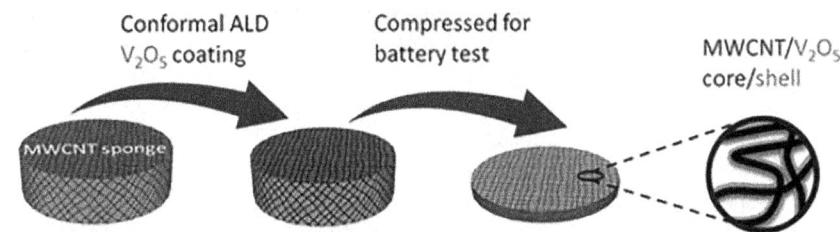

FIGURE 2.9 MWCNTs/V_2O_5 sponge using ALD thin film coating for power density Li-ion batteries. Reproduced with permission [77].

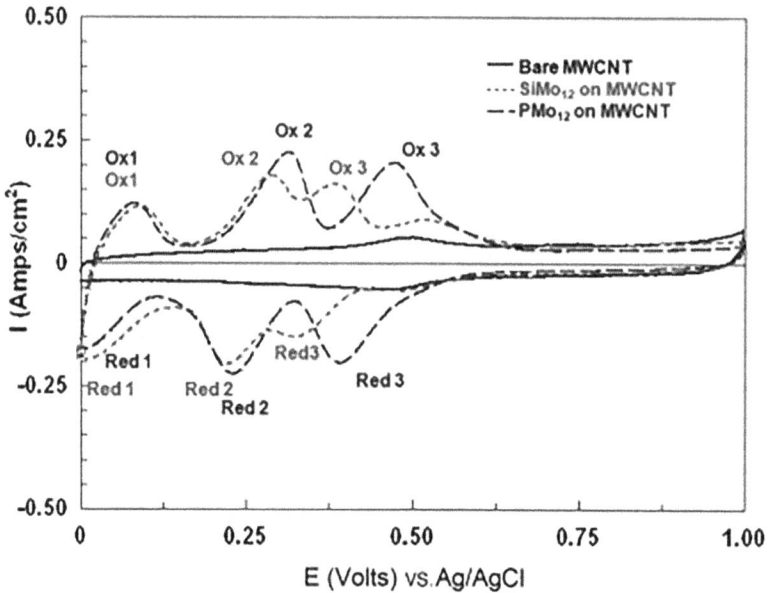

FIGURE 2.10   Electrochemical analyses of bare MWCNTs and POMs coated MWCNTs in 1 M H_2SO_4. Reproduced with permission [80].

may successfully lead to the fabrication of nanostructured electrodes for battery and sensor applications. In another study, Akter et al. [80] demonstrated different pseudocapacitive polyoxometalates (POMs) modified MWCNTs for electrochemical applications. As shown in Figure 2.10, MWCNTs coating results in a significant increase in the stored charge and capacitance for POMs dip coating on the surface of MWCNTs due to a synergistic effect. Table 2.8 shows various carbon electrodes, which have been studied for electrochemical applications.

2.4.7   CATALYST

In heterogeneous catalysis, CNTs can be used in many reactions due to their excellent chemical and physical properties. Carbon nanomaterials can be tailored in different shapes, and they are cheaper than some common supports such as Al_2O_3 or SiO_2. Wang et al. [89] prepared 3D graphite oxide-exfoliated CNTs as a promising catalyst support for the Methanol Oxidation Reaction (MOR). The results suggest that the 3D catalysts show a distinct hydrogen adsorption/desorption region, which results in superior activities toward methanol electrooxidation. Zhang et al. [90] used CNTs structures as catalyst supports for Proton Exchange Membrane (PEM) fuel cells. It was found that addition of N_2 atoms into the CNTs backbone leads to a direct increase in the overall performance of the fuel cell (increased power density).

Table 2.9 represents examples of catalytic studies using the carbon-supported catalysts in different reactions.

TABLE 2.8

Electrochemical Performance of Various Carbon Materials

Material and Method	Surface Area (m² g⁻¹)	Cell Voltage (kV)	Reported Capacitance (F g⁻¹)	Electrolyte	References
CNT/carbon nanofibers (CNFs)	122.9	9.0	464.2	KOH	[81]
Polypyrrole (PPY)-MWCNTs	10.0	20.0	390	Na_2SO_4	[82]
MnO_x/CNT	100	5.0	400	$KMnO_4$	[83]
CNTs/activated carbon (AC)	2879	1.0	141	KOH	[84]
Graphene/MWCNTs/ MnO_2	High surface area	40	126	Na_2SO_4	[85]
CNTs/mesoporous carbon	300.0	3.0	60.2	KOH	[86]
CNTs/aluminum foil (Al)	340.0	2.5	170.0	EMITFSI in ACN	[87]
Polyaniline (PANI)-CNT	Large surface area	1.0	1744.0	H_2SO_4	[88]

TABLE 2.9

Lists of Metal-Supported CNTs Catalysts Tested in Different Reactions

Reaction	Catalysts	Comments	References
Ethanol oxidation	PtSn/MWCNTs	Excellent catalytic performance of the Pt/ MWCNTs electrode due to strong interaction between the metal alloy and the functionalized MWCNTs.	[91]
CO_2 methanation	NiCe/CNTs	High dispersity of Ce on the large surface of CNTs resulted in 83.8% conversion of CO_2 and almost 100% selectivity of methane.	[92]
Fischer-Tropsch synthesis	Co/CNTs, Co/ Carbon spheres (CS)	Co/CNT exhibited higher selectivity to lighter hydrocarbons due to the particle size of cobalt and the cobalt-CNTs interactions.	[93]
Formic acid electrooxidation	Pd-Co/CNTs, Pd/CNTs	The interaction between Pd and Co improves the oxidation reaction as the accumulation of poisoning-intermediates is prevented.	[94]
Glucose microfluidic fuel cell	Au/MWCNTs	The proper dispersion of Au nanoparticles, metal-support interaction and minimum corrosion effects lead to improvement in the electrocatalytic reaction.	[95]
Oxygen reduction reaction	Tungsten nitride decorated CNTs hybrid supported Pt (Pt-WN/CNTs-M)	Significantly improved performance due to the synergistic interaction and effect of CNT functionalization.	[96]

2.5 CHALLENGES AND OPPORTUNITIES

CNTs have many advances in composites, sensors, solar cells and electrochemical devices. The amazing mechanical, thermal and electronic properties of CNTs result in a very promising glimpse into the future of medicine. The challenge lies in the innovative ways to produce CNTs on a large scale, as the process of synthesizing CNTs continuously needs improvement in terms of efficiency of the process.

REFERENCES

[1] Iijima S. Helical microtubules of graphitic carbon. Nature 1991;354:56–8.
[2] Iijima S, Ichihashi T. Single-shell carbon nanotubes of 1-nm diameter. Nature 1993;363:603–5.
[3] Prasek J, Drbohlavova J, Chomoucka J, Hubalek J, Jasek O, Adam V, et al. Methods for carbon nanotubes synthesis – review. J Mater Chem 2011;21:15872–84. https://doi.org/10.1039/C1JM12254A.
[4] Gupta N, Gupta SM, Sharma SK. Carbon nanotubes: Synthesis, properties and engineering applications. Carbon Lett 2019;29:419–47. https://doi.org/10.1007/S42823-019-00068-2/FIGURES/14.
[5] Jia X, Wei F. Advances in production and applications of carbon nanotubes. Single-Walled Carbon Nanotub Prep Prop Appl 2019:299–333.
[6] Hou P, Zhang F, Zhang L, Liu C, Cheng H. Synthesis of carbon nanotubes by floating catalyst chemical vapor deposition and their applications. Adv Funct Mater 2022;32:2108541.
[7] Dong L, Park JG, Leonhardt BE, Zhang S, Liang R. Continuous synthesis of double-walled carbon nanotubes with water-assisted floating catalyst chemical vapor deposition. Nanomaterials 2020;10:365.
[8] Modekwe HU, Mamo M, Moothi K, Daramola MO. Synthesis of bimetallic NiMo/MgO catalyst for catalytic conversion of waste plastics (polypropylene) to carbon nanotubes (CNTs) via chemical vapour deposition method. Mater Today Proc 2021;38:549–52. https://doi.org/10.1016/j.matpr.2020.02.398.
[9] Yadav MD, Dasgupta K, Patwardhan AW, Kaushal A, Joshi JB. Kinetic study of single-walled carbon nanotube synthesis by thermocatalytic decomposition of methane using floating catalyst chemical vapour deposition. Chem Eng Sci 2019;196:91–103. https://doi.org/10.1016/j.ces.2018.10.050.
[10] Kumar U, Yadav BC. Synthesis of carbon nanotubes by direct liquid injection chemical vapor deposition method and its relevance for developing an ultra-sensitive room temperature based CO2 sensor. J Taiwan Inst Chem Eng 2019;96:652–63. https://doi.org/10.1016/j.jtice.2019.01.002.
[11] Ahmad S, Liao Y, Hussain A, Zhang Q, Ding E-X, Jiang H, et al. Systematic investigation of the catalyst composition effects on single-walled carbon nanotubes synthesis in floating-catalyst CVD. Carbon N Y 2019;149:318–27. https://doi.org/10.1016/j.carbon.2019.04.026.
[12] Ibrahim H, Buhari S, Abdulrahman S, Lawal A, Abdulkareem A, Muriana R, et al. Optimisation of synthesis parameters for Co-Mo/MgO catalyst yield in MWCNTs production. Niger J Technol Dev 2023;20.
[13] Wanjeri VWO, Gbashi S, Ngila JC, Njobeh P, Mamo MA, Ndungu PG. Chemical vapour deposition of MWCNT on silica coated Fe3O4 and use of response surface methodology for optimizing the extraction of organophosphorus pesticides from water. Int J Anal Chem 2019;2019.
[14] Ribeiro H, Schnitzler MC, da Silva WM, Santos AP. Purification of carbon nanotubes produced by the electric arc-discharge method. Surfaces and Interfaces 2021;26:101389. https://doi.org/10.1016/j.surfin.2021.101389.

[15] Maria KH, Mieno T. Synthesis of single-walled carbon nanotubes by low-frequency bipolar pulsed arc discharge method. Vacuum 2015;113:11–8. https://doi.org/10.1016/j. vacuum.2014.11.025.

[16] Sharma R, Sharma AK, Sharma V. Synthesis of carbon nanotubes by arc-discharge and chemical vapor deposition method with analysis of its morphology, dispersion and functionalization characteristics. Cogent Eng 2015;2:1094017.

[17] Raniszewski G, Wiak S, Pietrzak L, Szymanski L, Kolacinski Z. Influence of plasma jet temperature profiles in arc discharge methods of carbon nanotubes synthesis. Nanomaterials 2017;7:50.

[18] Hosseini AA, Allahyari M, Besheli SD. Synthesis of carbon nanotubes, nano fibbers and nano union by electric arc discharge method using NaCl accuse as solution and Fe and Ni particles and catalysts. IJEST 2012;1:217–29.

[19] Huang L, Wu B, Chen J, Xue Y, Liu Y, Kajiura H, et al. Synthesis of single-walled carbon nanotubes by an arc-discharge method using selenium as a promoter. Carbon N Y 2011;49:4792–800. https://doi.org/10.1016/j.carbon.2011.06.091.

[20] A JB, M J, D RR, Haridoss P. Synthesis of thin bundled single walled carbon nanotubes and nanohorn hybrids by arc discharge technique in open air atmosphere. Diam Relat Mater 2015;55:12–5. https://doi.org/10.1016/j.diamond.2015.02.004.

[21] Joseph Berkmans A, Ramakrishnan S, Jain G, Haridoss P. Aligning carbon nanotubes, synthesized using the arc discharge technique, during and after synthesis. Carbon N Y 2013;55:185–95. https://doi.org/10.1016/j.carbon.2012.12.025.

[22] Chrzanowska J, Hoffman J, Małolepszy A, Mazurkiewicz M, Kowalewski TA, Szymanski Z, et al. Synthesis of carbon nanotubes by the laser ablation method: Effect of laser wavelength. Phys Status Solidi 2015;252:1860–7.

[23] Ismail RA, Mohsin MH, Ali AK, Hassoon KI, Erten-Ela S. Preparation and characterization of carbon nanotubes by pulsed laser ablation in water for optoelectronic application. Phys E Low-Dimensional Syst Nanostructures 2020;119:113997. https://doi. org/10.1016/j.physe.2020.113997.

[24] Ismail RA, Mohsin MH, Ali AK, Hassoon KI, Erten-Ela S. Preparation and characterization of carbon nanotubes by pulsed laser ablation in water for optoelectronic application. Phys E Low-Dimensional Syst Nanostructures 2020;119:113997. https://doi. org/10.1016/j.physe.2020.113997.

[25] Mostafa AM, Mwafy EA, Awwad NS, Ibrahium HA. Synthesis of multi-walled carbon nanotubes decorated with silver metallic nanoparticles as a catalytic degradable material via pulsed laser ablation in liquid media. Colloids Surfaces A Physicochem Eng Asp 2021;626:126992. https://doi.org/10.1016/j.colsurfa.2021.126992.

[26] Altowyan AS, Toghan A, Ahmed HA, Pashameah RA, Mwafy EA, Alrefaee SH, et al. Removal of methylene blue dye from aqueous solution using carbon nanotubes decorated by nickel oxide nanoparticles via pulsed laser ablation method. Radiat Phys Chem 2022;198:110268. https://doi.org/10.1016/j.radphyschem.2022.110268.

[27] Mostafa AM, Mwafy EA, Toghan A. ZnO nanoparticles decorated carbon nanotubes via pulsed laser ablation method for degradation of methylene blue dyes. Colloids Surfaces A Physicochem Eng Asp 2021;627:127204. https://doi.org/10.1016/j. colsurfa.2021.127204.

[28] Mwafy EA, Mostafa AM. Multi walled carbon nanotube decorated cadmium oxide nanoparticles via pulsed laser ablation in liquid media. Opt Laser Technol 2019;111:249–54. https://doi.org/10.1016/j.optlastec.2018.09.055.

[29] Wang X, Liu X, Licht G, Wang B, Licht S. Exploration of alkali cation variation on the synthesis of carbon nanotubes by electrolysis of CO2 in molten carbonates. J CO2 Util 2019;34:303–12. https://doi.org/10.1016/j.jcou.2019.07.007.

[30] Novoselova IA, Oliinyk NF, Volkov SV, Konchits AA, Yanchuk IB, Yefanov VS, et al. Electrolytic synthesis of carbon nanotubes from carbon dioxide in molten salts and their characterization. Phys E Low-Dimensional Syst Nanostructures 2008;40:2231–7. https://doi.org/10.1016/j.physe.2007.10.069.

[31] Schwandt C, Dimitrov AT, Fray DJ. High-yield synthesis of multi-walled carbon nanotubes from graphite by molten salt electrolysis. Carbon N Y 2012;50:1311–5. https://doi.org/10.1016/j.carbon.2011.10.054.

[32] Schwandt C, Dimitrov AT, Fray DJ. High-yield synthesis of multi-walled carbon nanotubes from graphite by molten salt electrolysis. Carbon N Y 2012;50:1311–5. https://doi.org/10.1016/j.carbon.2011.10.054.

[33] Veksha A, Bin Mohamed Amrad MZ, Chen WQ, Binte Mohamed DK, Tiwari SB, Lim T-T, et al. Tailoring Fe2O3–Al2O3 catalyst structure and activity via hydrothermal synthesis for carbon nanotubes and hydrogen production from polyolefin plastics. Chemosphere 2022;297:134148. https://doi.org/10.1016/j.chemosphere.2022.134148.

[34] Manafi S, Nadali H, Irani HR. Low temperature synthesis of multi-walled carbon nanotubes via a sonochemical/hydrothermal method. Mater Lett 2008;62:4175–6. https://doi.org/10.1016/j.matlet.2008.05.072.

[35] Hidalgo P, Navia R, Hunter R, Coronado G, Gonzalez M. Synthesis of carbon nanotubes using biochar as precursor material under microwave irradiation. J Environ Manage 2019;244:83–91. https://doi.org/10.1016/j.jenvman.2019.03.082.

[36] Esohe Omoriyekomwan J, Tahmasebi A, Zhang J, Yu J. Synthesis of super-long carbon nanotubes from cellulosic biomass under microwave radiation. Nanomaterials 2022;12:737.

[37] Dervishi E, Li Z, Xu Y, Saini V, Biris AR, Lupu D, et al. Carbon nanotubes: Synthesis, properties, and applications. Part Sci Technol 2009;27:107–25.

[38] Nurazzi NM, Sabaruddin FA, Harussani MM, Kamarudin SH, Rayung M, Asyraf MRM, et al. Mechanical performance and applications of cnts reinforced polymer composites – A review. Nanomaterials 2021;11:2186.

[39] Kumam P, Shah Z, Dawar A, Rasheed HU, Islam S. Entropy generation in MHD radiative flow of CNTs Casson nanofluid in rotating channels with heat source/sink. Math Probl Eng 2019;2019.

[40] Birch ME, Ruda-Eberenz TA, Chai M, Andrews R, Hatfield RL. Properties that influence the specific surface areas of carbon nanotubes and nanofibers. Ann Occup Hyg 2013;57:1148–66.

[41] El Achaby M, Qaiss A. Processing and properties of polyethylene reinforced by graphene nanosheets and carbon nanotubes. Mater Des 2013;44:81–9. https://doi.org/10.1016/j.matdes.2012.07.065.

[42] Wepasnick KA, Smith BA, Bitter JL, Howard Fairbrother D. Chemical and structural characterization of carbon nanotube surfaces. Anal Bioanal Chem 2010;396:1003–14. https://doi.org/10.1007/S00216-009-3332-5/FIGURES/7.

[43] Baudot C, Tan CM, Kong JC. FTIR spectroscopy as a tool for nano-material characterization. Infrared Phys Technol 2010;53:434–8.

[44] Dresselhaus MS, Jorio A, Hofmann M, Dresselhaus G, Saito R. Perspectives on carbon nanotubes and graphene Raman spectroscopy. Nano Lett 2010;10:751–8.

[45] Herrero-Latorre C, Álvarez-Méndez J, Barciela-García J, García-Martín S, Peña-Crecente RM. Characterization of carbon nanotubes and analytical methods for their determination in environmental and biological samples: A review. Anal Chim Acta 2015;853:77–94. https://doi.org/10.1016/J.ACA.2014.10.008.

[46] Zenkel C, Albuerne J, Emmler T, Boschetti-de-Fierro A, Helbig J, Abetz V. New strategies for the chemical characterization of multi-walled carbon nanotubes and their derivatives. Microchim Acta 2012;179:41–8.

[47] Shaari N, Tan SH, Mohamed AR. Synthesis and characterization of CNT/Ce-TiO2 nanocomposite for phenol degradation. J Rare Earths 2012;30:651–8. https://doi.org/10.1016/S1002-0721(12)60107-0.

[48] Ma L, Dong X, Chen M, Zhu L, Wang C, Yang F, et al. Fabrication and water treatment application of carbon nanotubes (CNTs)-based composite membranes: A review. Membranes (Basel) 2017;7:16.

[49] Hao B, Ma PC. Chapter 3 – Carbon nanotubes for defect monitoring in fiber-reinforced polymer composites. In: Peng H, Li Q, Chen TBT-IA of CN, editors. Micro Nano Technol., Boston: Elsevier; 2017, p. 71–99. https://doi.org/10.1016/B978-0-323-41481-4.00003-4.

[50] Spinelli G, Lamberti P, Tucci V, Vertuccio L, Guadagno L. Experimental and theoretical study on piezoresistive properties of a structural resin reinforced with carbon nanotubes for strain sensing and damage monitoring. Compos Part B Eng 2018;145:90–9. https://doi.org/10.1016/j.compositesb.2018.03.025.

[51] Bregar T, An D, Gharavian S, Burda M, Durazo-Cardenas I, Thakur VK, et al. Carbon nanotube embedded adhesives for real-time monitoring of adhesion failure in high performance adhesively bonded joints. Sci Rep 2020;10:16833.

[52] Jiang K. Chapter 4 – Carbon nanotubes for displaying. In: Peng H, Li Q, Chen TBT-IA of CN, editors. Micro Nano Technol., Boston: Elsevier; 2017, p. 101–27. https://doi.org/10.1016/B978-0-323-41481-4.00004-6.

[53] Ma Z, Gao Y, Cao H. The effect of chemically modified multi-walled carbon nanotubes on the electro-optical properties of a twisted nematic liquid crystal display mode. Crystals 2022;12:1482.

[54] Pagidi S, Manda R, Bhattacharyya SS, Cho KJ, Kim TH, Lim YJ, et al. Superior electro-optics of nano-phase encapsulated liquid crystals utilizing functionalized carbon nanotubes. Compos Part B Eng 2019;164:675–82. https://doi.org/10.1016/j.compositesb.2019.01.091.

[55] Hao H, Liu P, Tang J, Cai Q, Fan S. Secondary electron emission in a triode carbon nanotube field emission display and its influence on the image quality. Carbon N Y 2012;50:4203–8. https://doi.org/10.1016/j.carbon.2012.04.070.

[56] Zaporotskova IV., Boroznina NP, Parkhomenko YN, Kozhitov LV. Carbon nanotubes: Sensor properties. A review. Mod Electron Mater 2016;2:95–105. https://doi.org/10.1016/J.MOEM.2017.02.002.

[57] Sayago I, Santos H, Horrillo MC, Aleixandre M, Fernández MJ, Terrado E, et al. Carbon nanotube networks as gas sensors for NO2 detection. Talanta 2008;77:758–64. https://doi.org/10.1016/j.talanta.2008.07.025.

[58] Moon S, Vuong NM, Lee D, Kim D, Lee H, Kim D, et al. Co3O4–SWCNT composites for H2S gas sensor application. Sensors Actuators B Chem 2016;222:166–72. https://doi.org/10.1016/j.snb.2015.08.072.

[59] Verma AL, Saxena S, Saini GSS, Gaur V, Jain VK. Hydrogen peroxide vapor sensor using metal-phthalocyanine functionalized carbon nanotubes. Thin Solid Films 2011;519:8144–8. https://doi.org/10.1016/j.tsf.2011.06.034.

[60] Ju S, Lee JM, Jung Y, Lee E, Lee W, Kim S-J. Highly sensitive hydrogen gas sensors using single-walled carbon nanotubes grafted with Pd nanoparticles. Sensors Actuators B Chem 2010;146:122–8. https://doi.org/10.1016/j.snb.2010.01.055.

[61] Shivananju BN, Yamdagni S, Fazuldeen R, Sarin Kumar AK, Hegde GM, Varma MM, et al. CO2 sensing at room temperature using carbon nanotubes coated core fiber Bragg grating. Rev Sci Instrum 2013;84.

[62] Chimowa G, Tshabalala ZP, Akande AA, Bepete G, Mwakikunga B, Ray SS, et al. Improving methane gas sensing properties of multi-walled carbon nanotubes by vanadium oxide filling. Sensors Actuators B Chem 2017;247:11–8. https://doi.org/10.1016/j.snb.2017.02.167.

[63] Young SJ, Lin ZD. Ethanol gas sensors based on multi-wall carbon nanotubes on oxidized Si substrate. Microsyst Technol 2018;24:55–8.

[64] Nguyen LQ, Phan PQ, Duong HN, Nguyen CD, Nguyen LH. Enhancement of NH3 gas sensitivity at room temperature by carbon nanotube-based sensor coated with Co nanoparticles. Sensors 2013;13:1754–62.

[65] Alharbi ND, Shahnawaze Ansari M, Salah N, Khayyat SA, Khan ZH. Zinc oxide-multi walled carbon nanotubes nanocomposites for carbon monoxide gas sensor application. J Nanosci Nanotechnol 2016;16:439–47.

[66] Sharma AK, Mahajan A, Bedi RK, Kumar S, Debnath AK, Aswal DK. Non-covalently anchored multi-walled carbon nanotubes with hexa-decafluorinated zinc phthalocyanine as ppb level chemiresistive chlorine sensor. Appl Surf Sci 2018;427:202–9. https://doi. org/10.1016/j.apsusc.2017.08.040.

[67] Wang J. Near infrared optical biosensor based on peptide functionalized single-walled carbon nanotubes hybrids for 2,4,6-trinitrotoluene (TNT) explosive detection. Anal Biochem 2018;550:49–53. https://doi.org/10.1016/j.ab.2018.04.011.

[68] Jeon I, Matsuo Y, Maruyama S. Single-walled carbon nanotubes in solar cells. Single-Walled Carbon Nanotub Prep Prop Appl 2018:271–98.

[69] Cui K, Chiba T, Omiya S, Thurakitseree T, Zhao P, Fujii S, et al. Self-assembled micro-honeycomb network of single-walled carbon nanotubes for solar cells. J Phys Chem Lett 2013;4:2571–6. https://doi.org/10.1021/jz401242a.

[70] Wang X, Li Z, Xu W, Kulkarni SA, Batabyal SK, Zhang S, et al. TiO2 nanotube arrays based flexible perovskite solar cells with transparent carbon nanotube electrode. Nano Energy 2015;11:728–35. https://doi.org/10.1016/j.nanoen.2014.11.042.

[71] Li X, Guard LM, Jiang J, Sakimoto K, Huang J-S, Wu J, et al. Controlled doping of carbon nanotubes with metallocenes for application in hybrid carbon nanotube/si solar cells. Nano Lett 2014;14:3388–94. https://doi.org/10.1021/nl500894h.

[72] Wang X, Li Z, Xu W, Kulkarni SA, Batabyal SK, Zhang S, et al. TiO2 nanotube arrays based flexible perovskite solar cells with transparent carbon nanotube electrode. Nano Energy 2015;11:728–35. https://doi.org/10.1016/j.nanoen.2014.11.042.

[73] Landi BJ, Ganter MJ, Cress CD, DiLeo RA, Raffaelle RP. Carbon nanotubes for lithium ion batteries. Energy Environ Sci 2009;2:638–54.

[74] Wang S, Chen F, Zhuang G, Wei K, Chen T, Zhang X, et al. Synthesis of an all-carbon conjugated polymeric segment of carbon nanotubes and its application for lithium-ion batteries. Nano Res 2023:1–6.

[75] Zou D, Ma X, Liu X, Zheng P, Hu Y. Thermal performance enhancement of composite phase change materials (PCM) using graphene and carbon nanotubes as additives for the potential application in lithium-ion power battery. Int J Heat Mass Transf 2018;120:33–41. https://doi.org/10.1016/j.ijheatmasstransfer.2017.12.024.

[76] Zhu S, Liu J, Sun J. Growth of ultrathin SnO2 on carbon nanotubes by atomic layer deposition and their application in lithium ion battery anodes. Appl Surf Sci 2019;484:600–9. https://doi.org/10.1016/j.apsusc.2019.04.163.

[77] Chen X, Zhu H, Chen Y-C, Shang Y, Cao A, Hu L, et al. MWCNT/V2O5 core/shell sponge for high areal capacity and power density Li-ion cathodes. ACS Nano 2012;6:7948–55.

[78] Gu W, Yushin G. Review of nanostructured carbon materials for electrochemical capacitor applications: Advantages and limitations of activated carbon, carbide-derived carbon, zeolite-templated carbon, carbon aerogels, carbon nanotubes, onion-like carbon, and graphene. Wiley Interdiscip Rev Energy Environ 2014;3:424–73.

[79] Lee SW, Kim J, Chen S, Hammond PT, Shao-Horn Y. Carbon nanotube/manganese oxide ultrathin film electrodes for electrochemical capacitors. ACS Nano 2010;4:3889–96. https://doi.org/10.1021/nn100681d.

[80] Akter T, Hu K, Lian K. Investigations of multilayer polyoxometalates-modified carbon nanotubes for electrochemical capacitors. Electrochim Acta 2011;56:4966–71. https:// doi.org/10.1016/j.electacta.2011.03.127.

[81] Kshetri T, Thanh TD, Singh SB, Kim NH, Lee JH. Hierarchical material of carbon nanotubes grown on carbon nanofibers for high performance electrochemical capacitor. Chem Eng J 2018;345:39–47. https://doi.org/10.1016/j.cej.2018.03.143.

[82] Li X, Zhitomirsky I. Electrodeposition of polypyrrole–carbon nanotube composites for electrochemical supercapacitors. J Power Sources 2013;221:49–56. https://doi. org/10.1016/j.jpowsour.2012.08.017.

[83] Zhao D-D, Yang Z, Kong ES-W, Xu C-L, Zhang Y-F. Carbon nanotube arrays supported manganese oxide and its application in electrochemical capacitors. J Solid State Electrochem 2011;15:1235–42.

[84] Geng X, Li L, Li F. Carbon nanotubes/activated carbon hybrid with ultrahigh surface area for electrochemical capacitors. Electrochim Acta 2015;168:25–31. https://doi.org/10.1016/j.electacta.2015.03.220.

[85] Deng L, Hao Z, Wang J, Zhu G, Kang L, Liu Z-H, et al. Preparation and capacitance of graphene/multiwall carbon nanotubes/MnO_2 hybrid material for high-performance asymmetrical electrochemical capacitor. Electrochim Acta 2013;89:191–8. https://doi.org/10.1016/j.electacta.2012.10.106.

[86] Qian X, Lv Y, Li W, Xia Y, Zhao D. Multiwall carbon nanotube@ mesoporous carbon with core-shell configuration: A well-designed composite-structure toward electrochemical capacitor application. J Mater Chem 2011;21:13025–31.

[87] Vignal T, Banet P, Pinault M, Lafourcade R, Descarpentries J, Darchy L, et al. Electropolymerized poly(3-methylthiophene) onto high density vertically aligned carbon nanotubes directly grown on aluminum substrate: Application to electrochemical capacitors. Electrochim Acta 2020;350:136377. https://doi.org/10.1016/j.electacta.2020.136377.

[88] Bavio MA, Acosta GG, Kessler T. Synthesis and characterization of polyaniline and polyaniline – Carbon nanotubes nanostructures for electrochemical supercapacitors. J Power Sources 2014;245:475–81. https://doi.org/10.1016/j.jpowsour.2013.06.119.

[89] Wang H, Kakade BA, Tamaki T, Yamaguchi T. Synthesis of 3D graphite oxide-exfoliated carbon nanotube carbon composite and its application as catalyst support for fuel cells. J Power Sources 2014;260:338–48. https://doi.org/10.1016/j.jpowsour.2014.03.014.

[90] Zhang W, Sherrell P, Minett AI, Razal JM, Chen J. Carbon nanotube architectures as catalyst supports for proton exchange membrane fuel cells. Energy Environ Sci 2010;3:1286–93.

[91] Thomas JE, Bonesi AR, Moreno MS, Visintin A, Castro Luna AM, Triaca WE. Carbon nanotubes as catalyst supports for ethanol oxidation. Int J Hydrogen Energy 2010;35:11681–6. https://doi.org/10.1016/j.ijhydene.2010.08.109.

[92] Wang W, Chu W, Wang N, Yang W, Jiang C. Mesoporous nickel catalyst supported on multi-walled carbon nanotubes for carbon dioxide methanation. Int J Hydrogen Energy 2016;41:967–75. https://doi.org/10.1016/j.ijhydene.2015.11.133.

[93] Xiong H, Motchelaho MAM, Moyo M, Jewell LL, Coville NJ. Correlating the preparation and performance of cobalt catalysts supported on carbon nanotubes and carbon spheres in the Fischer–Tropsch synthesis. J Catal 2011;278:26–40. https://doi.org/10.1016/j.jcat.2010.11.010.

[94] Morales-Acosta D, Ledesma-Garcia J, Godinez LA, Rodríguez HG, Álvarez-Contreras L, Arriaga LG. Development of Pd and Pd–Co catalysts supported on multi-walled carbon nanotubes for formic acid oxidation. J Power Sources 2010;195:461–5. https://doi.org/10.1016/j.jpowsour.2009.08.014.

[95] Guerra-Balcázar M, Cuevas-Muñiz FM, Castaneda F, Ortega R, Álvarez-Contreras L, Ledesma-García J, et al. Carbon nanotubes as catalyst support in a glucose microfluidic fuel cell in basic media. Electrochim Acta 2011;56:8758–62. https://doi.org/10.1016/j.electacta.2011.07.099.

[96] Jing S, Luo L, Yin S, Huang F, Jia Y, Wei Y, et al. Tungsten nitride decorated carbon nanotubes hybrid as efficient catalyst supports for oxygen reduction reaction. Appl Catal B Environ 2014;147:897–903. https://doi.org/10.1016/j.apcatb.2013.10.026.

3 Functionalization and Characterization of Carbon Nanotubes

Arash Yahyazadeh and Alivia Mukherjee

3.1 INTRODUCTION

In 1991, Iijima envisioned the existence of carbon nanotubes (CNTs) through a high-resolution transmission electron microscope (HR-TEM) and investigated the fundamental properties of the tubes [1]. According to a widely known definition, CNTs are hollow, graphitic materials including single-walled carbon nanotubes (SWCNT) and multi-walled carbon nanotubes (MWCNT). SWCNT have diameters between 1 and 10 nm, and MWCNT have diameters between 5 and 200 nm with the application of supercapacitors, transparent films, and nanocomposites [2]. Furthermore, an increase in the production of CNTs would permit fabrication processes of microelectronic circuits [3]. The undesirable material properties of CNTs such as the hydrophobicity and chemical inertness limits their application into new, functional electronic devices. Moreover, CNTs have been receiving the spotlight as a catalyst support, owing to their mesoporous structure, high electrical conductivity, and thermal stabilities.

Different forms of MWCNTs or SWCNTs tend to aggregate, and various functionalization approaches are promising means of overcoming this problem. Here, we provide an overview of the major developments concerning characterization techniques and functionalization of CNTs.

3.2 CHARACTERIZATION OF CNTS

Characterization of CNTs for detecting the properties of unfunctionalized and functionalized CNTs is of the utmost importance. In the following section, we highlight recent developments in the characterization of CNTs.

3.3 BULK CHARACTERIZATION

3.3.1 NUCLEAR MAGNETIC RESONANCE (NMR) SPECTROSCOPY

In a study by Shen et al., Polyethyleneimine (PEI)-modified CNTs were used for biomedical applications and characterized by NMR [4]. The amine group of PEI negatively charged MWCNTs and improved the biocompatibility of the materials. Shown in Figure 3.1 are the covalent linkages between MWCNTs and PEI, together

DOI: 10.1201/9781003376071-3

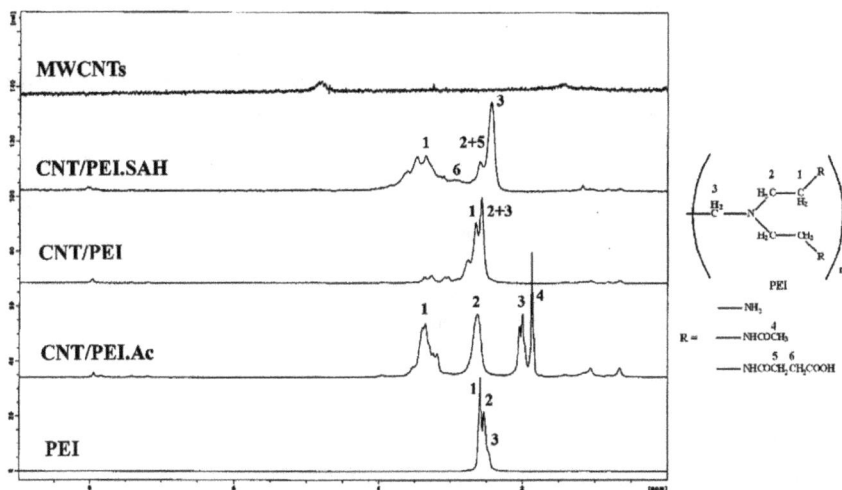

FIGURE 3.1 NMR spectroscopy of the PEI-modified MWCNTs. Reproduced with permission [4].

with the converting of the PEI amine groups to acetyl, confirmed by NMR spectroscopy. Castillo et al. [5] studied the functionalization of SWCNTs with chitosan and folic acid using NMR and confirmed that free folic acid diffused rapidly to SWCNTs. The signals at 1.90 and 6.37 ppm in the NMR spectrum were assigned to the protons of the methyl groups and the pteridine moiety proton from the folate, respectively. Table 3.1 summarizes some of the published works on solid-state NMR analysis of CNTs in the scientific literature.

3.4 FOURIER-TRANSFORM INFRARED (FTIR) AND RAMAN SPECTROSCOPY

Commonly, IR and FT-IR have been employed to evaluate surface modifications by employing different functional groups onto the CNT sidewalls [10]. An example of this application is the investigation of MWCNTs recovered after the separation of unleaded gasoline using FT-IR spectroscopy [11]. In fact, CNTs can remove various pollutants from water, owing to the aliphatic C-H and unsaturated C-H stretching in the IR spectrum. In a study by Hsan et al. [12], chitosan grafted MWCNTs (CSMWCNTs) as adsorbents for the capture and efficient catalysts for chemical fixation of CO_2. The presence of O-H and C-H bonds in f-MWCNTs indicates that pristine MWCNTs was successfully functionalized. Characteristic absorption bands at around 1739 cm^{-1} and 3404 cm^{-1} are attributed to carboxylic and hydroxyl groups, respectively (Figure 3.2).

Xin et al. [13] found that CNTs and polyethylene (PE) composites have synergistic influence on enhancing the high temperature performance of asphalt binder and can postpone the aging process. FT-IR analysis was employed to identify the chemical structure of asphalt binders, and it demonstrated the intensity of the absorption

TABLE 3.1

Details of Some Reported NMR Spectroscopy Analysis to Study the Structure and Properties of Functionalized CNT Nanocomposites

Sample	Application	Key Findings	References
CNT modified by acid	Synthesis of polysulfide copolymer nanocomposites.	Molecular weight can be calculated using NMR technique.	[6]
Alumina-based nanocomposites reinforced with CNT	The DWCNT-alumina nanocomposites improved toughening than the 5vol.% SWCNT-Al$_2$O$_3$ system.	NMR showed that the alumina remains 6-coordinated and no chemical bonding occurred at the CNT/alumina composite interface.	[7]
Covalently functionalized CNTs	Evaluate concentration, dispersion, and aggregation of functionalized CNTs in water solutions.	Low Field NMR showed that water molecules interact in three different ways with CNTs.	[8]
Polyethylene glycol-linked conjugate of CNTs with Mangiferin	The effective and safer delivery of phytochemicals to the brain cancerous cells.	NMR spectroscopy revealed the peak of methylene proton of polyethylene glycol at 3.42 ppm and the peak of tetrahydropyran at 4.92 ppm.	[9]

bands at 1601.3 cm^{-1} and 721.3 cm^{-1} for PE modified asphalt. Yin et al. [14] prepared a series of Nano-MgAl$_2$O$_4$ modified CNTs composite materials as efficient nanoadsorbent for perfluorooctanoic (PFOA) removal. It was found that surface-modified CNTs display a weak band in the 1300–1700 cm^{-1} range and confirm MgAl$_2$O$_4$ formation. Phosphate adsorption on Mg-loaded MWCNTs and characterization of adsorbent were investigated by Jiang et al. [15]. The FT-IR results confirmed that PO$_4^{3-}$ was adsorbed through reacting with magnesium oxide according to the peak at 3690 cm^{-1}. Besides, MWCNTs dispersed magnesium oxide into nanoflakes structures, and Mg@CNT nanocomposite obtained 87% of the equilibrium adsorption capacity within a day. The kinetic and isotherm results suggested that the presence of sulfate and carbonate reduced the phosphate adsorption capacity. In another study, CuZn/CNTs catalysts were employed in methanol steam reforming for H$_2$ production [16]. Furthermore, based on FT-IR spectra, the peaks at 1710 and 1195 cm^{-1} are related to C-O and C=O stretching vibrations, which improved dispersion of the treated CNTs in water. Also, according to the results, the perfect performance of the promoted CuZn/CNTs catalysts leads to considering CNTs as a proper catalyst support. The use of Raman spectroscopy for the characterization of advanced materials and devices such as CNTs has been presented [17]. Raman spectroscopy is a non-contact and non-invasive technique with a high energy resolution to serve as the relevant metric of a CNT-based sensor. As shown in Fig 3.3, Raman spectroscopy has now become a technique to investigate the physicochemical characterization of carbon materials. Briefly, the Raman spectra for CNT displayed two bands at 1345 and

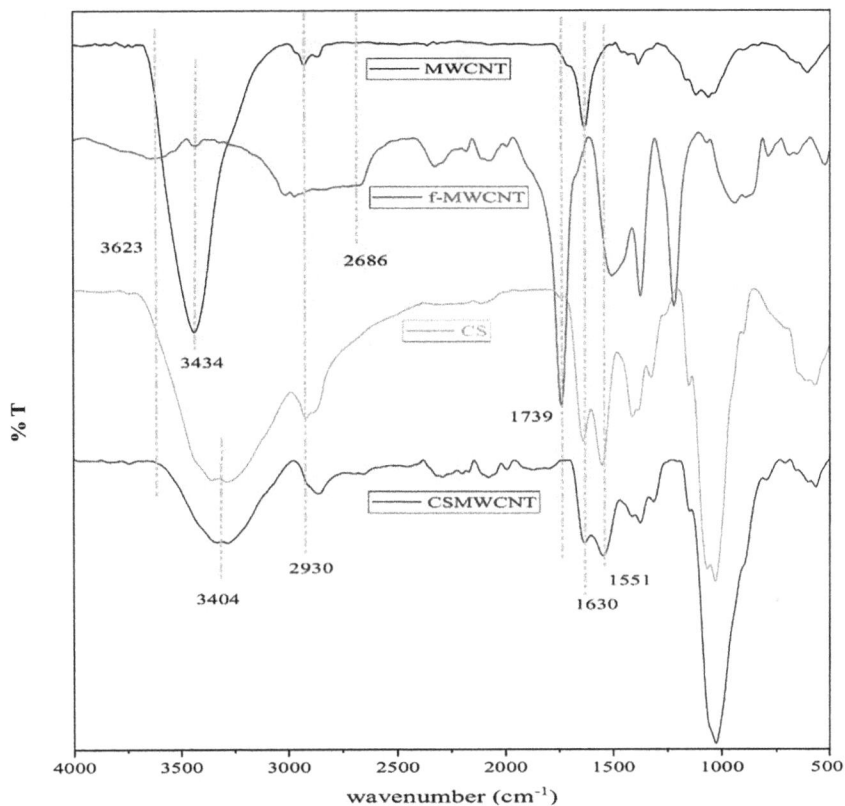

FIGURE 3.2 FTIR spectrum of fresh MWCNTs, functionalized MWCNTs (f-MWCNTs), Chitosan (CS), and CSMWCNTs. Reproduced with permission [12].

1576 cm^{-1} corresponding to amorphous carbon layers and the graphitic structures, respectively [18] (Figure 3.3).

The structure and surface characterization of CNTs and graphene nanoflakes (GNFs) through Raman spectroscopy have been investigated by Chernyal et al. [19]. It was found that the strong broadening of the D-band at ~ 1350 cm^{-1} in the structure of GNFs makes it close to that of soot and coal (Figure 3.4).

Akbarzadeh et al. [20] applied highly effective polyaniline-grafted CNTs to enhance active protective functioning in a silane coating. It was reported that the intensity of the Raman appeared at 2684 cm^{-1} related to the number of graphene sheet that comprise CNTs surfaces. A summary of Raman spectroscopy characterization applied for the determination of CNTs is presented in Table 3.2.

3.5 SURFACE CHARACTERIZATION

3.5.1 X-ray Photoelectron Spectroscopy (XPS)

XPS method is employed to understand chemical state information and bonding nature of dopant for some kinds of carbon materials. Basiuk et al. [26] reported

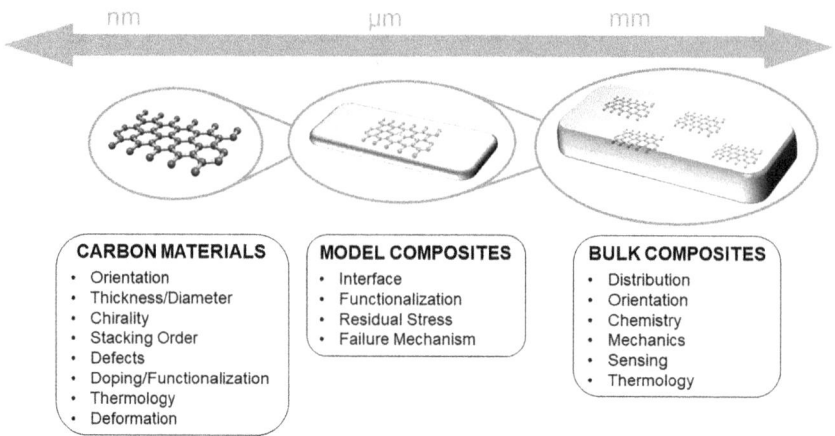

FIGURE 3.3 Applications of Raman spectroscopy to characterize carbon composites. Reproduced with permission [18].

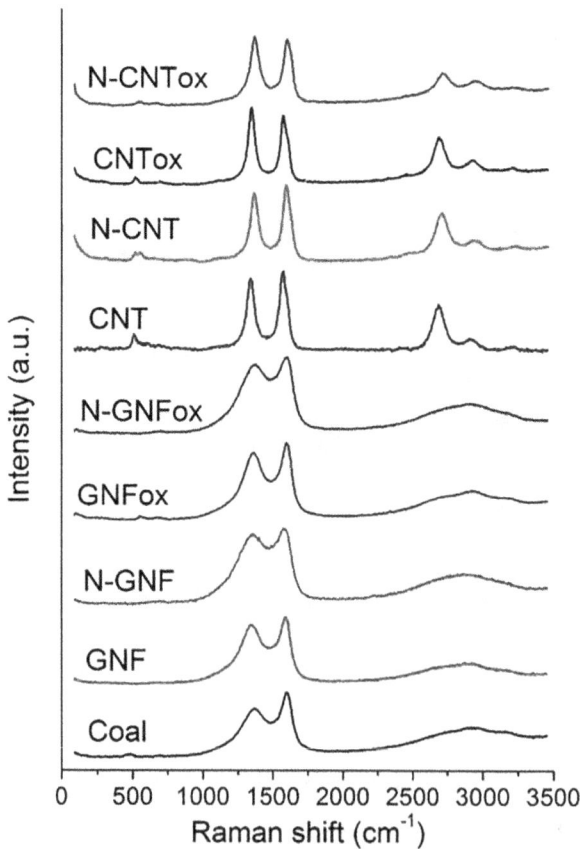

FIGURE 3.4 The Raman bands of CNTs and GNFs, together with oxidized and N-doped derivatives. Reproduced with permission [19].

TABLE 3.2

Review of Raman Scattering Spectroscopy for CNTs Characterization

CNT Type (Synthesis Method)	Sample	Analytical Characteristics	Outline	References
MWCNTs (CVD)	MWCNTs decorated with silver nanoparticles	Raman Spectroscopy using Laser spot size from 1 to 300 mm and optimized spectral range of 400–1100 nm	The deposition of silver in the MWCNTs leads to structural defects in the carbon wall and changing the intensity of the bands	[21]
MWCNTs (-)	Poly (lactic acid)/ CNTs nanocomposite	Raman microscope spectrometer using 514 nm wavelength edge laser physics	The presence of MWCNTs in the composites leads to higher frequencies, especially for the G band	[22]
MWCNTs (CVD)	CNT reinforced magnesium matrix composites	Raman Spectroscopy using spectral range of 0–3750 cm^{-1}	The Raman spectra analysis confirmed that the dimeters of the CNTs were between 5–15 nm	[23]
MWCNTs (-)	CNTs/polylactic acid (PLA) composites	Raman spectra using 633 nm excitation line via microscope operating with an objective	The content of CNTs in the composite affects I_D/I_G intensity ratio due to the degree of crystallinity	[24]
MWCNTs (CVD)	MWCNTs for removal of 4-tert-octylphenol	Raman microscope using spectral range of 1000–3500 cm^{-1}	Incomplete transformation of hydrocarbons is responsible for D+D' peaks	[25]

experimental investigation of SWNT noncovalently functionalized with 3d metal phthalocyanines using XPS. The deconvolution results of XPS spectroscopy exhibit shake-up and multiple splitting corresponding to the incorporation of nitrogen atoms in the crystalline surfaces. As shown in Figure 3.5, XPS high resolution spectra reveal multiple peaks near the signal, owing to the Me 2p (where Me = Co, Ni, Cu and Zn) core level. In line with the results of DFT calculations, the increase of spectral deconvolution for N 1s and Me 2p peaks is related to phthalocyanine deposition onto SWNTs.

CNTs, due to their hollow structure and strong adsorption capacity, can be widely used to remove different contaminants. Gu et al. [27] investigated phosphate adsorption from solution using a highly effective Zr-CNTs adsorbent. The XPS spectral measurements of Zr-CNTs confirmed that phosphate had been adsorbed on the surface of MWCNTs and formed chemical complexation between zirconium and phosphate. This is also proved by the asymmetric O 1s spectrum and the increase in the

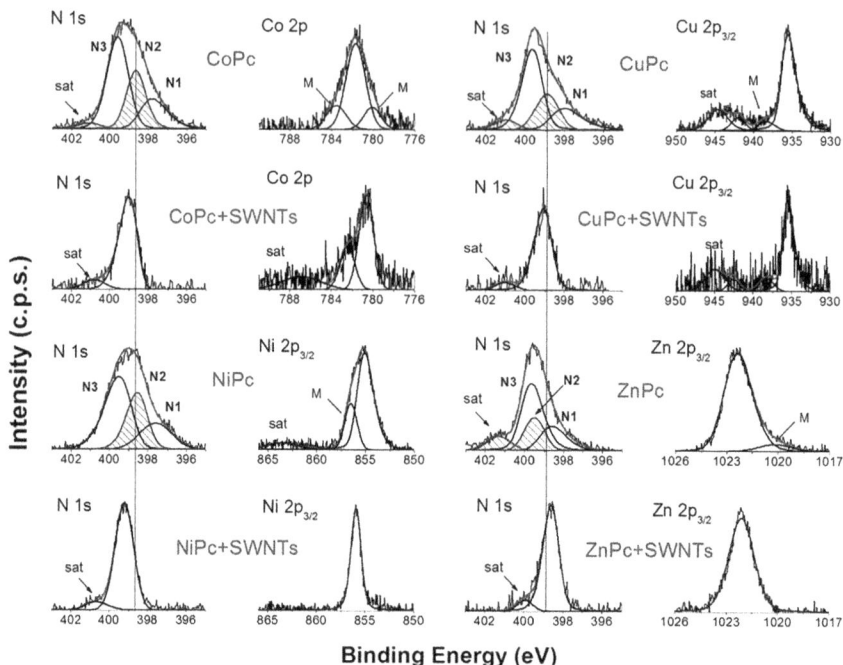

FIGURE 3.5 XPS high resolution of the transition metal spectra corresponding N 1s core levels. Reproduced with permission [26].

quantity of -OH from 49.3 to 68.7%. According to another study, CNTs was anchored onto chitosan sponge to adsorb fluoride from fertilizer industry effluent [28]. A summary of XPS spectroscopic technique employed to evaluate surface functionalization of the nanotube has been given in Table 3.3.

3.6 CHEMISTRY TITRATION

According to the method suggested by Chen et al. [35], chemical titration shows the content of the corresponding functional groups on the CNT samples. The surface of CNTs was functionalized using different oxidants and used in the preparation of Pd-Pt/CNTs catalysts. The results show that the surface of CNT-KMnO$_4$ is the most hydrophilic and possesses the highest absorption peaks from IR spectra. Furthermore, Kim et al. [36] employed one-pot titration for the simple elucidation of CNTs surfaces, and potentially, replaced the other carbon materials. The titration method provided the concentrations of surface acidic groups on functionalized CNTs based on the modified Henderson-Hasselbalch (H-H) equation. It was concluded that the one-pot titration methodology could separate the boundary between the indirect and direct titration methods for highly oxidized CNT or graphene oxides. Zhang and coworkers [37] have determined characterization of functionalized MWCNTs

TABLE 3.3

Structure, Purity, and Chemical Reactivity of CNTs by XPS

CNT Type (Synthesis Method)	Sample	Analytical Characteristics	Outline	References
CNTs ()	CNTs/ferrihydrite catalyst (CNTs/Fh)	C1 s peak with bind energy of 284.80 eV was employed as calibration standard	XPS results showed that rapid transport of electrons from iron during the surface reactions is negligible and CNTs can increase the electron transfer.	[29]
CNTs ()	Magnetite decorated CNTs/reduced graphene oxide nanosheets	XPS proved the formation of Fe_3O_4, NH_2-Fe_3O_4, CNTs, CNTs-COOH, and Fe_3O_4-CNTs	XPS showed that the successfully aggregation of Fe_3O_3 onto the surface of CNTs.	[30]
MWCNTs ()	Three dimensional porous MXene/CNTs microspheres (C-MCM)	XPS spectroscopy was employed to analyze the chemical structure of MXene/C-CNTs microspheres (MCM)	High-resolution XPS for the C 1s spectrum showed four peaks at 282.7, 284, 285.3, and 288.4 eV were assigned to C-Ti, C-C, C-O and O-C=O, respectively.	[31]
CNTs (CVD)	Ruthenium oxide nanorods prepared CNTs grown carbon cloth (RuO_2-CNTs-CC)	XPS characterizations reveals metal distributions of the RuO_2-CNTs-CC	The deconvoluted XPS spectra of C 1s confirms the presence of noticeable fragility in the CNT structure.	[32]
SWCNTs	Biosensor based on CNTs field-effect transistor (CNT-FET)	XPS was employed to confirm the absence or presence of 1-pyrenebutanoic acid succinimidyl ester (PBASE) on the CNT surface	XPS results suggest the PBASE is immobilized on the surface of CNTs.	[33]
Commercialized CNTs	Ba-Ru/CNTs catalyst for ammonia synthesis	XPS technique was employed to shed light on the role of CNTs support and the electron status of the Ru catalyst	XPS results reveal that Ru is electron deficient for CNT support and the easier reduction behavior of defective support.	[34]

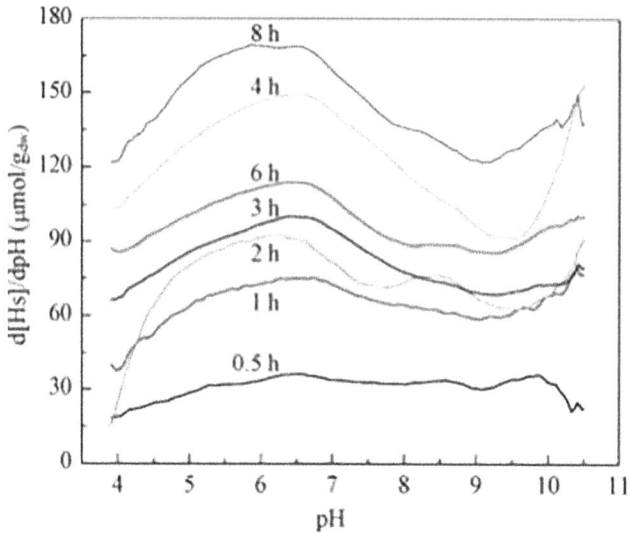

FIGURE 3.6 Titration curves of different chemical oxidants on the surface of CNT. Reproduced with permission [37].

by potentiometric proton titration. Furthermore, NEXAFS edge spectra has been used for the BrØnsted sites over the surface of CNT. It is worth noting that the acid-base titration curves of the treated MWCNT show five major peaks and have higher concentration of carboxyl groups. As shown in Figure 3.6, the titration tests of the functionalized CNTs using various oxidants exhibited different behavior.

Qi et al. [38] found encouraging results for using a chemical titration method to measure active oxygen functional groups on the surface of CNTs. It was found that the number of active sites determined by chemical titration with PH can also be measured through XPS or GC concentration analysis. In another study, a simple, recoverable titration method was used for determination of amine-functionalized CNTs [39]. Titration method confirmed the effects of CO_2 absorption and the dilution of titrant on amine-functionalized CNTs. In a study by Vennerberg et al. [40], MWCNTs were studied in terms of oxidation behavior using an optimized acid-base (Boehm) titration method. Chemical titration was subjected to different O_3 treatment times, and the results suggest that hydroxyl groups act as an intermediate species. Titration measurements of the acidic functional groups indicate that fluidization of the MWCNTs in ozone produces homogeneous apparent oxidation. Bortolamiol et al. [41] reported a quantitative investigation of the modifications induced in double-walled CNTs (DWNTs) by chemical titrations. When oxidation purification is employed to the DWNTs, the titration of the acidic functions is carried out under a dilute hydrochloric acid and an excess of NaOH. The results indicated that surface functionalization reduced by 60% after washing with 0.01 M NaOH.

3.7 TEMPERATURE-PROGRAMMED DESORPTION (TPD) AND REDUCTION (TPR)

The reducibility and basicity of the CNTs supported catalysts can be characterized by H_2-TPR and CO_2-TPD analysis and correlated with the catalytic activity [42]. Ronero-Saez et al. [43] employed nickel-ZrO_2 catalyst supported on CNTs for catalytic performance during CO_2 methanation, using sequential and co-impregnation methods. H_2-TPR analysis suggested that the presence of zirconia effectively promoted the reduction of NiO. On the other hand, in the catalyst synthesized by co-impregnation, nickel was mostly blocked by zirconia and not accessible to hydrogen in the catalyst surface. Furthermore, CO_2-TPD measurements of the CNT supported catalysts showed a greater CO_2 adsorption for the catalyst prepared by sequential impregnation.

Zhang et al. [44] used in situ, supported MnO_x-CeO_x on CNTs for the catalytic reduction of nitric oxide using ammonia. The reducibility results confirmed that there is a strong interaction between MnO_x, CeO_x, and CNTs, which can affect the activity of the samples in the low-temperature, selective catalytic performance. NH_3-TPD has been employed to measure the surface acid amount, suggesting that the amount of BrØnsted acid sites is intensified.

Saleh et al. [45] showed CNTs as a co-support with alumina for MoNi catalysts for their performance in the hydrodesulfurization (HDS) of dibenzothiophene. Figure 3.7 shows NH_3-TPD and H_2-TPR techniques of both the AlMoNi and AlCNTMoNi catalysts, which confirms the acidity sites and the nanoparticle interaction along with support materials.

A summary of the techniques, including TPR and TPD, which were used for characterization of CNTs and CNT-related materials, is presented in Table 3.4.

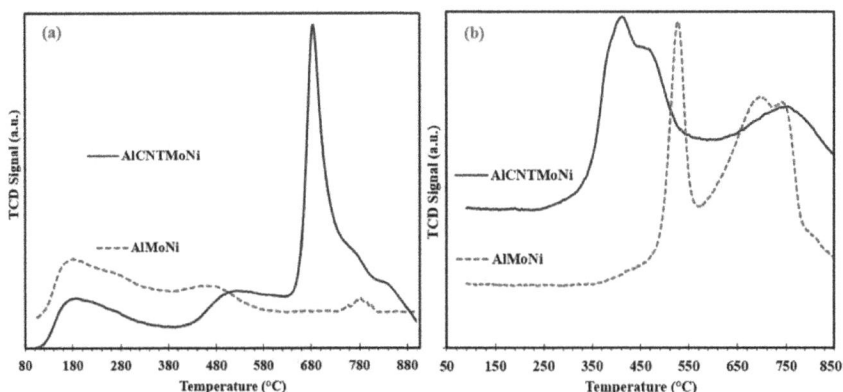

FIGURE 3.7 NH_3-TPD (a) and H_2-TPR (b) characterization techniques for the AlMoNi and AlCNTMoNi catalysts. Reproduced with permission [45].

TABLE 3.4

TPD-TPR Techniques Used for Nanotube Characterization

Technique	CNT Type	Sample/Application	Analytical Characteristics	Outlines	References
NH_3-TPD	MWCNTs	CNTs templated modified with Co and Ni/Methanol to light olefins (MTO) reaction	NH_3-TPD equipped with a thermal conductivity detector (TCD) was conducted in a flow of He by increasing temperature from 100 to 900 °C	NH_3-TPD profiles indicate the presence of acid sites with two different strengths.	[46]
H_2-TPR	MWCNTs	Pd doped CNTs/Hydrogen storage material	H_2-TPR tests were carried out up to 880 °C	TPR experiments indicated that H_2 uptake capacity was a function of annealing temperature.	[47]
NH_3-TPD	MWCNTs	Cobalt loaded CNTs catalyst/Fischer-Tropsch synthesis	NH_3-TPD experiments were conducted using 5%NH_3/95%He with a flowing rate of 20mL/min	The NH_3-TPD patterns show the decomposition peaks at 252 °C, 440 °C and 695 °C respectively, which might be assigned to the oxygen-containing groups on the CNTs.	[48]
H_2-TPR	MWCNTs	V_2O_5-CeO_x/TiO_2-CNTs catalysts were prepared by sol-gel method/ Reduction of NO_x with NH_3	H_2-TPR experiment was conducted at a heating rate of 10 °C/min using a stream of 5% hydrogen in argon	H_2-TPR profiles confirm that the large number of chemisorbed oxygen introduced via the addition of CeO_x	[49]
NH_3-TPD	MWCNTs	CNTs supported CeO_2 catalysts coated with SiO_2/Catalytic cracking of naphtha	NH_3-TPD measurements equipped with TCD detector and were exposed to ammonia for 60 min	The NH_3-TPD profiles for the catalysts confirm the decomposition of NH_4^+ ions formed over strong BrØnsted sites.	[50]
CH_4-TPD	CNTs	CeO_2 supported Co, Fe, and Ni catalysts/Catalytic decomposition of methane (CDM)	CH_4-TPD analysis used nitrogen at a ramping rate of 5 °C/min at temperatures of 50–900 °C	The TPD profiles of methane for CDM catalysts confirm the possible oxidation of the surface carbon by the residual oxygen on the catalysts	[51]

3.8   GAS ADSORPTION ISOTHERM

The nitrogen adsorption isotherms of different purities of CNTs were employed for the characterization, including the pore structure and pore size distributions (PSDs) [52]. Oktaviano et al. [53] proposed the nitrogen adsorption isotherm as fast way to determine carbonaceous deposits and amorphous carbon showing low graphitization structures. It was found that purified MWCNTs and drilled structure MWCNTs (DMWCNTs) are similar in type IV, indicating the mesoporous characteristic of both samples. Additionally, the pore size distribution analysis determined by the BJH method confirmed that fresh MWCNTs have different pore sizes, compared to purified MWCNTs and DMWCNTs, due to amorphous carbon layers. In another study, Sawant et al. [54] explored effect of in-situ boron doped CNTs (BCNTs) for hydrogen adsorption. Based on the nitrogen adsorption-desorption isotherms, the specific surface area and pore volume were calculated 901 m^2g^{-1} and 0.198 ccg^{-1}, respectively. It can be concluded that H_2 adsorption on BCNTs follows pseudo second-order kinetics. Table 3.5 reveals the detailed description of the published literature on structural characterization of CNTs for the adsorption of different compounds.

3.9   ELECTRON ENERGY LOSS SPECTROSCOPY (EELS) AND NEAR-EDGE X-RAY ABSORPTION FINE STRUCTURE (NEXAFS)

The NEXAFS in three modes, including Auger electron yield (AEY), partial electron yield (PEY), and total electron yield (TEY), provides electronic and structural information about chemical functionalities [60]. Sivkov et al. [61] reported that the presence of CO and water-dissociation products on the surface of carbon materials affects the electronic structure of all samples, which was confirmed by NEXAFS and XPS. The NEXAFS spectroscopy of MWCNTs in energy region 286–291 eV exhibited additional structures aligned with the transitions of the C 1s electrons to the C-O, C-O-C, and C=O surface functional groups. Combination of x-ray photoelectron (XP) and NEXAFS spectroscopies to measure the covalent functionalization of SWCNTs with nitrene was studied by Nickl et al. [62]. According to NEXAFS analysis, all C K-edge spectra have resonance features for C 1s π^* (C=C) between 284 and 285 eV. It was found that the distinguishing characteristic at 288.0 eV can correspond to C=N transition, and it can be employed to compare nondestructive covalent functionalization for various nanomaterials. MWCNTs surface was functionalized by vacuum-ultraviolet photochemical reactions, and comparative NEXAFS spectra suggests corrections due to chemical reactions (Figure 3.8) [63].

It has been found that charge-transfer behavior across carbon nanotube-quantum dot (QD) heterostructures strongly depends on QD sizes and QD coverage densities [64]. More specially, NEXAFS and Raman spectroscopies, coupled with electrical transport measurements, have been employed to complement the optical data. In another study, Winter et al. [65] prepared poly(dimethylsiloxane)/poly(methyl methacrylate)/MWCNT composites and investigated effects of adding nanofillers of different diameters. Additionally, NEXAFS spectroscopy results confirm phase separation within the bulk of the micrometer-sized graphitic balls and enhance CNT-polymer interactions. Moreover, nanostructured manganese oxide supported

TABLE 3.5
Summary of CNTs Characterization to Study the Textural Properties

Sample	Textural Properties S_{BET}: Specific Surface Area (m²/g) V_{total}: Total Pore Volume (cm³/g)	Applications	Remarks	References
MWCNTs impregnated with polyethyleneimine (MWCNT/PEI)	S_{BET} =146; V_{total} = 1.586	CO_2 capture	The CO_2 adsorption isotherms suggest that interaction between the adsorbents and CO_2 molecules is chemisorption.	[55]
Mg-metal-organic framework (MOF)-74-carbon composites	S_{BET} =1660–1720; V_{total} = 0.49	CO_2 adsorption	The introduction of a small fraction of CNT and graphene oxide (GO) resulted in additional microporosity inside MOFs and improved the uptake of CO_2 adsorption.	[56]
Functionalized MWCNTs	S_{BET} =100.93–119.04; V_{total} = 1.59–27.35	CO_2 adsorption	The attachment of oxygen functionalities leads to increase in CO_2 uptake capacity from 0.0874 to 0.3274 mmol/g.	[57]
Activated MWCNTs (A-MWCNTs)	S_{BET} =327–495; V_{total} = 0.881–1.186	The hydrogen storage capacity	Chemical activation leads to the formation of mesopores and increase the hydrogen storage capacity of the A-MWCNTs.	[58]
Amino modified MWCNT sorbents	S_{BET} =74–121; V_{total} = 0.34–0.63	CO_2 adsorption	Functionalities highly influence the CO_2 adsorption-separation properties of the MWCNT samples.	[59]

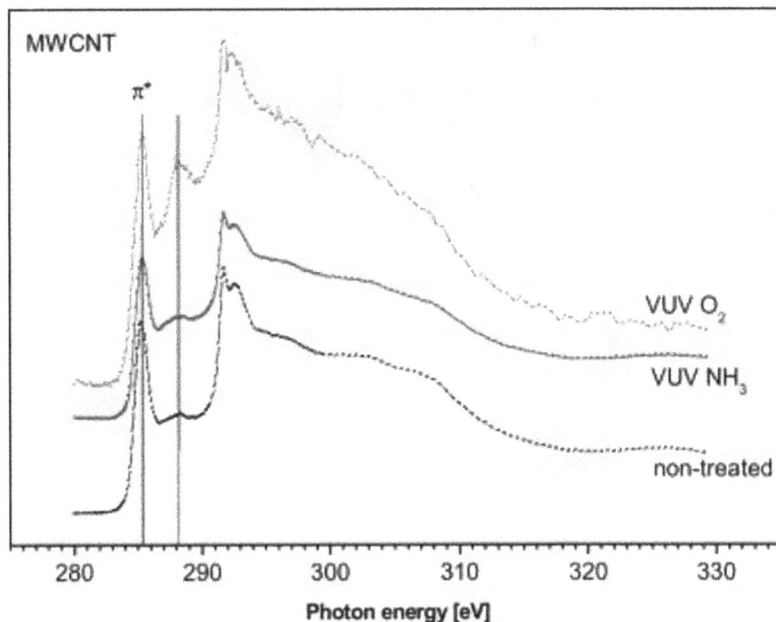

FIGURE 3.8 Comparative NEXAFS spectra of pristine and treated MWCNTs in either NH_3 or O_2. Reproduced with permission [63].

on CNTs using two typical preparation techniques has been considered for efficient water splitting [66]. In situ NEXAFS spectroscopy in the total electron yield mode determines the change in the oxidation state of manganese in the electrode material with variation of the pre-treatment temperature.

3.10 IMAGING METHODS

Transmission electron microscopy (TEM) and scanning electron microscopy (SEM) methods have traditionally been reported for detailed CNT characterization. For instance, Azzam et al. [67] used TEM and field emission-SEM (FE-SEM) using energy dispersive X-ray spectroscopy (EDX) to assess the morphologies of the prepared CNTs and fabricated nanocomposites (Figure 3.9). It was found that the basic structure of the prepared CNTs seems to be bamboo-like, and the crystal structure of the distributed AgNPs cannot be easily detected.

Say et al. [23] reported the strength of the CNT-reinforced magnesium matrix composites prepared by the powder metallurgy (PM) technique. Microstructural observations were performed by SEM-TEM, and the results showed CNTs with a range of diameters from 15 to 20 nm with a few μm lengths. It can be concluded that the CNT ropes ratio in the matrix was raised, increasing the CNT content. A summary of imaging techniques which have been used for characterization of CNTs and CNT-related materials is presented in Table 3.6.

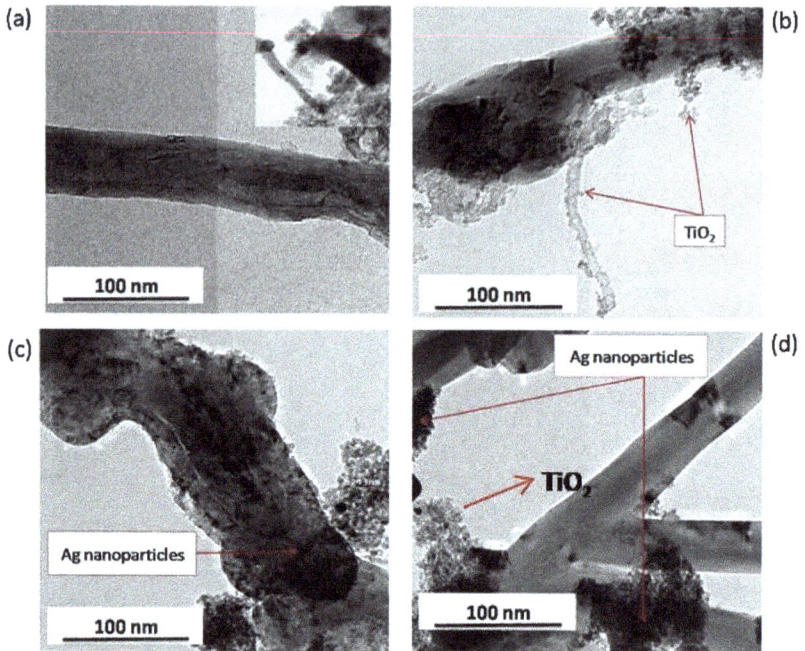

FIGURE 3.9 TEM photographs of (a) CNTs, (b) TiO$_2$@CNTs, (c) TiO$_2$@CNTs/AgNPs, and (d) TiO$_2$@CNTs/AgNPs/Cationic surfactant (C10) nanocomposites. Reproduced with permission [67].

TABLE 3.6

Summary of the SEM/TEM Micrographs Application for Characterization of CNTs

Characterization Techniques	Used for Studying	Notes	References
SEM	The morphology of nanomaterials	The addition of CNT into the nanocomposites leads to better dispersion.	[68]
TEM	Morphological characteristics of 3D printed thermoelectric polyurethane (TPU)/ MWCNTs filaments	The TEM images reveal MWCNTs have been sufficiently disentangled.	[69]
SEM combined energy dispersive X-ray analysis (EDAX)	The surface morphology of Gd@ZnO-MWCNTs nanocomposite	SEM images indicating the well dispersed ZnO nanoparticles on the cylindrical surface of MWCNTs.	[70]
High resolution TEM (HRTEM)	The morphology of the fresh catalysts and the CNTs on the surfaces of the spent catalysts	The TEM image confirms the tip growth mechanism of the CNTs with inner dimeters around 5 nm.	[71]
FE-SEM	The morphologies of ZnO/ CNT nanocomposites	SEM images indicating the diameters of CNTs and ZnO nanoparticles are 40–50 nm and 50 nm, respectively.	[72]

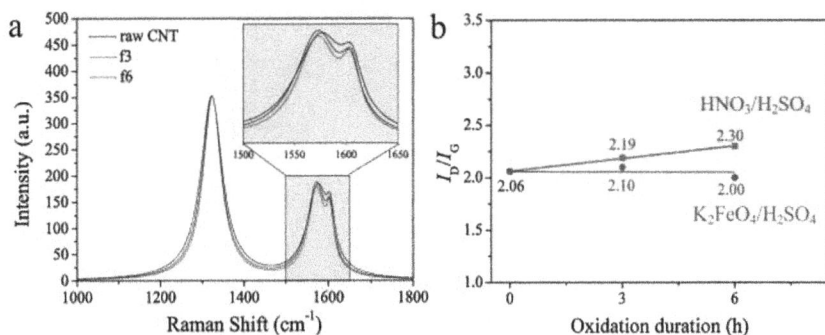

FIGURE 3.10 (a) The Raman bands of fresh CNTs and acid-treated CNTs; (b) effect of HNO_3/H_2SO_4 and K_2FeO_4/H_2SO_4 oxidations. Reproduced with permission [75].

3.11 CNTS FUNCTIONALIZATION

3.11.1 COVALENT FUNCTIONALIZATION

Covalent functionalization of CNTs that allows the introduction of different functional groups investigated by Mann et al. [73]. The authors showed the covalent functionalization of CNTs with peptides and proteins to tailor both the nanotube's photophysical properties together with their surface property. Covalent functionalization leads to increased solubility in aqueous environments, which opens up a realistic chance for applications in (bio)photonics, biosensing, and biomedicine. Li and collaborators explored surface charge and cellular processing of covalently functionalized MWCNTs in a predictive toxicological model and concluded that surface charge plays a vital role in the structure-activity relationships [74]. FTIR spectra of functionalized MWCNTs confirmed the presence of carboxylic groups at 1730 cm^{-1} as well as O-H stretching vibrations at 3430 cm^{-1}. According to Zhang et al. [75], nondestructive covalent CNT functionalization, carried out using selective oxidation of the original, defects with K_2FeO_4. As shown in Figure 3.10, the remarkable increase in the relative intensity of flaw site to graphite band is assumed as a signal for successful covalent functionalization.

Later, Rebelo et al. [76] investigated a comparative Raman study of covalently functionalized MWCNTs by diazonium chemistry and oxidation reactions. It was observed that covalent functionalization results in the formation of smaller sp^2 domains and can introduce core-shell structures. The resulting Raman spectra of original MWCNTs showed an intense and broad D band, in comparison to oxidized materials, depending on the aniline used.

3.12 NON-COVALENT FUNCTIONALIZATION

3.12.1 POLYMER FUNCTIONALIZED CNTs

Non-covalent surface modification of CNTs with polymers enhances the bio-affinity of CNTs and preserves their intrinsic properties. Making composite materials from conjugated polymers and CNTs can be achieved by establishing π-π interactions.

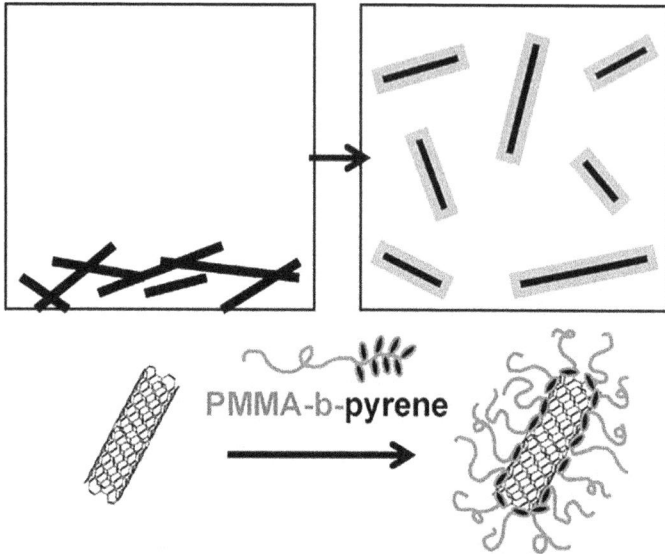

FIGURE 3.11   Block copolymers benefit, compared to symmetric random copolymers for non-covalent CNT functionalization. Reproduced with permission [79].

Polymers such as cellulose derivatives provide three-dimensional nanostructures [77]. Ponnamma et al. [78] non-covalently functionalized MWCNTs with surfactants of different charges to study homogenous dispersion of surfactant/MWCNTs in aqueous medium. Raman spectra of the samples indicates the de-bundling of CNTs and maximum level of dispersion for natural rubber latex (NRL)-surfactant/CNT nanocomposites. Based on all the results, anionic sodium dodecyl sulphate (SDS) improves the rate of MWCNT dispersion through electrostatic interaction. Meuer et al. [79] reported pyrene containing deblock copolymers for the non-covalent interaction of CNTs. As indicated in Figure 3.11, block copolymers with a short anchor are more efficient in dispersing CNTs copolymer.

3.13   SURFACE OXIDATION

3.13.1   FUNCTIONALIZE CNT SURFACES WITH METAL OR METAL OXIDE NANOPARTICLES

CNTs generally present p-type properties at ambient pressure, and the adsorption of reducing molecules renders CNTs ideal candidates in the gas-sensing field. Metal functionalization of the CNT surface can be achieved through different processes such as thermal evaporation, suppering, and electrochemical performance [80]. Tessonnier et al. [81] employed a novel and efficient method for the deposition of fine particles inside or outside of MWCNTs. Selective synthesis of Ni nanoparticles inside CNTs occurred due to the difference in the interface energies of organic-inorganic solutions with the CNT surface (Figure 3.12). It was shown that nickel functionalization of the CNT layer makes it possible to synthesize new materials for catalysis,

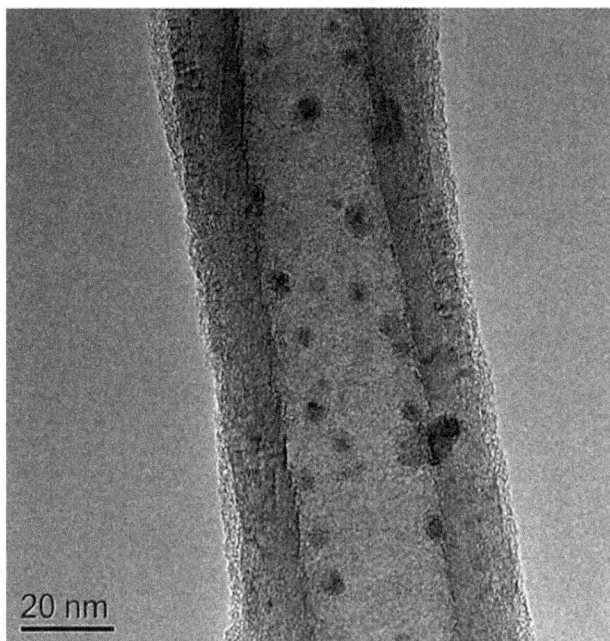

FIGURE 3.12   TEM image of the Ni nanoparticles penetrated inside the CNT. Reproduced with permission [83].

magnetism, and electronics. Moreover, the synthesis of metal (Co, Ru, Pd) catalysts supported on CNTs using the selective hydrogenation of glycerol has been investigated [82]. It was found that surface molecular reaction is an important factor to consider in improving the filling yield of CNTs. The catalytic performances of the Ru, Co, and Pd supported catalysts confirmed the remarkable influence of the confinement in CNTs support.

3.14   CONCLUDING REMARKS

CNTs are being extensively investigated by different analytical techniques to study changes in the CNTs' surface chemistry due to functionalization. The functionalized CNTs show the covalent and the non-covalent adsorption of functional groups, which lead to the practical applications of CNTs in many sectors. Also, the development of CNT-polymer composites due to unique intrinsic properties has been considered.

REFERENCES

[1] A.C. Dupuis, The catalyst in the CCVD of carbon nanotubes – a review, Prog. Mater. Sci. 50 (2005) 929–961. https://doi.org/10.1016/J.PMATSCI.2005.04.003.
[2] C. Herrero-Latorre, J. Álvarez-Méndez, J. Barciela-García, S. García-Martín, R.M. Peña-Crecente, Characterization of carbon nanotubes and analytical methods for their determination in environmental and biological samples: A review, Anal. Chim. Acta. 853 (2015) 77–94. https://doi.org/10.1016/J.ACA.2014.10.008.

[3] A. Yahyazadeh, B. Khoshandam, R.V. Kumar, An investigation into the role of substrates in the physical and electrochemical properties of carbon nanotubes prepared by chemical vapor deposition, Phys. B Condens. Matter. 562 (2019) 42–54. https://doi.org/10.1016/J.PHYSB.2019.03.010.

[4] M. Shen, S.H. Wang, X. Shi, X. Chen, Q. Huang, E.J. Petersen, R.A. Pinto, J.R.J. Baker, W.J.J. Weber, Polyethyleneimine-mediated functionalization of multiwalled carbon nanotubes: Synthesis, characterization, and in vitro toxicity assay, J. Phys. Chem. C. 113 (2009) 3150–3156. https://doi.org/10.1021/jp809323e.

[5] J.J. Castillo, M.H. Torres, D.R. Molina, J. Castillo-León, W.E. Svendsen, P. Escobar, F. Martínez O., Monitoring the functionalization of single-walled carbon nanotubes with chitosan and folic acid by two-dimensional diffusion-ordered NMR spectroscopy, Carbon N. Y. 50 (2012) 2691–2697. https://doi.org/10.1016/j.carbon.2012.02.010.

[6] M. Sheydaei, E. Alinia-Ahandani, Synthesis and characterization of methylene-xylene-based polysulfide block-copolymer/carbon nanotube nanocomposites via in situ polymerization method, J. Sulfur Chem. 41 (2020) 421–434.

[7] K.E. Thomson, D. Jiang, W. Yao, R.O. Ritchie, A.K. Mukherjee, Characterization and mechanical testing of alumina-based nanocomposites reinforced with niobium and/or carbon nanotubes fabricated by spark plasma sintering, Acta Mater. 60 (2012) 622–632. https://doi.org/10.1016/j.actamat.2011.10.002.

[8] D. Porrelli, M. Cok, M. Abrami, S. Bosi, M. Prato, M. Grassi, S. Paoletti, I. Donati, Evaluation of concentration and dispersion of functionalized carbon nanotubes in aqueous media by means of Low Field Nuclear Magnetic Resonance, Carbon N. Y. 113 (2017) 387–394. https://doi.org/10.1016/j.carbon.2016.11.025.

[9] P.J. Harsha, N. Thotakura, M. Kumar, S. Sharma, A. Mittal, R.K. Khurana, B. Singh, P. Negi, K. Raza, A novel PEGylated carbon nanotube conjugated mangiferin: An explorative nanomedicine for brain cancer cells, J. Drug Deliv. Sci. Technol. 53 (2019) 101186. https://doi.org/10.1016/j.jddst.2019.101186.

[10] A. Yahyazadeh, B. Khoshandam, Carbon nanotube synthesis via the catalytic chemical vapor deposition of methane in the presence of iron, molybdenum, and iron–molybdenum alloy thin layer catalysts, Results Phys. 7 (2017) 3826–3837. https://doi.org/10.1016/J.RINP.2017.10.001.

[11] D. Lico, D. Vuono, C. Siciliano, J.B. Nagy, P. De Luca, Removal of unleaded gasoline from water by multi-walled carbon nanotubes, J. Environ. Manage. 237 (2019) 636–643. https://doi.org/10.1016/j.jenvman.2019.02.062.

[12] N. Hsan, P.K. Dutta, S. Kumar, N. Das, J. Koh, Capture and chemical fixation of carbon dioxide by chitosan grafted multi-walled carbon nanotubes, J. CO2 Util. 41 (2020) 101237. https://doi.org/10.1016/j.jcou.2020.101237.

[13] X. Xin, Z. Yao, J. Shi, M. Liang, H. Jiang, J. Zhang, X. Zhang, K. Yao, Rheological properties, microstructure and aging resistance of asphalt modified with CNTs/PE composites, Constr. Build. Mater. 262 (2020) 120100. https://doi.org/10.1016/j.conbuildmat.2020.120100.

[14] S. Yin, J.F. López, J.J.C. Solís, M.S. Wong, D. Villagrán, Enhanced adsorption of PFOA with nano MgAl2O4@CNTs: Influence of pH and dosage, and environmental conditions, J. Hazard. Mater. Adv. 9 (2023) 100252. https://doi.org/10.1016/j.hazadv.2023.100252.

[15] S. Jiang, J. Wang, S. Qiao, J. Zhou, Phosphate recovery from aqueous solution through adsorption by magnesium modified multi-walled carbon nanotubes, Sci. Total Environ. 796 (2021) 148907. https://doi.org/10.1016/j.scitotenv.2021.148907.

[16] H. Shahsavar, M. Taghizadeh, A.D. Kiadehi, Effects of catalyst preparation route and promoters (Ce and Zr) on catalytic activity of CuZn/CNTs catalysts for hydrogen production from methanol steam reforming, Int. J. Hydrogen Energy. 46 (2021) 8906–8921. https://doi.org/10.1016/j.ijhydene.2021.01.010.

[17] A. Jorio, R. Saito, Raman spectroscopy for carbon nanotube applications, J. Appl. Phys. 129 (2021) 21102.

[18] Z. Li, L. Deng, I.A. Kinloch, R.J. Young, Raman spectroscopy of carbon materials and their composites: Graphene, nanotubes and fibres, Prog. Mater. Sci. 135 (2023) 101089. https://doi.org/10.1016/j.pmatsci.2023.101089.

[19] S.A. Chernyak, A.S. Ivanov, D.N. Stolbov, T.B. Egorova, K.I. Maslakov, Z. Shen, V.V. Lunin, S.V Savilov, N-doping and oxidation of carbon nanotubes and jellyfish-like graphene nanoflakes through the prism of Raman spectroscopy, Appl. Surf. Sci. 488 (2019) 51–60. https://doi.org/10.1016/j.apsusc.2019.05.243.

[20] S. Akbarzadeh, M. Ramezanzadeh, B. Ramezanzadeh, M. Mahdavian, R. Naderi, Fabrication of highly effective polyaniline grafted carbon nanotubes to induce active protective functioning in a silane coating, Ind. Eng. Chem. Res. 58 (2019) 20309–20322. https://doi.org/10.1021/acs.iecr.9b04217.

[21] L.M. Hoyos-Palacio, D.P. Cuesta Castro, I.C. Ortiz-Trujillo, L.E. Botero Palacio, B.J. Galeano Upegui, N.J. Escobar Mora, J.A. Carlos Cornelio, Compounds of carbon nanotubes decorated with silver nanoparticles via in-situ by chemical vapor deposition (CVD), J. Mater. Res. Technol. 8 (2019) 5893–5898. https://doi.org/10.1016/j.jmrt.2019.09.062.

[22] T. Batakliev, I. Petrova-Doycheva, V. Angelov, V. Georgiev, E. Ivanov, R. Kotsilkova, M. Casa, C. Cirillo, R. Adami, M. Sarno, Effects of graphene nanoplatelets and multiwall carbon nanotubes on the structure and mechanical properties of poly (lactic acid) composites: A comparative study, Appl. Sci. 9 (2019) 469.

[23] Y. Say, O. Guler, B. Dikici, Carbon nanotube (CNT) reinforced magnesium matrix composites: The effect of CNT ratio on their mechanical properties and corrosion resistance, Mater. Sci. Eng. A. 798 (2020) 139636. https://doi.org/10.1016/j.msea.2020.139636.

[24] X. Zhou, J. Deng, C. Fang, W. Lei, Y. Song, Z. Zhang, Z. Huang, Y. Li, Additive manufacturing of CNTs/PLA composites and the correlation between microstructure and functional properties, J. Mater. Sci. Technol. 60 (2021) 27–34. https://doi.org/10.1016/j.jmst.2020.04.038.

[25] Z.A. ALOthman, A.Y. Badjah, I. Ali, Facile synthesis and characterization of multi walled carbon nanotubes for fast and effective removal of 4-tert-octylphenol endocrine disruptor in water, J. Mol. Liq. 275 (2019) 41–48. https://doi.org/10.1016/j.molliq.2018.11.049.

[26] E. V Basiuk, L. Huerta, V.A. Basiuk, Noncovalent bonding of 3d metal(II) phthalocyanines with single-walled carbon nanotubes: A combined DFT and XPS study, Appl. Surf. Sci. 470 (2019) 622–630. https://doi.org/10.1016/j.apsusc.2018.11.159.

[27] Y. Gu, M. Yang, W. Wang, R. Han, Phosphate adsorption from solution by zirconium-loaded carbon nanotubes in batch mode, J. Chem. Eng. Data. 64 (2019) 2849–2858. https://doi.org/10.1021/acs.jced.9b00214.

[28] L.N. Affonso, J.L. Marques, V.V.C. Lima, J.O. Gonçalves, S.C. Barbosa, E.G. Primel, T.A.L. Burgo, G.L. Dotto, L.A.A. Pinto, T.R.S. Cadaval, Removal of fluoride from fertilizer industry effluent using carbon nanotubes stabilized in chitosan sponge, J. Hazard. Mater. 388 (2020) 122042. https://doi.org/10.1016/j.jhazmat.2020.122042.

[29] R. Zhu, Y. Zhu, H. Xian, L. Yan, H. Fu, G. Zhu, Y. Xi, J. Zhu, H. He, CNTs/ferrihydrite as a highly efficient heterogeneous Fenton catalyst for the degradation of bisphenol A: The important role of CNTs in accelerating Fe(III)/Fe(II) cycling, Appl. Catal. B Environ. 270 (2020) 118891. https://doi.org/10.1016/j.apcatb.2020.118891.

[30] C. Liang, P. Song, A. Ma, X. Shi, H. Gu, L. Wang, H. Qiu, J. Kong, J. Gu, Highly oriented three-dimensional structures of Fe3O4 decorated CNTs/reduced graphene oxide foam/epoxy nanocomposites against electromagnetic pollution, Compos. Sci. Technol. 181 (2019) 107683. https://doi.org/10.1016/j.compscitech.2019.107683.

[31] Y. Cui, F. Wu, J. Wang, Y. Wang, T. Shah, P. Liu, Q. Zhang, B. Zhang, Three dimensional porous MXene/CNTs microspheres: Preparation, characterization and microwave absorbing properties, Compos. Part A Appl. Sci. Manuf. 145 (2021) 106378. https://doi.org/10.1016/j.compositesa.2021.106378.

[32] S. Asim, M.S. Javed, S. Hussain, M. Rana, F. Iram, D. Lv, M. Hashim, M. Saleem, M. Khalid, R. Jawaria, Z. Ullah, N. Gull, RuO2 nanorods decorated CNTs grown carbon cloth as a free standing electrode for supercapacitor and lithium ion batteries, Electrochim. Acta. 326 (2019) 135009. https://doi.org/10.1016/j.electacta.2019.135009.

[33] M.A. Zamzami, G. Rabbani, A. Ahmad, A.A. Basalah, W.H. Al-Sabban, S. Nate Ahn, H. Choudhry, Carbon nanotube field-effect transistor (CNT-FET)-based biosensor for rapid detection of SARS-CoV-2 (COVID-19) surface spike protein S1, Bioelectrochemistry. 143 (2022) 107982. https://doi.org/10.1016/j.bioelechem.2021.107982.

[34] Y. Ma, G. Lan, W. Fu, Y. Lai, W. Han, H. Tang, H. Liu, Y. Li, Role of surface defects of carbon nanotubes on catalytic performance of barium promoted ruthenium catalyst for ammonia synthesis, J. Energy Chem. 41 (2020) 79–86. https://doi.org/10.1016/j.jechem.2019.04.016.

[35] J. Chen, Q. Chen, Q. Ma, Influence of surface functionalization via chemical oxidation on the properties of carbon nanotubes, J. Colloid Interface Sci. 370 (2012) 32–38. https://doi.org/10.1016/j.jcis.2011.12.073.

[36] Y.S. Kim, C.R. Park, One-pot titration methodology for the characterization of surface acidic groups on functionalized carbon nanotubes, Carbon N. Y. 96 (2016) 729–741. https://doi.org/10.1016/j.carbon.2015.08.078.

[37] Z. Zhang, L. Pfefferle, G.L. Haller, Comparing characterization of functionalized multi-walled carbon nanotubes by potentiometric proton titration, NEXAFS, and XPS, Chinese J. Catal. 35 (2014) 856–863. https://doi.org/10.1016/S1872-2067(14)60123-6.

[38] W. Qi, W. Liu, B. Zhang, X. Gu, X. Guo, D. Su, Oxidative dehydrogenation on nanocarbon: Identification and quantification of active sites by chemical titration, Angew. Chemie Int. Ed. 52 (2013) 14224–14228.

[39] E. Moaseri, M. Baniadam, M. Maghrebi, M. Karimi, A simple recoverable titration method for quantitative characterization of amine-functionalized carbon nanotubes, Chem. Phys. Lett. 555 (2013) 164–167. https://doi.org/10.1016/j.cplett.2012.10.064.

[40] D.C. Vennerberg, R.L. Quirino, Y. Jang, M.R. Kessler, Oxidation behavior of multi-walled carbon nanotubes fluidized with ozone, ACS Appl. Mater. Interfaces. 6 (2014) 1835–1842. https://doi.org/10.1021/am4048305.

[41] T. Bortolamiol, P. Lukanov, A.-M. Galibert, B. Soula, P. Lonchambon, L. Datas, E. Flahaut, Double-walled carbon nanotubes: Quantitative purification assessment, balance between purification and degradation and solution filling as an evidence of opening, Carbon N. Y. 78 (2014) 79–90.

[42] A. Yahyazadeh, V.B. Borugadda, A.K. Dalai, L. Zhang, Optimization of olefins' yield in Fischer-Tropsch synthesis using carbon nanotubes supported iron catalyst with potassium and molybdenum promoters, Appl. Catal. A Gen. 643 (2022) 118759. https://doi.org/10.1016/J.APCATA.2022.118759.

[43] M. Romero-Sáez, A.B. Dongil, N. Benito, R. Espinoza-González, N. Escalona, F. Gracia, CO2 methanation over nickel-ZrO2 catalyst supported on carbon nanotubes: A comparison between two impregnation strategies, Appl. Catal. B Environ. 237 (2018) 817–825. https://doi.org/10.1016/j.apcatb.2018.06.045.

[44] D. Zhang, L. Zhang, L. Shi, C. Fang, H. Li, R. Gao, L. Huang, J. Zhang, In situ supported MnO x–CeO x on carbon nanotubes for the low-temperature selective catalytic reduction of NO with NH 3, Nanoscale. 5 (2013) 1127–1136.

[45] T.A. Saleh, Carbon nanotube-incorporated alumina as a support for MoNi catalysts for the efficient hydrodesulfurization of thiophenes, Chem. Eng. J. 404 (2021) 126987. https://doi.org/10.1016/j.cej.2020.126987.

[46] A.Z. Varzaneh, J. Towfighi, S. Sahebdelfar, Carbon nanotube templated synthesis of metal containing hierarchical SAPO-34 catalysts: Impact of the preparation method and metal avidities in the MTO reaction, Microporous Mesoporous Mater. 236 (2016) 1–12. https://doi.org/10.1016/j.micromeso.2016.08.027.

[47] E. Erünal, F. Ulusal, M.Y. Aslan, B. Güzel, D. Üner, Enhancement of hydrogen storage capacity of multi-walled carbon nanotubes with palladium doping prepared through supercritical CO2 deposition method, Int. J. Hydrogen Energy. 43 (2018) 10755–10764. https://doi.org/10.1016/j.ijhydene.2017.12.058.

[48] C. Xing, G. Yang, D. Wang, C. Zeng, Y. Jin, R. Yang, Y. Suehiro, N. Tsubaki, Controllable encapsulation of cobalt clusters inside carbon nanotubes as effective catalysts for Fischer–Tropsch synthesis, Catal. Today. 215 (2013) 24–28. https://doi.org/10.1016/j.cattod.2013.02.018.

[49] Q. Li, X. Hou, H. Yang, Z. Ma, J. Zheng, F. Liu, X. Zhang, Z. Yuan, Promotional effect of CeOX for NO reduction over V2O5/TiO2-carbon nanotube composites, J. Mol. Catal. A Chem. 356 (2012) 121–127. https://doi.org/10.1016/j.molcata.2012.01.004.

[50] K. Keyvanloo, A. Mohamadalizadeh, J. Towfighi, A novel CeO2 supported on carbon nanotubes coated with SiO2 catalyst for catalytic cracking of naphtha, Appl. Catal. A Gen. 417–418 (2012) 53–58. https://doi.org/10.1016/j.apcata.2011.12.024.

[51] M.-J. Kim, S. Joo Park, K. Duk Kim, W. Kim, S. Chan Nam, K. Seok Go, S. Goo Jeon, Fabrication of carbon nanotube with high purity and crystallinity by methane decomposition over ceria-supported catalysts, J. Ind. Eng. Chem. 119 (2023) 315–326. https://doi.org/10.1016/j.jiec.2022.11.050.

[52] F. Li, Y. Wang, D. Wang, F. Wei, Characterization of single-wall carbon nanotubes by N2 adsorption, Carbon N. Y. 42 (2004) 2375–2383. https://doi.org/10.1016/j.carbon.2004.02.025.

[53] H.S. Oktaviano, K. Yamada, K. Waki, Nano-drilled multiwalled carbon nanotubes: Characterizations and application for LIB anode materials, J. Mater. Chem. 22 (2012) 25167–25173.

[54] S. V Sawant, S. Banerjee, A.W. Patwardhan, J.B. Joshi, K. Dasgupta, Effect of in-situ boron doping on hydrogen adsorption properties of carbon nanotubes, Int. J. Hydrogen Energy. 44 (2019) 18193–18204. https://doi.org/10.1016/j.ijhydene.2019.05.029.

[55] M.-S. Lee, S.-Y. Lee, S.-J. Park, Preparation and characterization of multi-walled carbon nanotubes impregnated with polyethyleneimine for carbon dioxide capture, Int. J. Hydrogen Energy. 40 (2015) 3415–3421. https://doi.org/10.1016/j.ijhydene.2014.12.104.

[56] K. Kamal, D.I. Grekov, A.M. Shariff, M.A. Bustam, P. Pré, Improving textural properties of magnesium-based metal-organic framework for gas adsorption by carbon doping, Microporous Mesoporous Mater. 323 (2021) 111246. https://doi.org/10.1016/j.micromeso.2021.111246.

[57] A. Mukhtar, N. Mellon, S. Saqib, A. Khawar, S. Rafiq, S. Ullah, A.G. Al-Sehemi, M. Babar, M.A. Bustam, W.A. Khan, M.S. Tahir, CO2/CH4 adsorption over functionalized multi-walled carbon nanotubes; an experimental study, isotherms analysis, mechanism, and thermodynamics, Microporous Mesoporous Mater. 294 (2020) 109883. https://doi.org/10.1016/j.micromeso.2019.109883.

[58] S.-Y. Lee, S.-J. Park, Influence of the pore size in multi-walled carbon nanotubes on the hydrogen storage behaviors, J. Solid State Chem. 194 (2012) 307–312. https://doi.org/10.1016/j.jssc.2012.05.027.

[59] M.A.O. Lourenço, M. Fontana, P. Jagdale, C.F. Pirri, S. Bocchini, Improved CO2 adsorption properties through amine functionalization of multi-walled carbon nanotubes, Chem. Eng. J. 414 (2021) 128763. https://doi.org/10.1016/j.cej.2021.128763.

[60] A. Kuznetsova, I. Popova, J.T.J. Yates, M.J. Bronikowski, C.B. Huffman, J. Liu, R.E. Smalley, H.H. Hwu, J.G. Chen, Oxygen-containing functional groups on single-wall carbon nanotubes: NEXAFS and vibrational spectroscopic studies, J. Am. Chem. Soc. 123 (2001) 10699–10704. https://doi.org/10.1021/ja011021b.

[61] D.V. Sivkov, O.V. Petrova, S.V. Nekipelov, A.S. Vinogradov, R.N. Skandakov, K.A. Bakina, S.I. Isaenko, A.M. Ob'edkov, B.S. Kaverin, I. V Vilkov, Quantitative characterization of oxygen-containing groups on the surface of carbon materials: XPS and NEXAFS study, Appl. Sci. 12 (2022) 7744.

[62] P. Nickl, J. Radnik, W. Azab, I.S. Donskyi, Surface characterization of covalently functionalized carbon-based nanomaterials using comprehensive XP and NEX-AFS spectroscopies, Appl. Surf. Sci. 613 (2023) 155953. https://doi.org/10.1016/j.apsusc.2022.155953.

[63] P.-L. Girard-Lauriault, R. Illgen, J.-C. Ruiz, M.R. Wertheimer, W.E.S. Unger, Surface functionalization of graphite and carbon nanotubes by vacuum-ultraviolet photochemical reactions, Appl. Surf. Sci. 258 (2012) 8448–8454. https://doi.org/10.1016/j.apsusc.2012.03.012.

[64] L. Wang, J. Han, Y. Zhu, R. Zhou, C. Jaye, H. Liu, Z.-Q. Li, G.T. Taylor, D.A. Fischer, J. Appenzeller, S.S. Wong, Probing the dependence of electron transfer on size and coverage in carbon nanotube–quantum dot heterostructures, J. Phys. Chem. C. 119 (2015) 26327–26338. https://doi.org/10.1021/acs.jpcc.5b08681.

[65] A.D. Winter, E. Larios, F.M. Alamgir, C. Jaye, D. Fischer, E.M. Campo, Near-edge X-ray absorption fine structure studies of electrospun poly(dimethylsiloxane)/poly(methyl methacrylate)/multiwall carbon nanotube composites, Langmuir. 29 (2013) 15822–15830. https://doi.org/10.1021/la404312x.

[66] K. Mette, A. Bergmann, J. Tessonnier, M. Hävecker, L. Yao, T. Ressler, R. Schlögl, P. Strasser, M. Behrens, Nanostructured manganese oxide supported on carbon nanotubes for electrocatalytic water splitting, ChemCatChem. 4 (2012) 851–862.

[67] E.M.S. Azzam, N.A. Fathy, S.M. El-Khouly, R.M. Sami, Enhancement the photocatalytic degradation of methylene blue dye using fabricated CNTs/TiO2/AgNPs/Surfactant nanocomposites, J. Water Process Eng. 28 (2019) 311–321. https://doi.org/10.1016/j.jwpe.2019.02.016.

[68] H. Huang, Y. Chen, Z. Chen, J. Chen, Y. Hu, J.-J. Zhu, Electrochemical sensor based on Ce-MOF/carbon nanotube composite for the simultaneous discrimination of hydroquinone and catechol, J. Hazard. Mater. 416 (2021) 125895. https://doi.org/10.1016/j.jhazmat.2021.125895.

[69] L. Tzounis, M. Petousis, S. Grammatikos, N. Vidakis, 3D printed thermoelectric polyurethane/multiwalled carbon nanotube nanocomposites: A novel approach towards the fabrication of flexible and stretchable organic thermoelectrics, Materials (Basel). 13 (2020) 2879.

[70] S. Agrahari, A.K. Singh, R.K. Gautam, I. Tiwari, Fabrication of gadolinium decorated spherical zinc oxide attached on carbon nanotubes (Gd@ ZnO-MWCNTs) for electrochemical detection of a bisphenol derivative BPSIP in real sample matrices, J. Appl. Electrochem. 53 (2023) 345–358.

[71] J. Wang, B. Shen, M. Lan, D. Kang, C. Wu, Carbon nanotubes (CNTs) production from catalytic pyrolysis of waste plastics: The influence of catalyst and reaction pressure, Catal. Today. 351 (2020) 50–57. https://doi.org/10.1016/j.cattod.2019.01.058.

[72] N. Arsalani, S. Bazazi, M. Abuali, S. Jodeyri, A new method for preparing ZnO/CNT nanocomposites with enhanced photocatalytic degradation of malachite green under visible light, J. Photochem. Photobiol. A Chem. 389 (2020) 112207. https://doi.org/10.1016/j.jphotochem.2019.112207.

[73] F.A. Mann, N. Herrmann, F. Opazo, S. Kruss, Quantum defects as a toolbox for the covalent functionalization of carbon nanotubes with peptides and proteins, Angew. Chemie Int. Ed. 59 (2020) 17732–17738.

[74] R. Li, X. Wang, Z. Ji, B. Sun, H. Zhang, C.H. Chang, S. Lin, H. Meng, Y.-P. Liao, M. Wang, Z. Li, A.A. Hwang, T.-B. Song, R. Xu, Y. Yang, J.I. Zink, A.E. Nel, T. Xia, Surface charge and cellular processing of covalently functionalized multiwall carbon nanotubes determine pulmonary toxicity, ACS Nano. 7 (2013) 2352–2368. https://doi.org/10.1021/nn305567s.

[75] Z. Zhang, X. Xu, Nondestructive covalent functionalization of carbon nanotubes by selective oxidation of the original defects with K2FeO4, Appl. Surf. Sci. 346 (2015) 520–527. https://doi.org/10.1016/j.apsusc.2015.04.026.

[76] S.L.H. Rebelo, A. Guedes, M.E. Szefczyk, A.M. Pereira, J.P. Araújo, C. Freire, Progress in the Raman spectra analysis of covalently functionalized multiwalled carbon nanotubes: Unraveling disorder in graphitic materials, Phys. Chem. Chem. Phys. 18 (2016) 12784–12796.

[77] Y. Zhou, Y. Fang, R.P. Ramasamy, Non-covalent functionalization of carbon nanotubes for electrochemical biosensor development, Sensors. 19 (2019) 392. https://doi.org/10.3390/S19020392.

[78] D. Ponnamma, S.H. Sung, J.S. Hong, K.H. Ahn, K.T. Varughese, S. Thomas, Influence of non-covalent functionalization of carbon nanotubes on the rheological behavior of natural rubber latex nanocomposites, Eur. Polym. J. 53 (2014) 147–159. https://doi.org/10.1016/j.eurpolymj.2014.01.025.

[79] S. Meuer, L. Braun, R. Zentel, Pyrene containing polymers for the non-covalent functionalization of carbon nanotubes, Macromol. Chem. Phys. 210 (2009) 1528–1535.

[80] A. Abdelhalim, A. Abdellah, G. Scarpa, P. Lugli, Metallic nanoparticles functionalizing carbon nanotube networks for gas sensing applications, Nanotechnology. 25 (2014) 55208.

[81] J.-P. Tessonnier, O. Ersen, G. Weinberg, C. Pham-Huu, D.S. Su, R. Schlögl, Selective deposition of metal nanoparticles inside or outside multiwalled carbon nanotubes, ACS Nano. 3 (2009) 2081–2089. https://doi.org/10.1021/nn900647q.

[82] T.T. Nguyen, P. Serp, Confinement of metal nanoparticles in carbon nanotubes, ChemCatChem. 5 (2013) 3595–3603. https://doi.org/10.1002/CCTC.201300527.

[83] J.-P. Tessonnier, O. Ersen, G. Weinberg, C. Pham-Huu, D.S. Su, R. Schlögl, Selective deposition of metal nanoparticles inside or outside multiwalled carbon nanotubes, ACS Nano. 3 (2009) 2081–2089. https://doi.org/10.1021/nn900647q.

4 Toxicology of Carbon Nanotubes

H. Harija, Anindya Nag, Andreas Richter, and Mehmet Ercan Altinsoy

4.1 INTRODUCTION

An interdisciplinary area of science and technology known as nanotechnology focuses on altering, researching, and using materials and devices on the nanoscale. It entails interaction with objects and elements that are typically 1 to 100 nanometres in size. A nanometre is a billionth of a metre, smaller than the diameter of a human hair – roughly 100,000 times smaller (1). Nanomaterials are substances with distinct features and structures on the nanoscale. Due to unique quantum and surface effects, material's behaviour at this scale can differ dramatically from that of its bulk counterparts (2). Some common types of nanomaterials include nanoparticles, nanotubes, nanowires, nanocomposites, nanofilms, and so on. Numerous sectors, including electronics, healthcare, energy, aerospace, and environmental restoration, use nanomaterials in numerous ways (3, 4). Nanomaterials provide several notable benefits, including improved mechanical properties, increased surface area for higher reactivity, better electrical and thermal conductivity, and precise medication administration in medicine. Though it raises important ethical and safety questions, the steady development of nanotechnology offers enticing possibilities. As a result, it is crucial to strike a balance between utilizing nanotechnology's potential benefits and guaranteeing its responsible growth, which calls for addressing any hazards to the environment and human health (5, 6). In recent years, numerous studies have been conducted on various types of nanomaterials due to extensive development in nanotechnology, with Carbon Nanotubes (CNTs) gaining significant attention (7–9).

Carbon atoms have the unusual ability to bond together in various ways, leading to different carbon allotropes; graphite, amorphous carbon, fullerenes and diamond are some of the well-defined allotropes of carbon. As a result, the physical characteristics of carbon are determined mainly by its allotropic form. Carbon is remarkably well maintained in its allotropic forms. CNTs, graphite, graphene, and diamond have recently been explored for their superior thermal conductivity at standard pressure and temperature. High tensile strength and elastic modulus make CNTs one of the stiffest and most vital materials (10–12). CNTs are nanostructures made entirely of carbon atoms arranged in a cylindrical pattern. They have a diameter of a few nanometres and a length ranging from a few nanometres to several centimetres. CNTs have several distinct properties that set them apart from other nanomaterials. One of the most significant advantages of CNTs is their high aspect ratio, which means

DOI: 10.1201/9781003376071-4

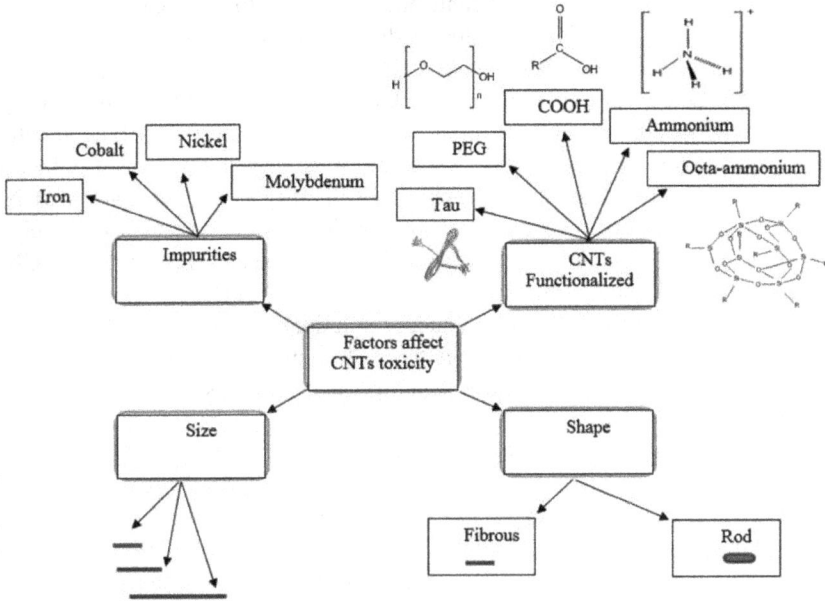

FIGURE 4.1 A list of the elements influencing the toxicity of CNTs (22).

they have a very high surface area-to-volume ratio. Based on this property, CNTs are suitable for energy storage, catalysis, and sensing applications (13–15). CNTs are ideal for composites and other structural materials because they possess high tensile strength and stiffness. The high electrical conductivity of CNTs makes them a valuable component in sensors and electronic devices. CNTs stand out from other nanomaterials due to their unique properties, making them highly versatile and helpful in various applications. However, concerns have been raised about their toxicity to human health and the environment (16–19). The toxicology of Carbon Nanotubes is an active area of study and much remains to be discovered. Notably, the complexity of the toxicity of CNTs is closely related to elements involving their size, shape, surface functionalization, and impurities. In contrast to their longer, agglomerated counterparts, shorter and more dispersed CNTs show decreased toxicity. Several factors significantly impact the health effects, including individual susceptibility, exposure length, and the magnitude of dosage (20, 21). Figure 4.1 illustrates a list of the elements influencing CNTs' toxicity (22).

As shown in Figure 4.1, the most frequently employed materials in CNTs synthesis are Co, Fe, Ni, and Mo. During CNTs' synthesis, these metals are used as catalysts to encourage CNTs' growth. CNTs are often produced in bulk, forming aggregates and clumps, as they are grown commercially. The size of agglomerates, the features of a particle's surface, and the presence of metallic impurities all affect the level of risk to a person's health. During the synthesis process, these dangers can particularly manifest through unintentional contact with skin – or inhalation

(22). Metal pollutants are essential contributors to CNTs toxicity, causing cell death via various pathways, including mitochondrial degeneration and oxidative stress. One of the techniques utilized to lessen CNTs cytotoxicity is CNTs purification. Enhancing functionalization can increase CNTs purity (23–25). To enable proper absorption, distribution, and elimination and to make the CNTs biocompatible while ensuring low toxicity, Carbon Nanotubes must be soluble in water to be used in biological systems. Due to their increased solubility in organic solvents, CNTs can be functionalized by covalent or non-covalent methods, which may help to achieve superior compatibility with physiological systems (26, 27). When CNTs are functionalized, the hydrophobicity decreases on the surface of graphene, making it easier for them to disperse in aqueous solutions. Additionally, it enables the graphene surface to assemble in order to create composite materials (26). There are two types of nanoparticles: those with high aspect ratio, such as nanowires and nanotubes; and those with low aspect ratio, such as nanospheres and nanotubes. Fibre-like (large aspect ratio) or clustered nanoparticles are generally known to be far more toxic and detrimental than rounded and ringed nanoparticles (27–29). Other critical elements affecting a CNT's toxicity are its length and diameter, which change during manufacturing. Studies have demonstrated that CNTs with longer lengths and larger diameters are more hazardous than those with smaller diameters. Compared to long CNTs with a length of 825 nm, short CNTs with 220 nm cause little to no irritation (23).

The functionalization of Carbon Nanotubes significantly impacts human health because it changes their morphological and physicochemical properties. CNTs are functionalized by attaching chemical groups or molecules to their surface or sidewalls. Functionalization is essential in order to tailor the properties of CNTs and make them suitable for specific applications. A Carbon Nanotube (CNT) is a tubular or rod-like shape with a diameter in the nanometre range. They belong to the group of carbon allotropes. These tubes may have two or more layers, with a diameter that ranges from 3 to 30 nm. The two primary varieties of Carbon Nanotubes are SWCNTs (Single-Walled Carbon Nanotubes) and MWCNTs (Multi-Walled Carbon Nanotubes) (30). The single layer of carbon atoms that make up Single-Walled Carbon Nanotubes (SWCNTs) are arranged in the shape of a tube. Depending on their structural makeup, they can be metallic or semiconducting, making them useful for different electronic applications (31). On the other hand, Multi-Walled Carbon Nanotubes (MWCNTs) comprise several layers of concentrically arranged cylindrical graphene sheets. Depending on the number of layers and how the graphene sheets are placed, MWCNTs can display various characteristics (32). Strong van der Waals forces cause SWCNTs to have a high surface area, which may result in the formation of bundles, and hence, diminish the efficiency of surface area. Due to their higher defect density and smaller sidewall surface area, MWCNTs have a lesser propensity to form bundles. Therefore, this functionalization strategy can help lessen the toxicity while distributing the CNTs without forming bundles. Compared to SWCNTs, they are typically less susceptible to defects and have better mechanical properties (33). Schematic diagrams of two varieties of CNTs and different applications of CNTs are shown in Figure 4.2 (a and b), respectively (34).

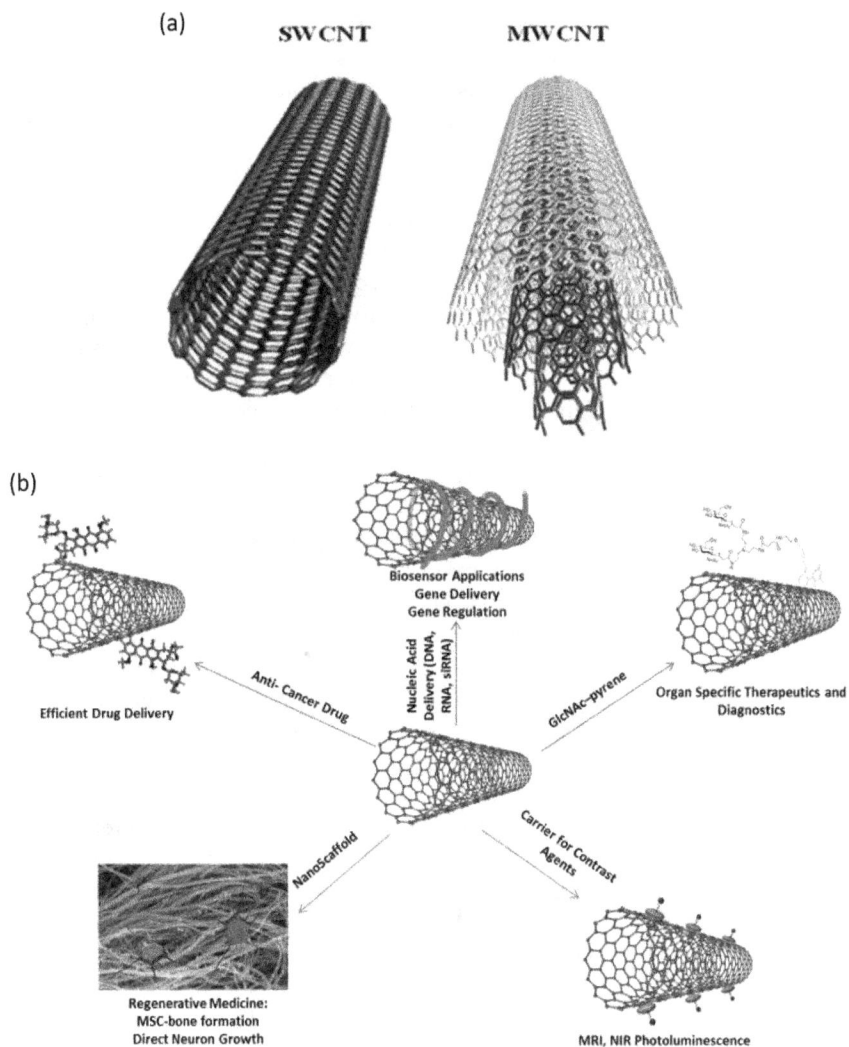

FIGURE 4.2 (a) Schematic diagrams of SWCNTs and MWCNTs. (b) Different applications of CNTs (34).

4.2 OVERVIEW OF TOXICOLOGY OF CNTs

Carbon Nanotubes (CNTs) are a notable and distinct class of advanced materials distinguished by various extraordinary and advantageous characteristics. The risk of higher exposure to the toxicity of CNTs is growing, as they are frequently being used in manufacturing and other applications. Therefore, a thorough toxicological investigation of CNTs is essential to understand the mechanisms associated with toxicity and to create effective plans to reduce the risk of exposure to these materials.

This section explores a collection of studies that focus on the toxicological aspects of CNTs, including their effects on the environment and human health.

The number of atoms, ions, or molecules that make up a nanoparticle is the main factor that determines its toxicity. Nanoparticles serve as the active ingredients in nanoformulations. The term "nanoformulations" refers to the creation and application of nanoparticles in numerous industries, particularly in the domains of medicine and drug delivery. In order to attain certain qualities and functions for various reasons, nanoparticles are manipulated and engineered at the nanoscale. Nano formulations enable targeted therapy with exact placement to particular cells or tissues, increase solubility, and improve medication delivery by encapsulating drugs within nanoparticles. Because of their small size, along with their ability to infiltrate or get absorbed into living systems, nanoparticles can spread much too quickly. Figure 4.3 illustrates the toxicity and the locations where the active components interact with the target (35).

Due to their natural tendency to aggregate, nanoparticles transfer a substantial amount of active chemicals to the site of action in the target organism. Their high dispersion rate also makes it faster for them to accumulate, as compared to other types of drug delivery formulations. Nanoscale-sized CNTs are easily inhalable, allowing easy entry into the respiratory system. The central and peripheral nervous systems, lymphatic system, and circulatory system are quickly affected as the CNTs enter the lungs, as is depicted in Figure 4.4. The chemical reactivity, surface properties, and the capacity to bind with the proteins found in the human body may affect how easily nanomaterials are transmitted throughout the body. They are deposited in various parts of the respiratory tract according to the size and physical structure of the particles. Following the deposition, the nanosized particles travel through various pathways and mechanisms to the extrapulmonary region before being directed to the site of the target organ (36).

Due to the rapidly expanding field of nanotechnology, humans are always at risk of being exposed to nanoparticles through ingestion, inhalation, absorption through skin, and injection into veins. Even though CNTs are an up-and-coming class of nanomaterials, little is known about how they might affect human health. Selvakumar et al. (2023) discussed the significance of the toxicity of MWCNTs and its relationship with biological systems. The capacity of a substance to harm or destroy cells is known as cytotoxicity (37). The main determinants of the toxicity of CNTs include length, stiffness, concentration, time of exposure, state of aggregation, purity, diameter, and functionalization. To lessen cytotoxicity, new varieties of Carbon Nanotubes have been developed to be used for biological applications. Cytotoxicity can express itself in various ways, depending on the type of cell and the foreign material entering the body. Apoptosis, oxidative stress, DNA damage, etc., are prevalent cytotoxic conditions. It is crucial to conduct cytotoxicity tests before compounds like medications, chemicals, and nanoparticles are released into the environment to determine their safety. Cytotoxicity is evaluated using a variety of in-vitro and in-vivo techniques (38), ranging from straightforward cell viability tests to more intricate models that replicate human tissues or organs (39). In-vitro studies utilize parts of an organism removed from its natural biological environment to provide a more thorough investigation than can be carried out with the organism per se. Instead of extracting a tissue

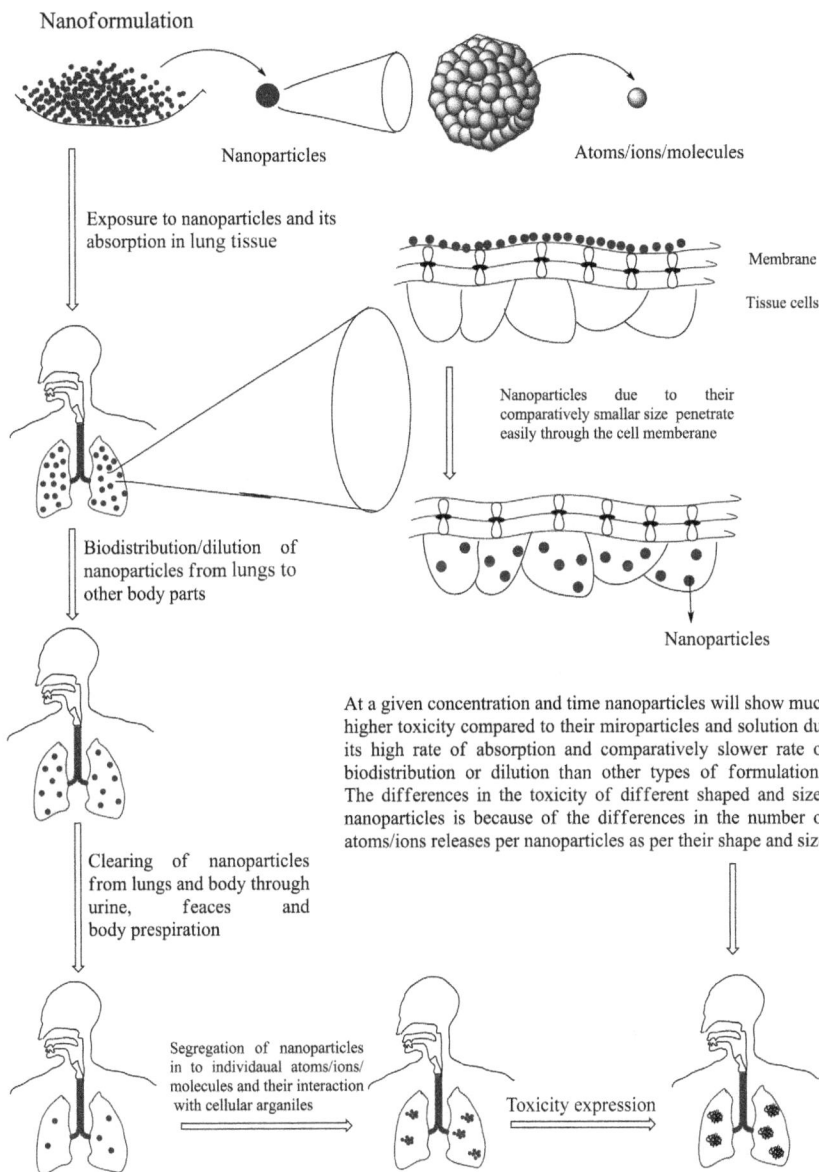

FIGURE 4.3 Schematic representation of the interaction of nanoparticles with the target (35).

or using a dead creature for experiments, in-vivo uses a living organism. The authors (37) compared CNTs with a clay nanomaterial known as Halloysite Nanotubes (HNTs). HNTs with tubular structures made from sheets of aluminosilicate kaolin are also popular because of their distinctive characteristics and potential applications (40, 41). HNTs are used in various catalytic studies, nanoelectronics, glass coatings, forensic sciences, etc. Due to their distinct physiochemical characteristics and ease

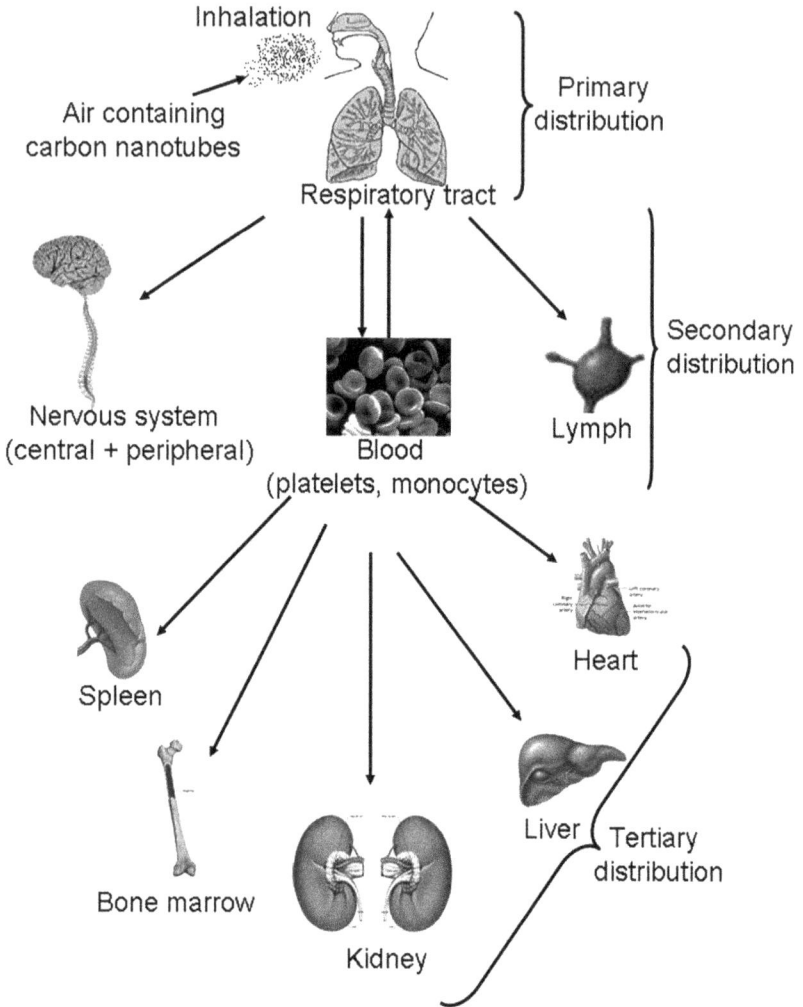

FIGURE 4.4 Levels of distribution of CNTs in the human body (36).

of forming hydrogen bonds with biomaterials, HNTs are more biocompatible and less toxic than CNTs, making them more suitable for various biomedical applications (42, 43).

The release of Carbon Nanotubes (CNTs) into water bodies is highly concerning, as it endangers the organisms in the aquatic ecosystem. CNTs consumed by aquatic species at lower trophic levels of the food chain are carried to organisms at higher trophic levels, ultimately affecting human health (44, 45). The work done by Gao et al. (2023) explains the impairment of multiple organs in fish induced by CNTs contamination. *Cyprinus carpio*, a large freshwater fish, was exposed to two different concentrations of MWCNTs for about four weeks. Based on the findings, it is stated that MWCNTs caused liver injuries and also induced endoplasmic reticulum stress

(ERS). Endoplasmic reticulum (ER) is a complex organelle present in the eukaryotic cells of plants and animals. It is essential for many cellular processes, especially the production, folding, alteration, and transport of proteins and lipids. The ER is a cytoplasmic network that extends from the nuclear membrane and is composed of membrane-enclosed tubules and sacs (cisternae). In endogenous or exogenous stimulation, the accumulation of unfolded proteins and calcium imbalance in the ER cause ER stress. Endoplasmic reticulum (ER) stress, sometimes referred to as the unfolded protein response (UPR), is a physiological response that happens when there is an imbalance between the requirement for appropriate folding of proteins and the ER's capacity to fold proteins. When unfolded or misfolded proteins build up in the ER, the stress response gets activated. The oxidative enzymatic pathways are activated by CNTs, which trigger the oxidative stress response. The authors suggest that the molecular activity of carbon nanomaterials in fish can be understood more conveniently, which has enormous theoretical significance in maintaining a healthy and sustainable aquatic environment (46).

Nanoparticle medicinal formulations have a wide range of potential applications because of their capacity to deliver pharmaceuticals with targeted and sustained release, increased solubility, and decreased adverse drug reactions to the body. Several effective nanometre-sized drug carriers have recently been created, and Carbon Nanotubes (CNT) have witnessed increasingly numerous applications in cancer treatment. Yan et al. (2019) reviewed the possible toxic effects of CNTs working as anti-tumour drug carriers. Extensive development of nanotechnology and their subsequent applications in the field of Oncology has resulted in the employment of CNTs as anti-cancer drugs. These nano molecules, however, have several disadvantages in clinical settings. CNTs, after dissociating from the drug molecule, are retained in the body in the process of delivering anti-cancer drugs through CNTs, which causes secondary harm. The authors suggest a method to address this problem by loading particular proteins on the surface of CNTs, which induces neutrophil Myeloperoxidase (MPO) release (47). Neutrophils, a kind of white blood cell involved in the immunological response, generate an enzyme called Myeloperoxidase (MPO). MPO is essential for neutrophil antimicrobial activity, notably for eradicating invasive pathogens (48). And the simulation of MPO causes CNTs to break down and eventually produce attenuating effects. In addition, dual drug-loading techniques work well to safeguard healthy tissues.

With their ability to adsorb harmful gases, heavy metals, and antibiotics, Carbon Nanomaterials (CNMs) play a significant part in the reduction of environmental pollution. As nanoparticles become more prevalent in our daily lives, environmental exposure to them is unavoidable; as a result, research on nanotoxicity is gaining interest. They are harmful to the living organisms and are hard to break down naturally. The toxicity of CNMs such as Carbon Nanotubes (CNTs), graphene (GRA), and fullerene (C60) towards living organisms at the cellular level was described by Peng et al. (2020). The fundamental component of graphite is graphene, a two-dimensional (2D) sheet of carbon atoms arranged in a honeycomb lattice (49). A schematic representation of the CNMs (50) is shown in Figure 4.5. SWCNTs and C60 are particularly harmful to cells because they cause apoptosis, also known as cell death (51, 52). C60 can harm the DNA of a cell and induce cytotoxicity (53, 54).

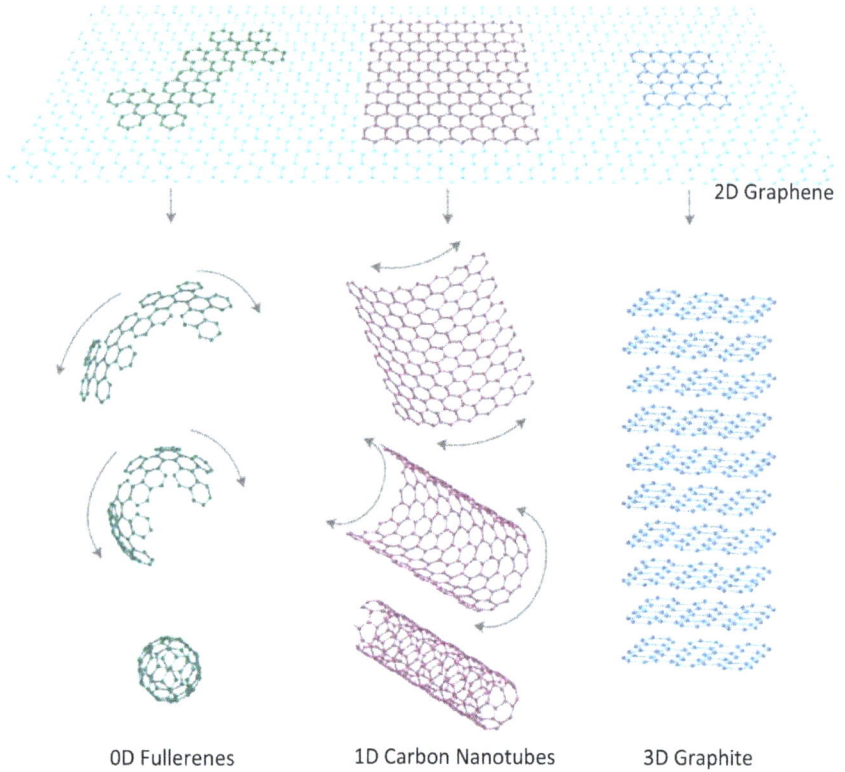

FIGURE 4.5 Schematic representation of CNMs (50).

The principle of the toxicity of C60s towards organisms is its high photosensitivity. C60, when exposed to light, reaches an excited state and forms fullerenes. CNTs are hazardous to aquatic organisms, as they do not dissolve in water, and rather, deposit in sediment (50). In the biological environment, they can easily pass through living things. Due to these characteristics, nanomaterials can damage DNA without having to be exposed to light, and consequently, induce oxidative stress, membrane irregularity, and formation of organic radicals.

The toxicity of CNMs is susceptible to changes, as their structure tends to change in response to environmental interactions. The design and length of CNMs affect their toxicity (55–57). Combining biodegradation and non-biodegradation techniques in the degradation of CNMs produces fewer pollutants while increasing the efficacy of the process (58). Additionally, conjugating biological molecules like antibodies, DNA, and proteins with CNTs may help to reduce toxicity (59). The toxic properties of CNTs are strongly proportional to their aggregation state and are usually the result of aggregation of stiffer, more solid, and larger CNTs (60, 61). Well-suspended CNTs are found to be less hazardous to in-vivo and in-vitro cell lines. Hence, the production of significantly less toxic and biocompatible CNTs can be predicted by adjusting all these variables. Thus, functionalization techniques using specific functional groups can reduce the toxicity of CNTs (60).

The study of in-vitro toxicity of CNTs is more practical than that of in-vivo toxicity, since it uses tissues from test animals, hence raising fewer ethical questions and being less expensive. However, in-vitro studies could not create toxicokinetic data, as they offer far less information on the reason or mechanism of cellular toxicity and death. In contrast, in-vivo tests are more complex and expensive (62–64). Therefore, extreme caution and meticulous evaluation are crucial for every experiment involving CNTs as a component to prevent a potential bias in the conclusions and interpretation of results of the cytotoxicity studies. Computational approaches might be considered the optimum tool to assess the toxicity of nanomaterials. One approach to toxicity assessment, called "in silico toxicology," makes use of computer simulation, visualization, and prediction tools to study, predict, and assess the toxicity of various compounds. In silico toxicity studies evaluate and forecast toxicity to ensure the best possible design of therapeutic drugs and to strengthen the industry-standard assays (65–68). Thus, combining computational modelling with the data obtained from in-vivo and in-vitro study models creates more precise estimates of potential hazards to humans. The possible health effects of Multi-Walled Carbon Nanotubes (MWCNTs) in agricultural applications (69–71) were thoroughly investigated by Rezaei Cherati, Sajedeh, et al. (2022). It is essential to look at how carbon-based nanomaterials can easily migrate through the food chain when introduced into plants. The authors fed CNT-contaminated tomatoes to mice as part of an extensive experiment to determine the dangers of Multi-Walled Carbon Nanotubes (MWCNTs) building up in the exposed parts of the tomato plants. The investigation revealed no toxicity response for all examined animal organs, and the presence of CNTs was determined to be minimal. The amount of CNTs absorbed by plant organs was quantified and visualized using microwave-induced heating (MIH) and radio frequency (RF) heating techniques. The plants contaminated with CNTs can be evaluated for potential dangers, using this thorough detection in conjunction with the evaluation of the in-vivo toxicity technique. These findings imply that there would be minimal damage after the consumption of agricultural plants fertilized with carbon-based nanomaterials as nano fertilizers (72).

4.2.1 Regulatory Aspects and Challenges

Health concerns have been found to be connected to all nanomaterials, including Carbon Nanotubes. Inflammatory responses in the lungs, upon inhalation, have been noted in various studies of pulmonary toxicity. Further, certain MWCNTs were found to have caused mesothelioma following intraperitoneal injection. The employment of CNTs, in general, has been criticized due mainly to these discoveries. The US Environmental Protection Agency ("EPA") and the European Commission ("EC, the executive body of the EU") have announced the use of CNTs as raw material for certain research and experiments. The CNTs used for the experiments will be regulated in the US and the EU. The quantity of compounds produced or imported annually must be registered with the European Union's Registration Evaluation Authorization and Restriction of Chemicals (REACH) committee when it exceeds the limit of one metric tonne. These raw materials include nanomaterials and PBT compounds, which are carcinogenic, bio-accumulative, and toxic (26, 73–75). There are no defined regulatory, safety, or handling standards regarding occupational exposure to personnel in manufacturing, production and maintenance, university researchers, R&D laboratory workers, and consumers. Due to the lack of any such precautionary regulations, there might be frequent instances of occupational exposure, some as ordinary as contact with pollutants from the manufacturing process. The dermal contact with spills or dust from the manufacturing process by secondary manufacturers employing CNTs in composites or embedded in other materials (76–79) has been found to have inflammatory responses.

Furthermore, studies on toxicological impact should be expanded beyond the hazards caused by inhalable pollutants to assess the risk of additional routes of exposure across the entire value chain. Moreover, the studies need further improvisation in order to manage nanomaterials effectively so that they ensure safe implementation in low-risk applications. The findings from these studies will provide information about the variables, including the particular fibre type, the material itself, and the complexities of the manufacturing process. Making a thorough, interdisciplinary framework which is compatible with standard sustainability evaluations across the life-cycle of the product could be a workable approach. This method would include relevant datasets to solve issues involving safety and its influence on the environment and human health. The intrinsic characteristics behind the toxicity of Carbon Nanotubes (CNTs) may be addressed through computational research on the relationships between the structure of CNTs and their functionality. Interestingly, compared to the extraction of resources and energy-intensive processing associated with conventional metals, minerals, and materials, CNTs might present a more sustainable alternative, requiring significantly fewer energy inputs and materials, in addition to having fewer adverse effects on the environment and society (80).

4.3   CONCLUSION

From the investigation of existing Carbon Nanotube (CNT) toxicology reviews, it was found that the toxicity of Carbon Nanotubes (CNTs) is influenced by elements comprising composition, structural arrangement, and surface changes. Despite the rapid development and continuous evolution of carbon nanomaterials, it is necessary to do additional toxicological research to determine how these materials affect human health. Inconsistencies in the discourse surrounding the toxicity of CNTs are reflected in the literature; while a few researches showcase increased toxicity, others highlight the absence of adverse effects. Nevertheless, a few studies have demonstrated that well-functionalized CNTs are non-toxic to animal cells. However, untreated or non-functionalized CNTs harm animal and human cells, even at lower concentrations. Notably, Carbon Nanotubes have gained acceptance as a highly effective delivery system for various synthetic and biological molecules, reiterating their status as effective drug-delivery techniques. Functionalization of CNTs for protein and nucleic acid delivery has been shown to improve safety and efficacy as well as to enhance formulation-related characteristics such as stability and solubility. It also exhibits benefits like minimal immunogenicity when compared to viral vectors. Different biomacromolecules (DNA, proteins) are joined to CNTs using functionalization. There are numerous therapeutic and diagnostic uses for gene delivery by functionalized CNTs that may be employed in biomedical areas to treat cancer, autoimmune diseases, and cardiovascular diseases. Regulations governing workplaces must prioritize the safety of workers, as the growing inclusion of CNTs in industrial applications may also lead to increased susceptibility towards its toxicity. This requires tackling the issue of keeping the amount of these minute pollutants in check, exacerbated by our incomplete knowledge of their nature, processes, and deleterious reactions caused by exposure to the same. Regarding occupational risk assessments, it is preferable to approach CNTs in a way similar to other bio-persistent fibres, following analogous control and assessment processes, until a more in-depth understanding is achieved. As a result, the ongoing study is critical in order to understand the mechanisms underlying the toxicity of CNTs, their interaction with biological systems, the possible harm they might do, and the best strategies to reduce the risks. This technique emphasizes a cautious and knowledgeable approach, allowing us to make well-informed choices. At the same time, we continue to learn more about the distinct issues that may entail in occupational settings involving Carbon Nanotubes.

ACKNOWLEDGEMENT

This work was supported by the Free State of Saxony and by the European Union (ESF Plus) by funding the research group "MultiMOD." This study was funded by the German Research Foundation (DFG, Deutsche Forschungsgemeinschaft) as part of Germany's Excellence Strategy – EXC 2050/1 – Project ID 390696704 – Cluster of Excellence "Centre for Tactile Internet with Human-in-the-Loop" (CeTI) of Technische Universität Dresden.

REFERENCES

1. Hulla J, Sahu S, Hayes A. Nanotechnology: History and future. Human & Experimental Toxicology. 2015;34(12):1318–21.
2. Roduner E. Size matters: Why nanomaterials are different. Chemical Society Reviews. 2006;35(7):583–92.
3. Bissessur R. Nanomaterials applications. In Polymer Science and Nanotechnology 2020 Jan 1 (pp. 435–53). Elsevier.
4. Mazari SA, Ali E, Abro R, Khan FS, Ahmed I, Ahmed M, Nizamuddin S, Siddiqui TH, Hossain N, Mubarak NM, Shah A. Nanomaterials: Applications, waste-handling, environmental toxicities, and future challenges–A review. Journal of Environmental Chemical Engineering. 2021 Apr 1;9(2):105028.
5. Hristozov D, Malsch I. Hazards and risks of engineered nanoparticles for the environment and human health. Sustainability. 2009 Nov 30;1(4):1161–94.
6. Kabir E, Kumar V, Kim KH, Yip AC, Sohn JR. Environmental impacts of nanomaterials. Journal of Environmental Management. 2018 Nov 1;225:261–71.
7. Gupta N, Gupta SM, Sharma SK. Carbon nanotubes: Synthesis, properties and engineering applications. Carbon Letters. 2019 Oct;29:419–47.
8. Anzar N, Hasan R, Tyagi M, Yadav N, Narang J. Carbon nanotube-A review on Synthesis, Properties and plethora of applications in the field of biomedical science. Sensors International. 2020 Jan 1;1:100003.
9. Norizan MN, Moklis MH, Demon SZ, Halim NA, Samsuri A, Mohamad IS, Knight VF, Abdullah N. Carbon nanotubes: Functionalization and their application in chemical sensors. RSC advances. 2020;10(71):43704–32.

10. Masroor S. Functionalized carbon nanotubes: An introduction. Functionalized Carbon Nanotubes for Biomedical Applications. 2023 Mar 7:1–20.
11. Dresselhaus MS, Avouris P. Introduction to carbon materials research. In Carbon Nanotubes: Synthesis, Structure, Properties, and Applications 2001 Mar 9 (pp. 1–9). Berlin, Heidelberg: Springer Berlin Heidelberg.
12. Dresselhaus MS, Endo M. Relation of carbon nanotubes to other carbon materials. In Mildred S. Dresselhaus, Gene Dresselhaus and Phaedon Avouris (Eds.), Carbon Nanotubes: Synthesis, Structure, Properties, and Applications 2001 (pp. 11–28). Berlin, Heidelberg: Springer.
13. Han T, Nag A, Mukhopadhyay SC, Xu Y. Carbon nanotubes and its gas-sensing applications: A review. Sensors and Actuators A: Physical. 2019 Jun 1;291:107–43.
14. Basheer BV, George JJ, Siengchin S, Parameswaranpillai J. Polymer grafted carbon nanotubes – Synthesis, properties, and applications: A review. Nanostructures & Nano-Objects. 2020 Apr 1;22:100429.
15. Ghalandari M, Maleki A, Haghighi A, Shadloo MS, Nazari MA, Tlili I. Applications of nanofluids containing carbon nanotubes in solar energy systems: A review. Journal of Molecular Liquids. 2020 Sep 1;313:113476.
16. Prajapati SK, Malaiya A, Kesharwani P, Soni D, Jain A. Biomedical applications and toxicities of carbon nanotubes. Drug and Chemical Toxicology. 2022 Jan 2;45(1):435–50.
17. Heller DA, Jena PV, Pasquali M, Kostarelos K, Delogu LG, Meidl RE, Rotkin SV, Scheinberg DA, Schwartz RE, Terrones M, Wang Y. Banning carbon nanotubes would be scientifically unjustified and damaging to innovation. Nature Nanotechnology. 2020 Mar;15(3):164–6.
18. Kobayashi N, Izumi H, Morimoto Y. Review of toxicity studies of carbon nanotubes. Journal of Occupational Health. 2017 Sep;59(5):394–407.
19. Aldosari H. Review of carbon nanotube toxicity and evaluation of possible implications to occupational and environmental health. Nano Hybrids and Composites. 2023 Aug 31;40:35–49.
20. Díez-Pascual AM. Chemical functionalization of carbon nanotubes with polymers: A brief overview. Macromolecules. 2021 Mar 30;1(2):64–83.
21. Rathinavel S, Priyadharshini K, Panda D. A review on carbon nanotube: An overview of synthesis, properties, functionalization, characterization, and the application. Materials Science and Engineering: B. 2021 Jun 1;268:115095.
22. Saleemi MA, Hosseini Fouladi M, Yong PV, Chinna K, Palanisamy NK, Wong EH. Toxicity of carbon nanotubes: Molecular mechanisms, signaling cascades, and remedies in biomedical applications. Chemical Research in Toxicology. 2020 Dec 15;34(1):24–46.
23. Madani SY, Mandel A, Seifalian AM. A concise review of carbon nanotube's toxicology. Nano Reviews. 2013 Jan 1;4(1):21521.
24. Pichardo S, Gutiérrez-Praena D, Puerto M, Sánchez E, Grilo A, Cameán AM, Jos Á. Oxidative stress responses to carboxylic acid functionalized single wall carbon nanotubes on the human intestinal cell line Caco-2. Toxicology in Vitro. 2012 Aug 1;26(5):672–7.
25. Donaldson K, Aitken R, Tran L, Stone V, Duffin R, Forrest G, Alexander A. Carbon nanotubes: A review of their properties in relation to pulmonary toxicology and workplace safety. Toxicological Sciences. 2006 Jul 1;92(1):5–22.
26. Agnihotri T, Shinde T, Gitte M, Paradia PK, Tekade RK, Jain A. Functionalized carbon nanotubes for gene therapy. Functionalized Carbon Nanotubes for Biomedical Applications. 2023 Mar 7:139–70.
27. Gomathi A, Hoseini SJ, Rao CN. Functionalization and solubilization of inorganic nanostructures and carbon nanotubes by employing organosilicon and organotin reagents. Journal of Materials Chemistry. 2009;19(7):988–95.
28. Kunzmann A, Andersson B, Thurnherr T, Krug H, Scheynius A, Fadeel B. Toxicology of engineered nanomaterials: Focus on biocompatibility, biodistribution, and biodegradation. Biochimica et Biophysica Acta (BBA)-General Subjects. 2011 Mar 1;1810(3):361–73.

29. Buzea C, Pacheco II, Robbie K. Nanomaterials and nanoparticles: Sources and toxicity. Biointerphases. 2007 Dec 1;2(4):MR17–71.

30. Hirlekar R, Yamagar M, Garse H, Vij M, Kadam V. Carbon nanotubes and its applications: A review. Asian Journal of Pharmaceutical and Clinical Research. 2009 Oct;2(4):17–27.

31. Jeon I, Xiang R, Shawky A, Matsuo Y, Maruyama S. Single-walled carbon nanotubes in emerging solar cells: Synthesis and electrode applications. Advanced Energy Materials. 2019 Jun;9(23):1801312.

32. Kukovecz Á, Kozma G, Kónya Z. Multi-walled carbon nanotubes. Springer Handbook of Nanomaterials. 2013:147–88.

33. Nag A, Alahi ME, Mukhopadhyay SC, Liu Z. Multi-walled carbon nanotubes-based sensors for strain sensing applications. Sensors. 2021 Feb 10;21(4):1261.

34. Alshehri R, Ilyas AM, Hasan A, Arnaout A, Ahmed F, Memic A. Carbon nanotubes in biomedical applications: Factors, mechanisms, and remedies of toxicity: Miniperspective. Journal of Medicinal Chemistry. 2016 Sep 22;59(18):8149–67.

35. Ali M. What function of nanoparticles is the primary factor for their hyper-toxicity? Advances in Colloid and Interface Science. 2023 Mar 12:102881.

36. Kayat J, Gajbhiye V, Tekade RK, Jain NK. Pulmonary toxicity of carbon nanotubes: A systematic report. Nanomedicine: Nanotechnology, Biology and Medicine. 2011 Feb 1;7(1):40–9.

37. Selvakumar S, Rajendiran T, Biswas K. Current advances on biomedical applications and toxicity of MWCNTs: A review. BioNanoScience. 2023 Apr 29:1–9.

38. Chiaretti M, Mazzanti G, Bosco S, Bellucci S, Cucina A, Le Foche F, Carru GA, Mastrangelo S, Di Sotto A, Masciangelo R, Chiaretti AM. Carbon nanotubes toxicology and effects on metabolism and immunological modification in vitro and in vivo. Journal of Physics: Condensed Matter. 2008 Nov 6;20(47):474203.

39. Lewinski N, Colvin V, Drezek R. Cytotoxicity of nanoparticles. Small. 2008 Jan 18;4(1):26–49.

40. Satish S, Tharmavaram M, Rawtani D. Halloysite nanotubes as a nature's boon for biomedical applications. Nanobiomedicine. 2019 Jul 12;6:1849543519863625.

41. Lvov Y, Aerov A, Fakhrullin R. Clay nanotube encapsulation for functional biocomposites. Advances in Colloid and Interface Science. 2014 May 1;207:189–98.

42. Liu M, He R, Yang J, Long Z, Huang B, Liu Y, Zhou C. Polysaccharide-halloysite nanotube composites for biomedical applications: A review. Clay Minerals. 2016 Jun 1;51(3):457–67.

43. Santos AC, Ferreira C, Veiga F, Ribeiro AJ, Panchal A, Lvov Y, Agarwal A. Halloysite clay nanotubes for life sciences applications: From drug encapsulation to bioscaffold. Advances in Colloid and Interface Science. 2018 Jul 1;257:58–70.

44. Azari MR, Mohammadian Y, Pourahmad J, Khodagholi F, Mehrabi Y. Additive toxicity of Co-exposure to pristine multi-walled carbon nanotubes and benzo α pyrene in lung cells. Environmental Research. 2020 Apr 1;183:109219.

45. Sayadi MH, Pavlaki MD, Martins R, Mansouri B, Tyler CR, Kharkan J, Shekari H. Bioaccumulation and toxicokinetics of zinc oxide nanoparticles (ZnO NPs) co-exposed with graphene nanosheets (GNs) in the blackfish (Capoeta fusca). Chemosphere. 2021 Apr 1;269:128689.

46. Gao X, Ma C, Wang H, Zhang C, Huang Y. Multi-walled carbon nanotube induced liver injuries possibly by promoting endoplasmic reticulum stress in Cyprinus carpio. Chemosphere. 2023 Jun 1;325:138383.

47. Yan H, Xue Z, Xie J, Dong Y, Ma Z, Sun X, Kebebe Borga D, Liu Z, Li J. Toxicity of carbon nanotubes as anti-tumor drug carriers. International Journal of Nanomedicine. 2019 Dec 31:10179–94.

48. Chen S, Chen H, Du Q, Shen J. Targeting myeloperoxidase (MPO) mediated oxidative stress and inflammation for reducing brain ischemia injury: Potential application of natural compounds. Frontiers in Physiology. 2020 May 19;11:433.

49. Peng Z, Liu X, Zhang W, Zeng Z, Liu Z, Zhang C, Liu Y, Shao B, Liang Q, Tang W, Yuan X. Advances in the application, toxicity, and degradation of carbon nanomaterials in environment: A review. Environment International. 2020 Jan 1;134:105298.
50. Lalwani G, D'Agati M, Khan AM, Sitharaman B. Toxicology of graphene-based nanomaterials. Advanced Drug Delivery Reviews. 2016 Oct 1;105:109–44.
51. Bottini M, Bruckner S, Nika K, Bottini N, Bellucci S, Magrini A, Bergamaschi A, Mustelin T. Multi-walled carbon nanotubes induce T lymphocyte apoptosis. Toxicology Letters. 2006 Jan 5;160(2):121–6.
52. Canapè C, Foillard S, Bonafè R, Maiocchi A, Doris E. Comparative assessment of the in vitro toxicity of some functionalized carbon nanotubes and fullerenes. RSC Advances. 2015;5(84):68446–53.
53. Sharma S, Naskar S, Kuotsu K. A review on carbon nanotubes: Influencing toxicity and emerging carrier for platinum based cytotoxic drug application. Journal of Drug Delivery Science and Technology. 2019 Jun 1;51:708–20.
54. Zhao X, Chang S, Long J, Li J, Li X, Cao Y. The toxicity of multi-walled carbon nanotubes (MWCNTs) to human endothelial cells: The influence of diameters of MWCNTs. Food and Chemical Toxicology. 2019 Apr 1;126:169–77.
55. OZTURK A. Toxicology of nanoscale materials used in water treatment. The Eurasia Proceedings of Science Technology Engineering and Mathematics. 2022 Dec 31;21:435–40.
56. Jiang T, Lin Y, Amadei CA, Gou N, Rahman SM, Lan J, Vecitis CD, Gu AZ. Comparative and mechanistic toxicity assessment of structure-dependent toxicity of carbon-based nanomaterials. Journal of Hazardous Materials. 2021 Sep 15;418:126282.
57. Liu Y, Zhao Y, Sun B, Chen C. Understanding the toxicity of carbon nanotubes. Accounts of Chemical Research. 2013 Mar 19;46(3):702–13.
58. Geim AK, Novoselov KS. The rise of graphene. Nature Materials. 2007 Mar;6(3):183–91.
59. Du J, Wang S, You H, Zhao X. Understanding the toxicity of carbon nanotubes in the environment is crucial to the control of nanomaterials in producing and processing and the assessment of health risk for human: A review. Environmental Toxicology and Pharmacology. 2013 Sep 1;36(2):451–62.
60. Montes-Fonseca SL, Orrantia-Borunda E, Aguilar-Elguezabal A, Horta CG, Talamás-Rohana P, Sánchez-Ramírez B. Cytotoxicity of functionalized carbon nanotubes in J774A macrophages. Nanomedicine: Nanotechnology, Biology and Medicine. 2012 Aug 1;8(6):853–9.
61. Allegri M, Perivoliotis DK, Bianchi MG, Chiu M, Pagliaro A, Koklioti MA, Trompeta AF, Bergamaschi E, Bussolati O, Charitidis CA. Toxicity determinants of multi-walled carbon nanotubes: The relationship between functionalization and agglomeration. Toxicology Reports. 2016 Jan 1;3:230–43.
62. Wang G, Zhang J, Dewilde AH, Pal AK, Bello D, Therrien JM, Braunhut SJ, Marx KA. Understanding and correcting for carbon nanotube interferences with a commercial LDH cytotoxicity assay. Toxicology. 2012 Sep 28;299(2–3):99–111.
63. Horie M, Kato H, Fujita K, Endoh S, Iwahashi H. In vitro evaluation of cellular response induced by manufactured nanoparticles. Chemical Research in Toxicology. 2012 Mar 19;25(3):605–19.
64. Sharifi S, Behzadi S, Laurent S, Forrest ML, Stroeve P, Mahmoudi M. Toxicity of nanomaterials. Chemical Society Reviews. 2012;41(6):2323–43.
65. Raies AB, Bajic VB. In silico toxicology: Computational methods for the prediction of chemical toxicity. Wiley Interdisciplinary Reviews: Computational Molecular Science. 2016 Mar;6(2):147–72.
66. Ying J, Zhang T, Tang M. Metal oxide nanomaterial QNAR models: Available structural descriptors and understanding of toxicity mechanisms. Nanomaterials. 2015 Oct 12;5(4):1620–37.

67. Puzyn T, Rasulev B, Gajewicz A, Hu X, Dasari TP, Michalkova A, Hwang HM, Toropov A, Leszczynska D, Leszczynski J. Using nano-QSAR to predict the Cytotoxicity of metal oxide nanoparticles. Nature Nanotechnology. 2011 Mar;6(3):175–8.
68. Toropova AP, Toropov AA, Veselinović AM, Veselinović JB, Leszczynska D, Leszczynski J. Monte Carlo–based quantitative structure-activity relationship models for toxicity of organic chemicals to Daphnia magna. Environmental Toxicology and Chemistry. 2016 Nov;35(11):2691–7.
69. Usman M, Farooq M, Wakeel A, Nawaz A, Cheema SA, ur Rehman H, Ashraf I, Sanaullah M. Nanotechnology in agriculture: Current status, challenges, and future opportunities. Science of the Total Environment. 2020 Jun 15;721:137778.
70. Mittal D, Kaur G, Singh P, Yadav K, Ali SA. Nanoparticle-based sustainable agriculture and food science: Recent advances and future outlook. Frontiers in Nanotechnology. 2020 Dec 4;2:579954.
71. Shang Y, Hasan MK, Ahammed GJ, Li M, Yin H, Zhou J. Applications of nanotechnology in plant growth and crop protection: A review. Molecules. 2019 Jul 13;24(14):2558.
72. Rezaei Cherati S, Anas M, Liu S, Shanmugam S, Pandey K, Angtuaco S, Shelton R, Khalfaoui AN, Alena SV, Porter E, Fite T. Comprehensive risk assessment of carbon nanotubes used for agricultural applications. ACS Nano. 2022 Jul 22;16(8):12061–72.
73. Amenta V, Aschberger K. Carbon nanotubes: Potential medical applications and safety concerns. Wiley Interdisciplinary Reviews: Nanomedicine and Nanobiotechnology. 2015 May;7(3):371–86.
74. Thomas J, Uvesato D, Bergkamo L, Herbat N. One bird with two stones-international regulation of carbon nanotubes. Client Alert. 2008, www.huntonak.com (accessed on 22/07/23).
75. Park HG, Yeo MK. Nanomaterial regulatory policy for human health and environment. Molecular & Cellular Toxicology. 2016 Sep;12:223–36.
76. Bergamaschi E, Garzaro G, Wilson Jones G, Buglisi M, Caniglia M, Godono A, Bosio D, Fenoglio I, Guseva Canu I. Occupational exposure to carbon nanotubes and carbon nanofibres: More than a cobweb. Nanomaterials. 2021 Mar 16;11(3):745.
77. Canu IG, Bateson TF, Bouvard V, Debia M, Dion C, Savolainen K, Yu IJ. Human exposure to carbon-based fibrous nanomaterials: A review. International Journal of Hygiene and Environmental Health. 2016 Mar 1;219(2):166–75.
78. Petersen EJ, Zhang L, Mattison NT, O'Carroll DM, Whelton AJ, Uddin N, Nguyen T, Huang Q, Henry TB, Holbrook RD, Chen KL. Potential release pathways, environmental fate, and ecological risks of carbon nanotubes. Environmental Science & Technology. 2011 Dec 1;45(23):9837–56.
79. Nowack B, David RM, Fissan H, Morris H, Shatkin JA, Stintz M, Zepp R, Brouwer D. Potential release scenarios for carbon nanotubes used in composites. Environment International. 2013 Sep 1;59:1–1.
80. Pasquali M, Mesters C. We can use carbon to decarbonize – and get hydrogen for free. Proceedings of the National Academy of Sciences. 2021 Aug 3;118(31):e2112089118.

5 Fabrication of Carbon Nanotubes-Based Sensors

Aniket Chakraborthy, Pratidhwani Biswal,
Anindya Nag, and Mehmet Ercan Altinsoy

5.1 INTRODUCTION

Flexible electronic devices are in high demand; they can be used for various purposes in wearable electronics, soft robots and implanted medical devices (1). Amongst them, wearable electronics play a prominent role. Wearable sensors, which can be mounted or worn on the skin to continuously and closely monitor the actions of an individual without obstructing or restricting the user's movements, are crucial for a number of upcoming applications, such as electronic skins (2, 3), real-time physiological information acquisition (4–6), human-machine interfaces (7–9) and energy harvesting (10–12). The decade before has brought about a significant advancement in intelligent, flexible, wearable technology, including flexible strain sensors (13) and batteries (14). The adaptable strain sensor, which has a high sensitivity and strain range, has a lot of prospective applications. Numerous theories, such as resistivity (15–19), capacitance (20–23), piezoelectricity (24–26) and inductance (27–29), have been used to build flexible strain sensors. The demand for these applications has grown significantly, which has also raised the standards for developing and effective completion of such sensors (30–33). Flexible or wearable electronics, a recent development (32, 33) that is well-matched with curved surfaces and moving parts, has interesting applications in large-area electronics. The potential for the creation of flexible electronics interfaces, for complex shapes and curved designs, has been accelerated by the rapid advancement in the design of extremely thin electronics and optoelectronics devices, biocompatible encapsulating surfaces, sensors and actuators (34–38). One of the most prevalent elements on earth is carbon, which is also one of the main components of carbon nanomaterials. Carbon nanomaterials exist in various nano allotropes with varying low-dimensional structures, containing Carbon Nanotubes (CNT) in fullerene/carbon quantum dots in zero-dimension, one-dimension and two-dimensional graphene (39–41).

For more than two decades, significant research has been carried out on CNTs, and CNTs have subsequently become the most commonly used nanomaterial. CNTs, made by corrugating a single layer or several layers of graphene nanoplatelets, have a distinctive one-dimensional, narrow, tube-like structure and display exceptional conductivity and physical properties (42–44). CNTs promise to advance a number

DOI: 10.1201/9781003376071-5

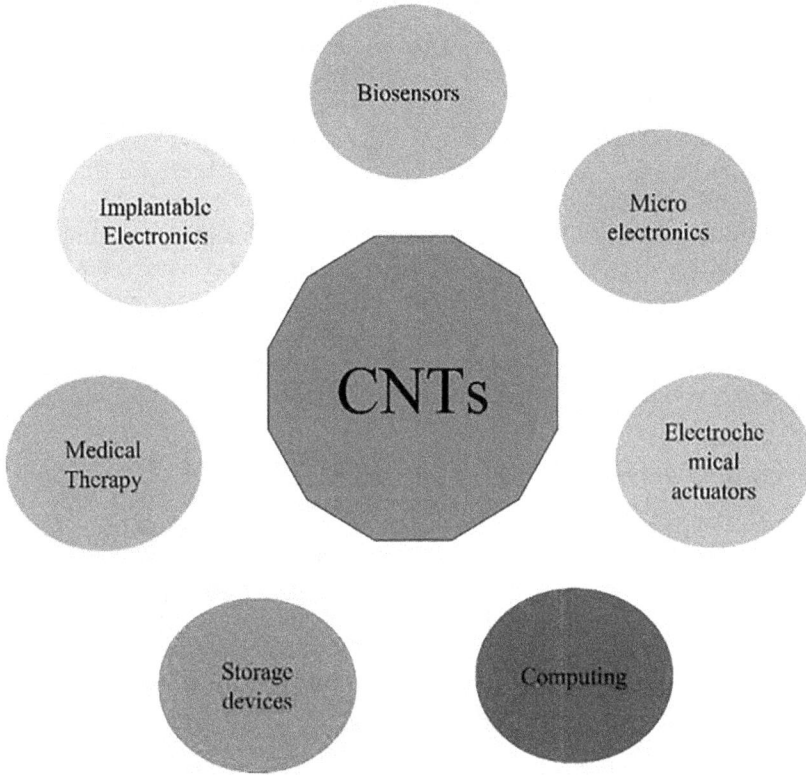

FIGURE 5.1 Application spectrum of Carbon Nanotubes depicting some of the common areas wherein the CNTs-based flexible sensors have been deployed.

of real-world technologies – prominently, in the sensing technology in the significant fields (45, 46). Hence, there is a tremendous amount of awareness focused on CNTs in many different disciplines. These characteristics include excessive, large surface-area-to-volume ratios and huge aspect-ratios, outstanding electrical conductivities, extreme chemical stability and fluorescent properties. CNTs can be utilized as electrochemical biosensors in a wide variety of fields and applications. Some of these include microelectronics, medical therapy and chemical sensing, as shown in Figure 5.1. The sensitivity of the CNTs was enhanced by doping with other components – or hybridizing with other materials. (47, 48). To be precise, the quick progress of synthetic chemistry and material science has led to an enormous number of polymeric materials being devised and employed as substrates to produce flexible devices, such as polydimethylsiloxane (PDMS) (49, 50), polyimide (PI) (51, 52), polyethylene terephthalate (PET) (53, 54) and polypropylene (55, 56). Researchers are currently investigating a variety of low-dimensional carbon nanomaterials and their related compounds, including graphene, reduced-graphene oxide and carbonized films (57, 58).

Conductive materials, which play a significant role to determine the functioning of strain sensors, typically include metallic materials like silver particles, silver nanowires and gold nanowires. It is also formed with carbon-based materials like carbon black, CNTs and graphene as well as intrinsically conducting polymers and metallic materials (19, 59). Strain sensors with high sensitivity and great stretchability can be fabricated by combining conductive materials with suitable elastic polymer materials, such as silicone rubber (60, 61) natural rubber (62, 63) and polyurethane (PU) (64, 65). Detections of peripheral signals such as strain, stress, temperature, light, moisture and chemical/biological species by conformal contact sensors could monitor real-time human activity conditions. Till date, numerous works have been reported on the fabrication of CNTs sensors. The sensors were developed using a number of different techniques. Here, a few of them are mentioned, including inkjet printing, 3D printing, laser ablation and screen printing.

5.2 CARBON NANOTUBES-BASED SENSORS

The formation of CNTs-based sensors has been carried out since its popularization in the 90s. Followed by the synthesis and optimization of the electrical, mechanical and thermal properties of CNTs, the flexible sensors are developed by integrating them using various printing techniques. These processes also include other nanomaterials and polymers that increase the overall effectiveness of the sensors in terms of performance.

5.2.1 SCREEN PRINTING

Labiano et al. (66) used traditional screen-printing techniques to develop textile antennas based on Carbon Nanotubes as smart wearables. CNTs/CNC ink was prepared for the screen-printing technique. The aqueous solution was equipped with CNC via sonication, and then, the CNTs were dispersed in the solution to get the ink composition. Two different antennas were designed and further named planar inverted cone antenna (PICA) and LOOP. PICA was selected, as it showed high capacity, better bandwidth signals, robustness and low power consumption. After investigation, it was found that the CNTs/CNC has better consistency beyond a week when compared with the pristine CNTs, which have very weak dispersion stability. Multiple layers of coating, done as a single layer, exhibits poor resistivity. As the layers increased, the resistance of the sheet decreased. For LOOP, designs with five layers were printed; using CNTs/CNC and CNTs/CNC in the ratio 1:1 and -0.5:1, respectively, lowered the resistance of the sheet from 11.5 ± 1.5 kW/sq. to 176 ± 37 W/sq. and from 3.7 ± 2.7 kW/sq. to 326 ± 78 W/sq., respectively. The fabric prototypes underwent testing in dynamic environment that simulated challenging real-world circumstances, including proximity to a human body and the effects of wrinkling, bending and fabric maintenance techniques, such as washing and ironing.

Jiang et al. (67) reported a fabric based on CNTs' conductive ink mixed with thermoplastic polyurethane (TPU) for screen-printing fabric electrodes which showed excellent mechanical properties. Carbon Nanotubes (CNTs) with high aspect ratio and conductivity are utilized as electrode fillers. Meanwhile, thermoplastic

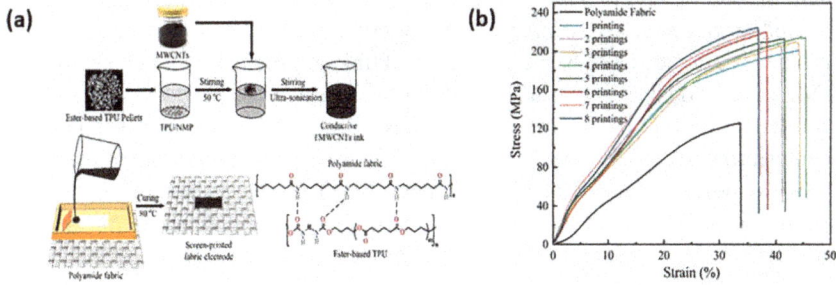

FIGURE 5.2 (a) shows the fabrication steps of the sensor, and (b) the tensile strength of the sensor with different cycles of printing (67).

polyurethane (TPU) possesses a diverse range of functional groups and a low thermal curing rate, making it suitable as a polymer matrix. Different weight ratios of TPU pellets were dissolved in NMP solvent to form a polymer matrix that was not very thick or thin. Later, MWCNTs were dissolved in the solution. A deformer – BYK052N – was used to remove the bubbles, which also affected the conductivity of the matrix. Conductivity entirely depends on the amount of CNTs in the matrix. Screen printing is one of the most promising methods for the fabrication of conductive textiles. Initially, the fabric was cleaned with ethanol and DI water to remove the impurities and then kept at 40°C to dry up. Using desktop screen printing, the conductive ink was used to print on the polyimide substrate, and the printing speed was maintained at 15 cm/s. The angle between the nozzle and the screen was 85°C. The fabric was cured at 80°C for 30 minutes.

The schematic step of the fabrication is shown in Figure 5.2 (a). As the concentration of the CNTs increased, the resistance of the matrix decreased. Owing to the fact that the CNTs could accelerate the conductive path formed in the screen-printing layer, drastically changing the impedance. The stress-strain curve revealed that the Young's modulus of the fabric increased. Figure 5.2 (b) demonstrates the tensile strength of the composite with varying number of printing layers on it. After the process was repeated 1,000 times, the average electrical conductivity of the material was found to be negligible, even after bending to5 mm and 1,000 cycles. The fabric underwent stretching, bending and folding formation, and no damage was observed in the conductivity and structure of the material. Tang et al. (68) used the screen-printing technique to develop organically integrated devices using CNTs and silver composite ink. The technique was used for the patterning of the electrodes. Ag/CNTs composite was developed to create an ideal ink composition. A thin, aluminium coating was thermally deposited to form the electrode layout of the device. A polymer layer was then applied on top, followed by a coating of cyclic olefin copolymer (COC) insulation. It was cured for 30 minutes at 130°C, and then Ag/CNTs ink was printed on it. Thermal evaporation was used to make a semiconductor layer on the electrode using C10-DNTT and PTCDI-C13. The printed electrodes formed a pattern with trapezoidal edges, allowing the semiconducting molecules to grow within the composite and ensuring a smooth extraction property. Even after repeated tensile stretching, the screen-printed Ag/CNTs showed a constant electrical conductance and accurately portrayed trapezoidal-shaped Ag/CNTs patterns. Ag/CNTs composite ink screen printing was used to create S/D electrodes and linkages for complementary inverters, NAND gates and NOR gates, among other organic integrated logic devices.

Similarly, in another article (69), Ag/CNT-based flexible thin films based on screen printing technology are mentioned. CNTs were dispersed in the aqueous solution, with sodium dodecylbenzene sulfonate (SDS) used as a dispersant for better dispersion of CNTs in the aqueous solution. Only Ag ink was screen printed, and another with the composite solution was screen printed on polyamide film, commonly known as Kapton film. The films were later sintered for 1 hour at a different temperature in the presence of nitrogen atmosphere. The performance of the Ag films was improved with the addition of CNTs fillers. The composite film could show excellent mechanical flexibility, even after 1,000 bending cycles with bending radii of 4 mm. The relative resistivity of the composite was 1.49, which was low compared to the 1.79 relative resistivity of the Ag film. The composite film showed an unchangeable electrical conductivity with the adhesion layer and remained strong, even after the accelerated ageing test with 500 cycles of thermal shocks.

Loghin et al. (70) reported an easy and scalable screen-printing technique using Ag and CNTs. Carboxymethyl cellulose (CMC) was utilized as a dispersant for better dissipation of CNTs in water. The CNTs solution was sprayed on a Kapton substrate, and interdigitated electrode patterns were screen printed with silver flakes. The resistive response of the sensor for NH_3 gas was measured for varying gas concentration. The sensor resistance increased when the sensor was kept inside the test chamber. To restore to its initial state, a direct current bias of 10 V was applied to the film which led to flow of current in the layer. The response of the sensors was noted as the concentration of gas was varied from 0 ppm to 80 ppm. It was found that the sensor has a linear response to the gas concentration. The result demonstrates an excellent sensing capability of the completely printed prototype that ranges between 0 and 22%.

Maddipatla et al. (71) developed a flexible, capacitive-based sensor to detect different pressures. Screen printing technique was used where PDMS was used as a dielectric material, and the electrodes were directly screen printed on the PDMS layer. Here, various loads were applied to the sensor, and the pressure was examined by its capacitive responses. Over the dynamic range of the sensor, its capacitive response towards pressure revealed a rise in capacitance from 6.49 pF to 7.02 pF. It was found that, at the maximum applied pressure of 337 kPa, the fabricated sensor gives an 8.2% shift in capacitance when compared to the normal base capacitance. The CNT-based pressure sensor also has a correlation coefficient of 0.9971 and a change in capacitance per kPa of 0.021%. Figure 5.3 (a and b) show the capacitance response of the sensor and the average percentage for change in capacitance. The outcomes demonstrate the viability of the CNT-based capacitive, pressurized sensor as an effective, adaptable and economical method of pressure monitoring.

Using a screen-printing approach, Pyo et al. (72) created a flexible tactile sensor based on a CNT-PDMS composite. Solvent evaporation was used to create the CNT-PDMS composite. CNTs were evenly distributed throughout the polymer matrix using a toluene solvent. The silicon wafer had an electrode layer put on it before the PI liquid was spin-coated on top of it. Thermal evaporation was then used to deposit

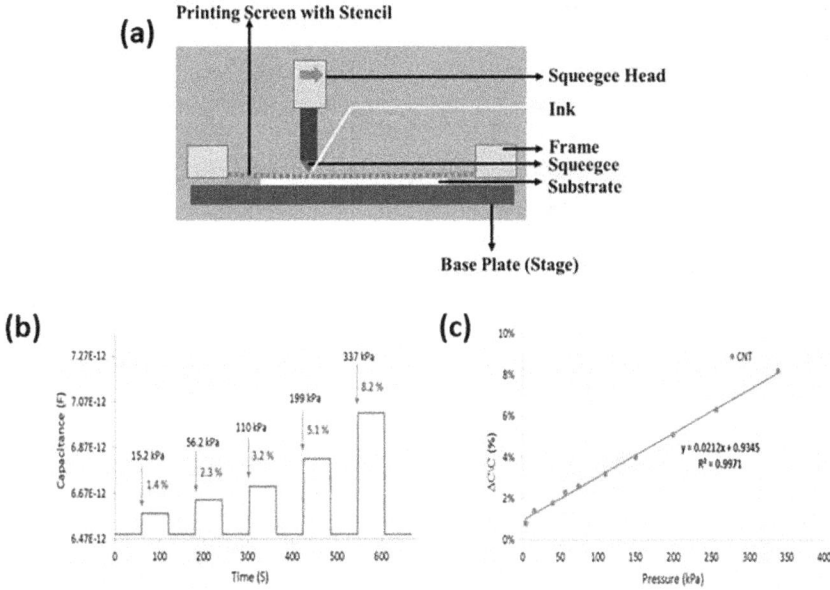

FIGURE 5.3 (a) Schematic image of screen-printing. (b) and (c) show the capacitance response of the sensor and the average percentage for change in capacitance values with respect to pressure (71).

the Cu/Au electrode layer. Later, screen printing of the CNT-PDMS composite was done on it. It was dried for 5 hours at 120°C. The manufactured sensor demonstrated significant piezoresistive properties for forces up to 2.0 N in the normal course and 0.5 N in the shear course. To ensure repeatability in load-sensing, extensive loading tests were conducted. The resistance did not show any significant change. The highest relative resistance change observed was 2.00% for a force of up to 2.0 N.

Turkani et al. (73) fabricated a five-layered negative temperature coefficient (NTC) thermistor using PET as a substrate. At first, the screen printing of the Ag electrodes was performed on the substrate; after that, it was cured at 130°C for 6 minutes. After printing the CNT-based active surface on the silver electrodes, the printing of the organic encapsulating layer was done and cured at 150°C for 5 minutes under UV radiation to improve the adhesion property of the encapsulated layer. An additional, organic, encapsulated layer was deposited on the previously mentioned active layer in order to eliminate pinholes. To finalize the five-layer assembly, a secondary, Ag-encapsulated layer was printed and dried at 130°C for 6 minutes. The performance of the sensor was studied by changing the temperature from -40°C to 100°C. An exponential decrease of the resistive response was found with an overall change of 53% along the temperature coefficient of resistance (TCR) of −0.4%/°C. The sensor also shows an improved performance of 0.34% and 0.1% change in relative humidity at constant temperatures of 30°C and 50°C. A response time of around 300 milliseconds and recovery time of 4 seconds was also estimated with an accuracy of ±5 °C. This sensor fabrication successfully demonstrated its significant potential for wearable and flexible electronics with comfortability and large-scale manufacturing capability.

Wu et al. (74) demonstrated a flexible temperature sensor made up of a combination of fake graphite (FG)/CNTs/PDMS. At first, the cyclohexane and PDMS were mixed by ultrasonication. The cyclohexane was added to reduce the agglomeration due to its miscibility with PDMS. Then, by varying the mass ratio of FG to CNTs, a homogeneous and aggregation-free dispersion was created by undergoing the process of ultrasonication for 30 min. After adding the PDMS curing agent, the mixture was kept in a vacuum oven to remove the additional cyclohexane. The prepared FG/CNTs/PDMS ink was printed on a PET substrate and dried at 100°C for 2 hours. The material's thinning due to shear and thixotropy characteristics were shown using FESEM and TGA. This low-cost, flexible sensor has exceptional temperature sensitive and stability, with a TCR value of 0.028 K-1 at a mass ratio of FG to CNTs close to 4:1.

Cao et al. (75) developed flexible, scalable, cost-effective and thin-film transistors with the help of separated single-wall CNTs (SWNCTs). The sensor was fabricated on a rigid Si/SiO_2 wafer and flexible PET substrate. Then, the layer functionalization was performed with poly-L-lysine and SWCNTs. Later, the printing of silver source, drain, barium titanate dielectric, and finally, silver gate was performed followed by etching of undesirable SWCNTs. The ink was spread on the screen, followed by drying in the oven at 140°C. The sensor shows excellent flexibility up to 7.67 cm^2 V^{-1} s^{-1}, an on/off ratio of 104 ~ 105, slight hysteresis and low operating voltage (< 10 V), in addition to prominent mechanical flexibility, with a bending radius of curvature of 3 mm and driving ability for organic LED. The sensor shows excellent movement up to 7.67 cm^2 V^{-1} s^{-1}.

Sloma et al. (76) demonstrated a screen-printed composite material based on CNTs and polyaniline (PANI) in the application of thermoelectric generators. Two different fabrication processes were applied for the sensor construction. Two separate solutions of HCl in ammonium persulphate (APS) and HCl with aniline were prepared followed by mixing, washing (with distilled water) and drying for 12 hours in a 50°C oven. To study the in-situ chemical polymerization, two different solutions were prepared: one with HCl, aniline and MWCNTs, and another, with HCl and APS, followed by their mixing, washing and drying at 50°C for 12 hours. The composite material was prepared by mixing the dried polyaniline with CNTs (40 wt. %) and PANI (60 wt. %). The dried product was printed on a rectangular template, maintaining the thickness at 50 ± 10 μm. Two other composite materials were also prepared by mixing PMMA polymer and PANI/CNTs, and a functional phase was prepared by in-situ polymerization but with low filler content. The sensor showed the best performance of σ = 405.45 S·m^{-1}, S = 15.4 μV·K^{-1}, and PF = 85.2 nW.m^{-1}K^{-2}.

Muhammad et al. (77) synthesized a modified, carbon-based electrode for electrochemical sensing and detection of thiamphenicol residue in milk. The fabrication of the sensor involves the carboxylation of carbon nanotubes with nitric acid, followed by refluxing with $HAuCl_4$. Further, the reduction of $HAuCl_4$ to Au nanoparticles was done with sodium citrate, and the final product was cured at 55°C for 2 hours. The sensor's performance was characterized by Cyclic Voltammetry (CV) and Differential Pulse Voltammetry (DPV), which illustrates a wide range of linear calibration of

0.1–30 µM with a detection limit of 0.003 µM. The sensor shows reproducibility, repeatability, stability, a lower detection limit and an extensive linear dynamic range for the detection of TAP in milk residue.

Hernández-Ibáñez et al. (78) developed an electrochemical lactate-based biosensor to detect the presence of lactose in embryonic cell cultures. The fabrication of the biosensor was done by mixing Multi-Walled Carbon Nanotubes (MWCNTs), FeMe and Chitosan (CS) in N, N'-dimethylformamide (DMF)/ethanol (1:3 v/v) solution through sonication, and then dropcasted onto graphitic working electrode. Then, the enzymatic solution prepared from Horse Radish Peroxidase (HRP), Bovine Serum Albumin (BSA) and Lactate Oxidase (LOx) dissolved in Phosphate Buffer Solution (PBS) was dropcasted on the previously mentioned layer, followed by drying the whole set-up under ultrahigh vacuum for 15 minutes. The characterization of the sensor was performed by CV and chronoamperometry (CA), which offers a linear range of 30.4–243.9 µM, a limit of detection of 22.6 µM and an excellent sensitivity of -3417 ± 131 µAM^{-1}. Moreover, the sensor exhibits high stability and excellent reproducibility, with a relative standard deviation of less than 3.8%, making it a potential candidate for lactate sensing application.

5.2.2 LASER ABLATION

The popularity of flexible sensors, owing to their excellent electrical, mechanical and thermal properties, has developed wide interest among researchers in order for them to explore different fabrication techniques (79). The selection of fabrication technique is highly dependent on characteristics of the conductive material as well as the substrate used. Among widely used laser-printed techniques, the laser ablation technique has become popular to fabricate flexible sensors. The reasons being – non-contact printing, the possibility of using various range of sensors, and utilization of many types of samples (80). Moreover, the quickness and the convenience of this technique act as advantages. Laser ablation makes it possible to use both polymer as well as metal-based materials (81). Additionally, the eco-friendly nature of the laser ablation technique is an additional advantage regarding environment (82). In this process, the surface of the target is subjected to a temperature higher than its boiling point and/or above the critical temperature during its interaction with the laser beam. After the laser-surface interaction, the vaporization starts. The temperature of the evaporated particles is further increased by the laser pulse by several tens of degrees kelvin, leading to formation of a plasma plume (83). Various methodologies adapted by different researchers for laser ablation fabrication of graphene and carbon nanotubes in flexible sensors are discussed in detail in this section.

Due to cost efficiency and high conductivity, transparent CNTs thin-films are widely used in applications involving electronic as well as optoelectronic devices. Jian et al. (82) fabricated the transparent, double-walled carbon nanotubes (DWCNTs) thin films using laser ablation technique to use in a matrix touch panel. PET was used as a substrate to form the sensors. This substrate was pre-deposited with 20–40 nm transparent DWNTs flexible thin-films. The application of a metal mask was performed upon the substrate. A pulsed Nd: YAG laser with a pulse width of 5–7 ns and a wavelength of 1,064 nm laser light was irradiated through the metal mask

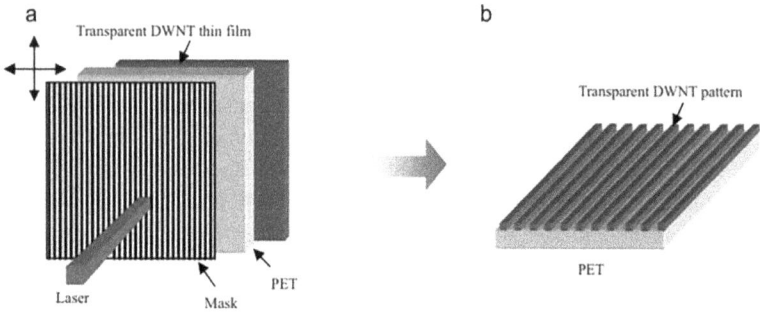

FIGURE 5.4 (a) Laser ablation of the DWCNTs-based flexible transparent matrix touch panel and (b) DWCNTs line patterns created by laser ablation after removal of the metal mask (82).

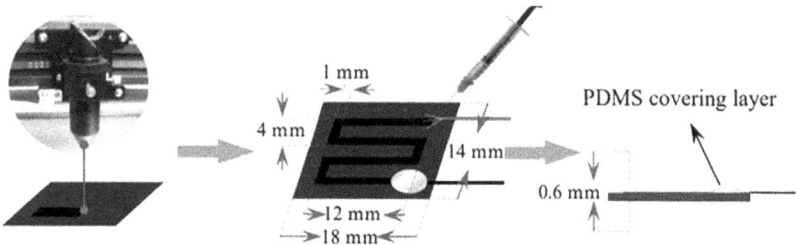

FIGURE 5.5 Laser ablation under variable power, followed by wire connection with conductive silver glue and packaging with PDMS (84).

subsequently on the DWCNTs thin layer. The defined area to be patterned on the PET substrate was entirely exposed to the laser light by simultaneous movement of the mask and the PET substrate. The absorption of laser energy by the DWNT thin film results in the evaporation of DWNTs from the PET substrate. The area protected with metal mask remains unaffected from laser ablation, and the laser-pattern is thus successfully applied. Figure 5.4 (82) shows the adapted laser ablation technique and the final laser patterned structure obtained.

Huang et al. (84) performed laser ablation to fabricate composite strain sensors. The materials used in this process were MWCNTs synthesized through Chemical Vapour Deposition (CVD) method as well as PDMS and a curing agent. Firstly, under continuous stirring for 10 minutes, the MWCNTs was dispersed into the PDMS with a planetary mixer. Further, a curing agent was added while constantly stirring for 5 minutes. The acquired mixture was then processed and cured with a home-made coating equipment at 60°C for 3 hours to create a composite film having a thickness of approximately 0.3 mm. The resultant composite film was then sliced into a size that would be appropriate for additional laser ablation processing. To construct a conducting layer on the surface of the MWCNTs/PDMS composite film, the surface of the produced film was treated using a CO_2 laser. Figure 5.5 shows the fabrication steps followed to form the MWCNTs/PDMS-based sensors.

Chrzanowska et al. (83) reported that, for a high-yielding synthesis of SWCNTs, a double-pulse laser technique was employed. The results indicate that an increase in the laser fluence may significantly alter the growth circumstances. The results also showed that the influence of UV laser radiation is low, as compared to that of infrared radiation, and the properties of the developed CNTs completely depend on laser fluence. At 355 nm laser wavelength, excellent SWCNTs could be produced at fluences F1 43J/cm^2; however, at 1,064 nm wavelength, good results could be produced at fluences F1–6J/cm^2, though the production yield fell at F > 3J/cm^2. The amount of SWCNTs particles was lowest at fluence F141 J cm^2 with a wavelength of 1,064 nm, and the large amount of CNTs particles had a diameter of 1.3 nm. With rising fluence, a wider range of SWCNTs diameters is covered by the population distribution.

Luisa Lascialfari et al. (85) demonstrated how nanohybrids could be created with ox-MWCNTs and Au NPs using a PLAL technique. In this study, a linker was employed to encourage the development of a reliable contact between the two parts. Although it may seem more time-consuming, this method enables more exact control over the chemical makeup of the nanostructured matrix and the composition of the nanohybrids. The tiny, quasi-monodispersed nanoparticles created with acetone as the solvent, and a 532 nm laser gave the greatest results. With this particular type of Au, NPs, both molecule 1 and molecule 2 linkers, were found to be effective. The nanostructured material carbon matrix was covered better due to greater solubility of molecule 2.

Naik et al. (86) showed a novel ZnS/Au/f-MWCNTs nanomaterial developed using a wet chemical procedure after having been assisted by pulsed laser technology. Using PLA on Zn aimed in DMSO – that is, served as both a solvent and a source of sulphur – ZnS nanospheres were produced. It was mentioned that neither any surfactant nor any reducing agents were utilized for the synthesis. In comparison to pure ZnS, the altered electrode made of ZnS, Au and f-MWCNTs demonstrated a considerable oxidative electrochemical reaction to 4-NP, which was explained by the beneficial synergistic interactions across the electrode components. Additionally, the increased electrochemical sensing performance of the composite was a result of strong catalytic activity of the individual material, low charge barrier and large electrochemically active surface area (ECSA). The improved composite electrode demonstrated remarkable sensitivity of 0.8084µA µM^{-1}cm^{-2} at pH 5 with a linear range of 10–150 µM and a LOD of 30 nM. The composite electrode's outstanding selectivity and sensitivity were shown by the interference and stability analyses. As a result, the present investigation explores the establishment of a novel, composite-based electrochemical sensor that might be used to detect harmful contaminants at the nanomolar level in contaminated water samples.

Araromi et al. (87) demonstrated the use of PDMS-carbon to cast electrodes with lasered patterns where oxygen plasma activation was used to establish strong, long-lasting adhesion to a silicone elastomer membrane. It shows the adaptability of the technology to construct a range of transducer designs with feature sizes under 200 m over the considerable regions. Even with short exposure time of 11 seconds, there is an excellent adherence between the electrode and elastomer membranes. Figure 5.6 (87) illustrates the fabrication technique along with certain characteristics of these flexible sensors.

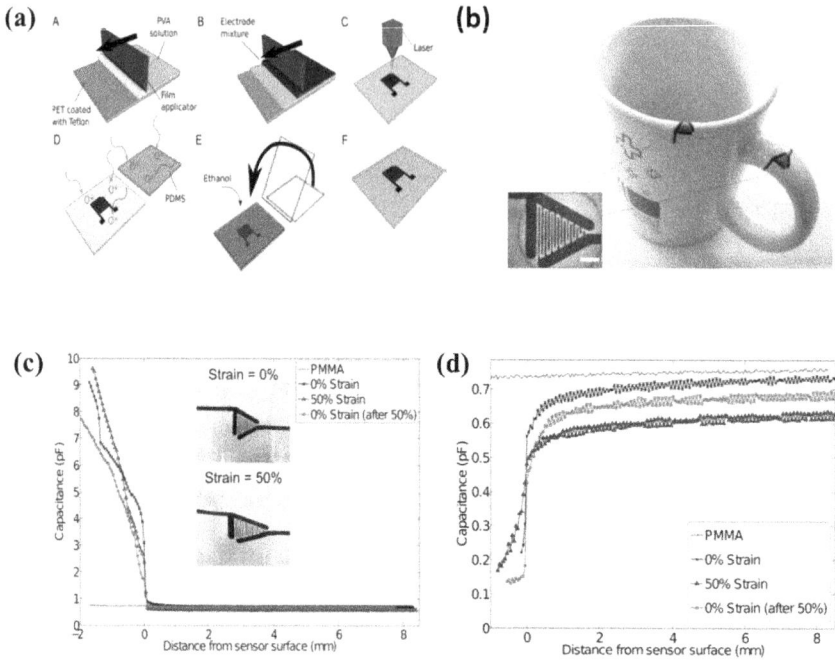

FIGURE 5.6   (a) Schematic for the fabrication of the electrodes. (b) Sensor was attached to the coffee mug. (c and d) Capacitance of the sensor at different strain levels (87).

Inset of Figure 5.6(c) shows the sensor's characteristics at 0 and 50% uniaxial strains in a perpendicular direction of the lines of the IDEs. PDMS balloon-encased high-permittivity insulators and change in impedance as a function of separation from grounded conductors were measured using LCR meter at a measuring frequency of 1 kHz [Figure 5.6 (b), (c)]. The initial value of capacitance drops by around 20% when the sensor was extended by 50%, but the ultimate values of capacitance are continued with both DI water and grounded ionic solution. The sensor has enhanced the possibility of developing thin, sharp, flexible touch-sensor arrays.

Tabassum et al. (88) described the production of MWCNT/Ta$_2$O$_5$ (MT) core-shell nanomaterials in a liquid environment using pulsed laser ablation techniques. Examining the absorption spectra of MT core-shell nanocomposite, fabricated in various refractive index solutions, was necessary to understand the characteristics of Refractive Index (RI) sensing. Figure 5.7 (88) shows the synthesis of the sensor with the experimental setup for measurements.

An 8.34 nm red shift in the sharp absorbing wavelength was observed while the refractive index of the synthesizing solution was changed from 1.33 to 1.39. In order to create Core-Shell (CS) nanostructures with the best possible sensitivity to RI sensing, the laser's operating properties, such as its laser fluence and ablation time, were modified (b). The RI sensitivity of the CS material is 240.58 nm/RIU at a RI value of 1.33, which is much better than the earlier reported values. Additionally,

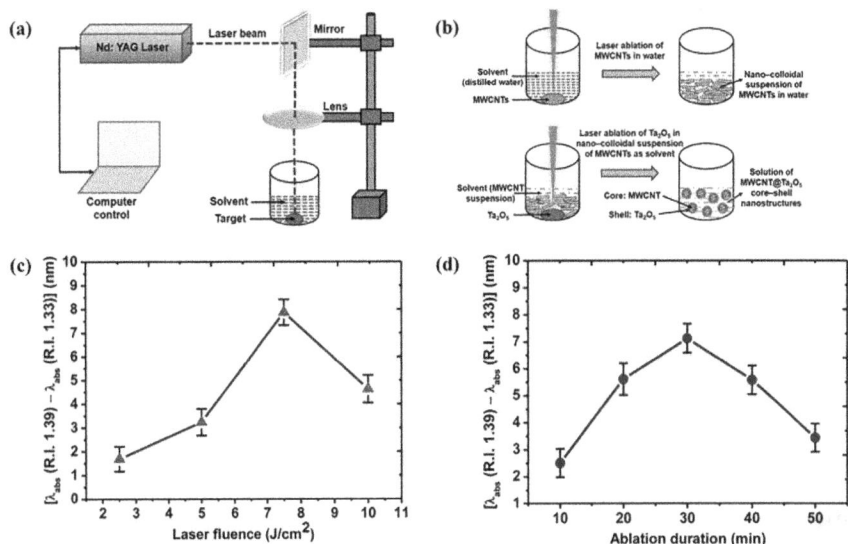

FIGURE 5.7 (a) Schematic representation of the experimental setup. (b) Steps for the synthesis of MWCNTs and Ta_2O_5. (c and d) Resistance due to laser fluence and ablation duration, respectively (88).

1.62×10^{-3} RIU of LOD was obtained, which is comparable to similar RI sensing systems. The sensing technique used in this study demonstrated a wide range of operation – specifically, RI from 1.33 to 1.39. This wide range supports its suitability for tracking (RI) in various scientific and technical applications. The highest RI sensitivity (240.58 nm/RIU) was obtained at a RI of 1.33, which also appears to be the RI of water, among other important biological fluids, demonstrating a promising future for the sensing scheme in the field of biosensing.

Shintake et al. (89) developed extremely flexible strain sensors using silicon elastomers and a composite of elastomers that were filled with carbon black. It was reported that, with the use of laser ablation, the used fabrication technique could be tailored to produce planar forms of sensor for a range of applications. The usage of the composite allowed the sensors to function at 500% load. The film-casting, layered synthesis technique enabled the development of multilayered sensors on a large scale, through an inexpensive batch-production process. The resulting sensors had either capacitance or resistance strain sensing capabilities. These sensors successfully completed over 10,000 testing cycles. The capacitive sensor had excellent repeatability, little hysteresis and good linearity (R2 = 0.9995). Its performance remained consistent, even at 50% strain rates. Not only do they provide guidance on selecting the most suitable sensor for a specific application, but they also highlight the excellent performance of highly flexible strain sensors constructed of elastomers filled with conductive carbon black (CB). The aforementioned CB-filled elastomer composite sensors, in particular, can be used with a variety of sensors like soft robotic systems and wearable gadgets.

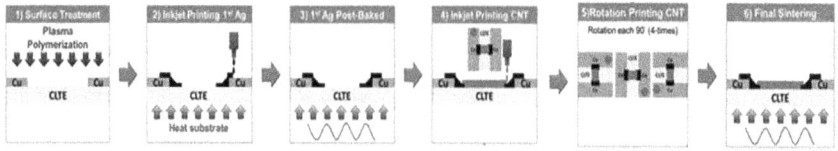

FIGURE 5.8   Use of inkjet printing technique to develop CNTs/Ag-based T flexible sensors (91).

5.2.3 INKJET PRINTING

Among various printing technologies to fabricate graphene-based flexible sensors, inkjet printing has been described (90) as a low-cost as well as rapid-printing method. No requirement of either pre-treatment or post-treatment, the possibility of simultaneous usage of multiple materials and accurate patterning are the advantages of this method. During the inkjet printing of CNTs, it is important to consider dispersion, surface tension and viscosity of the ink. Prior to carrying out the printing, a usable ink is prepared by the dispersion of Carbon Nanotubes in the liquid, followed by centrifugation, to differentiate the agglomerates and avoid the clogging of the printer nozzle. Afterwards, the prepared ink can be fed into an inkjet cartridge for printing.

Chen et al. (91) used a piezoelectric inkjet printer for printing the strain sensor on ceramic filled, woven-glass-reinforced PTFE composites substrates. A carbon nanotube ink (CNT-22, lab311, Ltd.) was used. Constant frequency of 1 kHz was used to print the silver and CNTs inks. Firstly, the surface of CLTE surface was treated with plasma polymerization at 5 watts for 5 seconds. Using 10 pL drops and a droplet spacing (DS) of 20 μm for five passes at 60°C, the first silver film was printed onto the CLTE substrate. This was followed by drying and baking for 30 minutes at 100°C. Several layers of CNTs films were printed, with a pattern rotation in every 90 degrees. Finally, sintering was carried out at 150°C for 3 hours to improve the conductivity. Figure 5.8 represents the described fabrication process (91).

Gas sensors play a vital role in both industrial and domestic applications. Due to excellent properties, such as chemical stability and higher sensitivity, CNTs are widely used as gas sensors. Alshammari et al. (92) fabricated CNTs-based sensors using inkjet printing. Two different inks – Ag nanoparticles ink as interdigitated electrodes and CNTs ink – were used in layers, as shown in Figure 5.9 (92). In the first step, Ag nanoparticles ink was used for printing on the PET substrates at a constant temperature of 60°C. Followed by printing, sintering was performed for a duration of 30 minutes at a temperature of 150°C. For the next step, CNTs inks were mixed with a little quantity of ethanol (20–30 vol. %) in order to adjust their viscosity to 5–7 cPs and surface tension to 28–33 mN/m. Subsequently, the regions between electrodes were printed with CNTs.

Michelis et al. (93) worked on MWCNTs strain sensors that were printed on sheets of ethylene tetrafluoroethylene (EFTE). The materials used were 0.125 mm thick foil of ETFE as substrate, MWCNTs Graphistrength C100, solvents – Dichlorobenzene, Acetone and Methanol – and the surfactant Sodium Dodecyl Benzene Sulphonate (SDBS). Firstly, MWCNTs were dispersed in dichlorobenzene. This step was

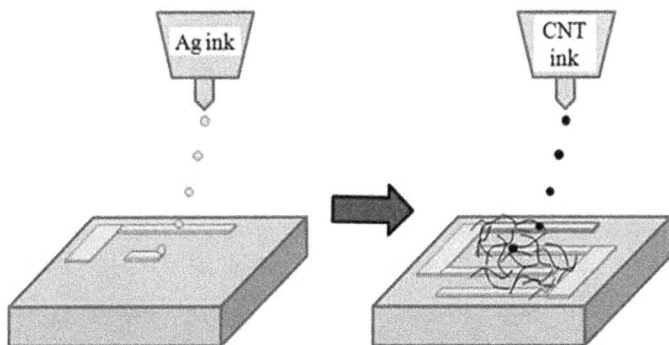

FIGURE 5.9   Inkjet printing using Ag ink and CNTs to form the electrodes (92).

followed by centrifugation for 4 hours at 10 kg. For homogenous deposition and better wettability, SDBS was added to the mixture. This resulted in the formation of a stable CNTs ink. Prior to CNTs printing, gold electrode was deposited on the substrates. During the process of printing, the substrate was heated 55°C. However, the cartridge was maintained at room temperature. The Dimatix Material Inkjet Printer 2800 with DMP-11601 cartridges was used to print the MWCNTs. It was possible to print multiple layers successively based on the property requirements.

J George et al. (94) presented an experimental proof-of-concept for an RF gas (VOC) sensor. Silver ink serves as the conductive element of the sensor, and PEDOT: PSS-MWCNTs serve as the sensitive portion. The first step in the process of fabrication was to apply a 2-micron thick layer of silver ink to the sensor's metallic component. Then, to remove the surplus solvent from it and improve its conductivity, it was heated at 60°C for 30 minutes. The PEDOT: PSS ink used to print the sensor component contains CNTs. It was then heated once more at 100°C to get rid of the extra solvents. After multiple temperature cycles that were required by the fabrication process, the characterization was completed. The relative permittivity at 2.45 GHz changed by 19% from the first reading, according to the measurements. Following that, this significant variance as well as the electrical conductivity of the printed components were taken into account when designing the gas sensor. Simulations were run by changing the conductivity of the sensitive component in order to estimate the behaviour of the resonator in presence of gas. Here, the goal was to measure the change in the conductivity of the sensitive element in response to changes in gas concentration. It was discovered that a change in concentrations between 0 ppm to 1,300 ppm causes a change in frequency of 6 MHz. By changing the conductivity of the sensitive material, simulations of the structure were performed.

Using inkjet printers to deposit CNTs and PEDOT: PSS-based inks on an adhesive polyamide fabric, the aim of this (95) work was to develop temperature sensors. Kuzubasoglu et al. suggested three distinct kinds of sensitive materials for resistive temperature sensors, including CNTs ink, PEDOT: PSS ink and CNTs/PEDOT: PSS composite ink. In order to obtain the optimal material, the dependability, sensitivity and flexibility of sensors manufactured with inks based on CNTs, PEDOT: PSS and

(a)

(b)

Wrist bending cycle

(c)

Bending cycles

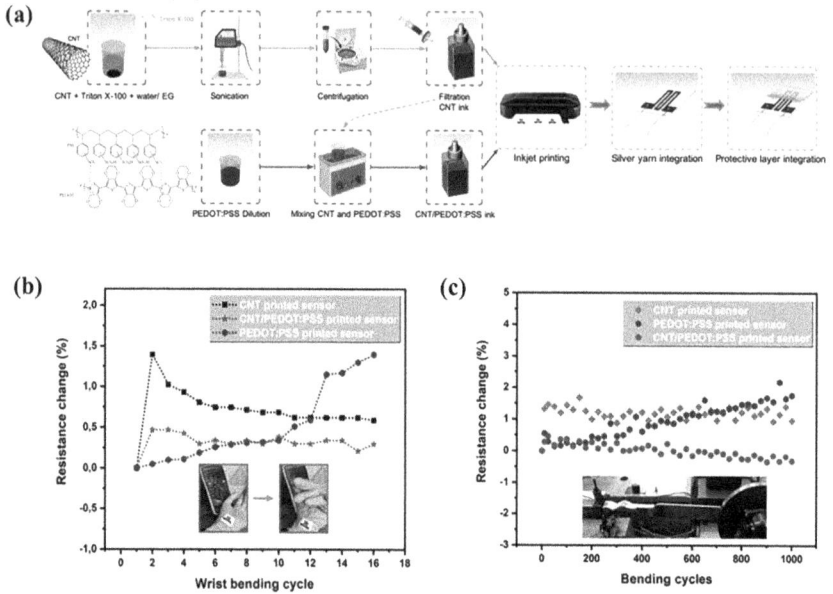

FIGURE 5.10 (a) Schematic representation of the fabrication process of the sensor. (b) Resistance changes during wrist movement. (c) Resistance changes during bending movement (95).

CNTs/PEDOT: PSS were tested. Figure 5.10 (95) shows the fabrication of the sensor. The most dependable performance was shown by the printed CNTs/PEDOT: PSS temperature sensor. The sensor's temperature sensitivity was -0.31 0.03%/°C, with a range of temperature from RT to 50°C. The resistance of the flattened sensor changes with time as a result of repetitive wrist bending, as seen in Figure 5.10(a). The highest values for the change in resistance for printed sensors made of CNTs, CNTs/PEDOT: PSS and PEDOT: PSS after 15 cycles were 0.6%, 0.3% and 1.4%, respectively. The respective results for the repeated bending tests are shown in Figure 5.10(c). It underwent repetitive bending tests up to 1,000 cycles, with a bending radius of 6 mm. The testing process lasted for three seconds for each cycle of bending.

In the work done by Kao et al. (96), CNTs was used as a sensing film and silver as a conducting film to create an inkjet-printed strain sensor. Multi-pass printing and shape spinning were the processes employed to create a uniform and stable-resistance CNTs film based on the digital picture of the inkjet printing technique. Strain sensors were created by layering silver and CNTs films that had been printed. Poor mechanical and electrical strain sensing is caused by the asymmetrical carbon nanotube layer. Because of the uneven CNTs film, the low strain performances, such as non-linearity, a limited measurement range and low sensitivity, were seen in strain sensors with unidirectional printing and 180 rotations. For 10 cycles between 0 and 3,128 $\mu\varepsilon$ strain, the change in the relative resistance for 8-layer was 0.76%, and the same for 20-layer strain sensor was 1.05% for 90 spin cycles for each sensor. Because of the density and homogeneous surface of multiple printing, the GF of the

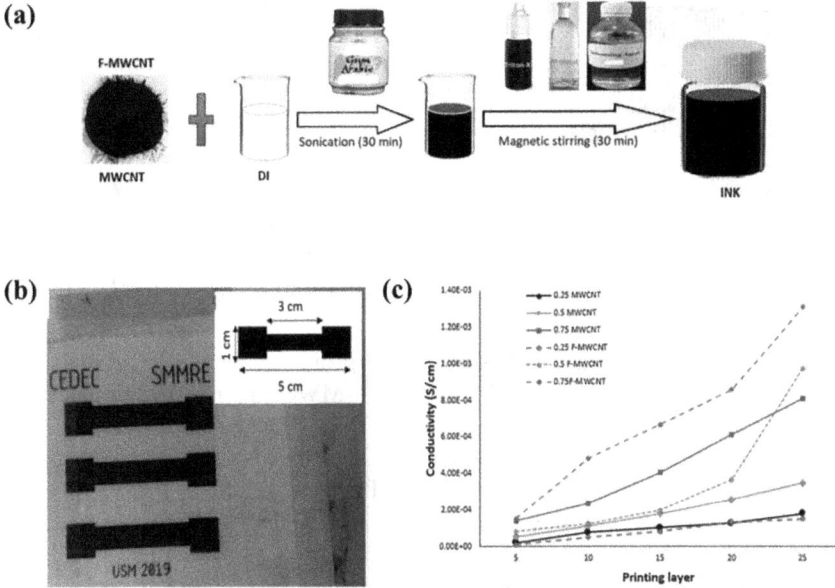

FIGURE 5.11 (a) Synthesis of the ink using MWCNTs. (b) Patterns printed on a PU substrate. (c) Conductivity of the material by number of layers (98).

20-layered sensor was higher. From 71 to 3,128, a linear and reversible behaviour of the strain sensors was observed with a 90-degree rotation. The CNT-based strain sensor was printed by inkjet printer and demonstrated that the sensor could be reversible, repeatable and linear. As a result, it was capable to categorize micro strains for scleroderma hardness differentiation and thin-film fracture detection.

As piezoresistive strain sensors, Ervasti et al. (97) demonstrated printed SWCNTs micro-patterns with stretchable, screen-printed Ag electrodes on a PDMS substrate. With a sheet resistance of 100 ohms, distinct nanoscale CNTs line patterns were deposited along with the flexible silver electrodes on the plasma-treated substrate. With a GF of 400, exceptional pressure sensitivity of 0.09 Pa^{-1} and convex bending deformation with a radius of less than 10 mm, the constructed structures exhibited excellent sensitivity to tensile strain. It was found that the fabricated sensor could be mounted on skin to monitor and identify cardiac pulses detected at the radial vein, when the finger was bent, and detect chest contraction or expansion during inhalation or exhalation.

Akindoyo et al. (98) showed how CNTs dispersions are ideal for inkjet printing using MWCNTs that have been dispersed in water to produce a nature friendly liquid conductive ink. Figure 5.11 (98) depicts the fabrication process of these sensors and their subsequent responses. In order to achieve the desired ink qualities, CNT quantities in the ink were varied between 0.25 to 0.75 wt%, while additives such as triton-x 100, polypropylene glycol and defoamer were incorporated. Figure 5.11(a) shows the steps for the preparation of the ink. The ink's fluidic characteristics, including viscosity, wetting, adsorption and fixation on the PU substrate, were successfully

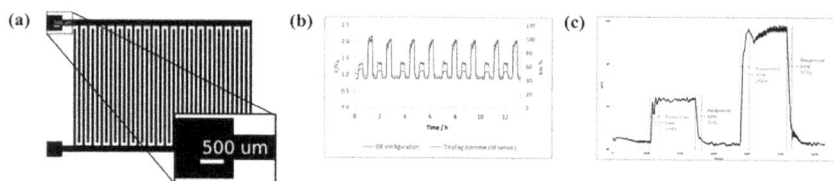

FIGURE 5.12 (a) Shows the image of the sensor design. (b) Shows the response of the sensor at different humidity levels. (c) Shows the response and recovery curve from the sensor (99).

accomplished. The material can be successfully printed using multiple rounds of the ink, and after 25 rounds for the MWCNTs and F-MWCNTs inks, the electrical conductivity can reach up till 8.11 9–10 S/cm and 1.31 9–10 S/cm, respectively. Before the experiments, the stretchable sensor was pasted on the bandage, which was printed on PU substrate. Additionally, the ink had been effectively printed using a professional desktop inkjet printer, enabling the utilization of low-cost, printable substrates and cartridges. The ink, therefore, exhibits significant promise for the environmentally viable and broad manufacturing of CNTs sensors by inkjet printing.

Gaspar et al. (99) used paper as both a flexible substrate and an active material that responds to the change in humidity. A Dimatix piezoelectric multi-nozzle inkjet printer was used to print using 10 pL ink cartridges. The drop separation was roughly 30 m. To optimize the evaporation of the solvent from the ejected droplets, the substrate temperature during printing was tuned to 60°C. The IDE arrangement used a 40-finger structure with 20 fingers for each electrode, with a line width of 200 m, a gap of 150 m and a length of 900 m, covering a total area of contact of 1 cm by 2 cm, as shown in Figure 5.12(a). When the sensor was exposed to different humidity level over a period of 12 hours at room temperature, the results showed consistent reproducibility of the structure shown in Figure 5.12 (99). The structure had a response time of about 5 minutes, but its recovery time for roughly 60 RH% was less than 4 minutes. The response time of the sensor was around 4 minutes, whereas the recovery time was less than 3 minutes for high humidity levels of 100 RH%.

5.2.4   3D PRINTING

3D printing, widely known as Additive manufacturing, is becoming popular because of its capacity to produce more sophisticated parts using computer designs. This enables flexible designing, better performance and weight reduction. 3D printing is being widely used in the field of nanomaterials, biomaterials, smart materials as well as functional materials (100). This printing technique involves layer-by-layer fabrication of a 3D object using a digital mode of file or design. It offers the possibility of using both powder and filament-based metals as well as plastics. Additionally, higher accuracy and cost effectiveness in fabrication of small parts act beneficiary for manufacturing of sensors (101).

Gnanasekaran et al. (102) printed a CNTs- and graphene-based polymer nanocomposite using desktop 3D printer. The first step was the preparation of CNTs and graphene dispersion. 50 mg of CNTs along with 100 ml of isopropanol was mixed separately with CNTs and graphene. Further, 2 hours of sonification was done in an ice bath (Bransonic 1510). This step was vital in order to prevent any temperature increase in the dispersions. The required amount of polymer nanocomposite polybutylene terephthalate (PBT) was added to the dispersions and sonicated. To produce a good dispersion free of any macroscale inhomogeneity and phase separation, isopropanol was evaporated at 90°C, using a water bath while the solution was being vigorously stirred. After 24 hours of drying, the molten, mixed composite was extruded using a DSM Xplore 15 ml mini extruder. To avoid warping during printing, double-sided tape was used as the printing substrate at a temperature of 70°C. The composite filaments were printed using an Ultimaker 2 3D printer. The squared monolayers measuring 3 cm×3 cm×0.3 mm were printed.

Strain sensors are a great part of automation industry. Based on the higher strain field requirements, flexible sensors are being highly favoured. Fused Filament Fabrication 3D printing process was employed to create efficient, flexible strain sensors made from CNTs- and GNPs-packed thermoplastic polyurethane (TPU) composites by Xiang et al. (101). The materials used were MWCNTs, GNPs, TPU particles and Dimethylformamide (DMF) reagent. Firstly, the CNTs and GNP dispersions were created with addition of DMF. Then, the sonification was done for 1 hour. Further, TPU particles were added and magnetically stirred for 2 hours. After 24 hours of drying in air oven at 80°C, the nanocomposite sheets were obtained. Dispersions with different CNTs:GNPs weight ratios, such as 7:1, 3:1, 1:1, 1:3 and 1:7, were prepared using DMF at defined proportions and drying the solution at 80°C for 24 hours. A desktop single-screw extruder was fed with the composite at 210°C for extrusion of filament. FFF using an ET-K1 desktop 3D printer was used for 3D printing of filaments onto a substrate. To ensure complete melting of the composite filament, the nozzle temperature was maintained at 220°C. To improve the adherence of the 0.1 mm thick first layer, filament was deposited at a rate of 20 mm/s onto a substrate. Throughout the process, the substrate was maintained at a temperature of 70°C. The printed samples were of dimension 50 mm x 10 mm x 1 mm. The entire procedure is schematically explained in Figure 5.13 (101).

Xiao et al. (103) proposed a flexible strain sensor by employing DLP 3D printing to fabricate a UV-curable MWCNTs/elastomer (MWCNTs/EA) composite. Aliphatic urethane diacrylate (AUD) and epoxy aliphatic acrylate (EAA) make up the highly elastic elastomer. The sensing device detected the highest strain of 60% and an overall strain of 0.01% with a MWCNTs mass proportion of 2%. In addition, the device was found to have a good mechanical reliability (10,000 cycles), sensitivity of 8.939, and predictable temperature and humidity dependences. The device also demonstrates how the Internet of Things may use NFC technology to perceive external stimuli and human movements. The study demonstrates the capability of a strain sensor array, created through dual-material printing on a DLP platform, to sense and detect distributions of environmental stimuli.

Mu et al. (104) demonstrated the use of the Digital Light Processing (DLP) 3D printing process to create detailed, conductive designs from components made up of MWCNTs, which were dissolved in a photocurable resin. The optimal quantity of MWCNTs was determined to be 0.3wt% with the printing constraints of 19.05m coat

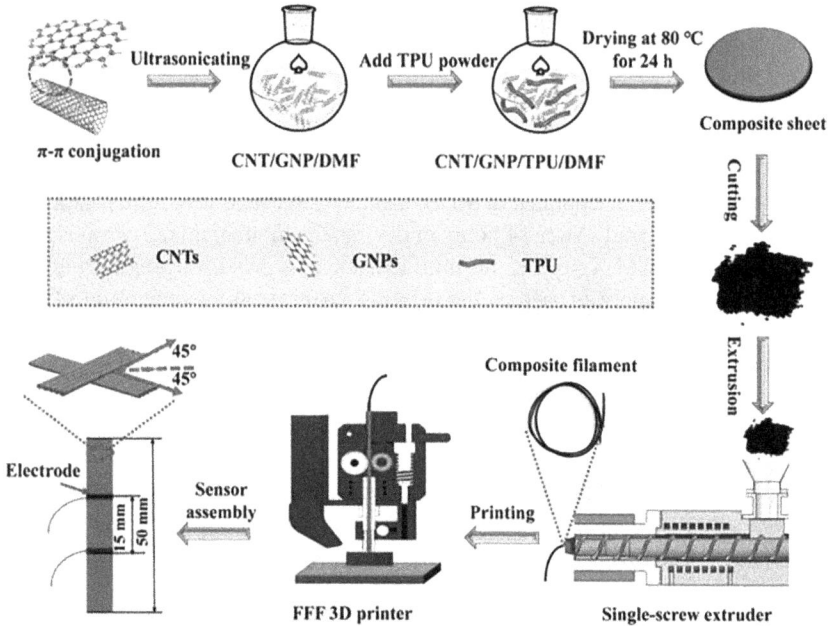

FIGURE 5.13   Schematic diagram of the use of 3D printed flexible strain sensor (101).

thickness and light radiations for a period of 40 seconds, taking into account both electrical conductivity and printability. The complicated printed designs offer decent electrical characteristics. The possible use of DLP-printed conductive complex structures includes stretchable spring circuits, partly conductive components, adaptive components with shape memory effects and hollow capacitive sensors. By balancing an easy, affordable and flexible manufacturing technique with the assuring ultimate functionality of the printed structures, these demonstrations suggest numerous possibilities in the development and synthesis of structural complicated components with customized electrical outputs.

Abshirini et al. (105) employed a 3D printing approach to develop strain sensors that are wearable and highly flexible on an electrically conductive composite substrate. Evaluating the printed sensors under continuous tensile pressure allowed researchers to assess the capacity of the sensors for strain sensing.

The experimental findings demonstrated the feasibility of fabricating a nanostructure with a usual GF of 13.01 in the range of 10–30% strain using a PDMS/MWCNTs preparation containing 1.5 wt.% amount of MWCNTs, printed using a needle with 0.41 mm diameter, and further, dried at 130°C. Improved MWCNT alignment in the nanocomposites, made possible by 3D printing, led to better performance of the sensors. The exceptional stretchability and the flexibility of the proposed sensors were demonstrated by their ability to withstand up to 146% tensile strain before breaking. In Figure 5.14 (105), it is shown that, initially, the piezoresistive response became more favourable as the pressure rate increased, representing better piezoresistive sensitivity at a better loading rate.

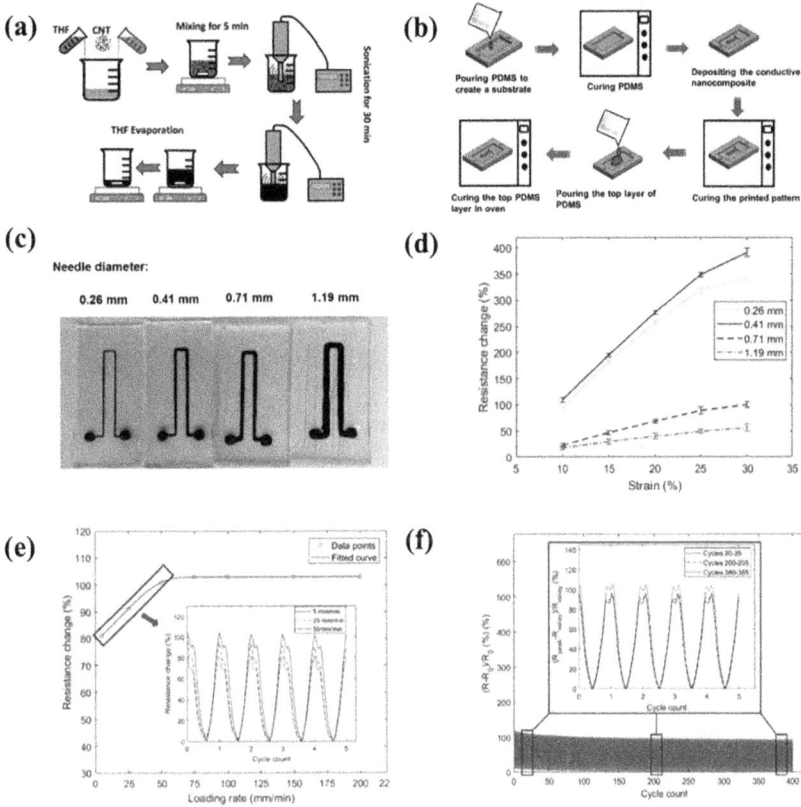

FIGURE 5.14 (a) Synthesis of the nanocomposite, (b) fabrication of the sensor, (c) effect on the sensor with various needle diameters, (d) resistance change under applied strain, (e) piezoresistive behaviour and resistance change under different loading, (f) durability test with 400 cycles (105).

However, the performance stabilized as the load rate reached 50 mm/min. In Figure 5.14(e), it can be seen that initially the piezoresistive response became more favourable as the input rate increased, representing better piezoresistive sensitivity at a higher input rate. However, the sensing performance became steady as the load rate reached 50 mm/min. An essential factor to check a system's robustness for a long-time sensing response is the strain sensor. Figure 5.10(d) demonstrates the piezoresistive behaviour of the sensor during a tensile fatigue test consisting of 400 cycles loaded up to a 10% strain. Wan et al. (106) provided a simple way to fabricate CNT-based 3D structures with electro-responsive, shape-changing capabilities. To make it printable, they first control the rheological and solvent evaporation qualities. In this combination, the shape-memory properties of PLMC and the electrical conductivity of CNTs are used. Within 16 seconds at a 25 V voltage, the sensor could display rapid electro-responsive, shape-changing behaviour. Subsequently, to create crosslinking networks, a hydrogen abstraction reaction was initiated using modified PLMC by

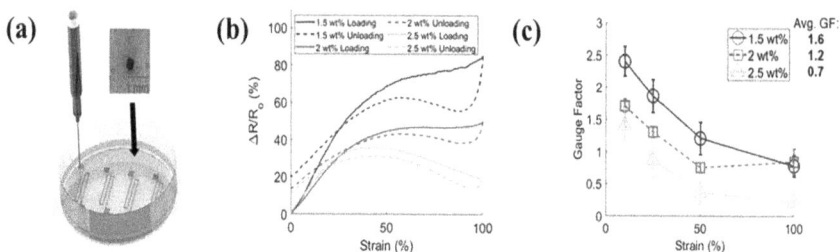

FIGURE 5.15 Pictorial representation (a) of the 3D printing, (b) of resistance change at different loadings, (c) of change in gauge factor at different wt.% ratios (107).

UV photo-initiators. The liquid sensors were printed out of cross-linked, CNT-based nanocomposites that could increase their strength and widen their liquid detecting range. The sensor has environmental adaptability due to the electro-responsive shape-memory behaviour, since the form can be changed, which was suitable for different environments. To observe the leakage of different solvents in different environments, the incorporation of multifunctional, CNT-based nanocomposites with shape-changing capabilities and customer-designed architectures may be of significant potential.

With the objective to develop a sensor with a broad strain-sensing range that may be mounted on the skin to detect human movements, Blake Herren et al. (107) developed ultra-stretchable strain sensors using an embedded 3D printing method. To determine the ideal quantity of nanoparticles that would produce the best sensitivity of the sensor, a nanocomposite consisting of silicone rubber and CNTs was mixed at various loadings. Step-sensing tests were run on the embedded sensor to determine the apparent sensing function when kept at various strains. Two tests were run to demonstrate the sensor's sensing behaviour under cyclic loading at maximum and minimum stresses and under increasing strains during one cycle. When loaded and unloaded, the sensor's relative resistance changed, and this change was astonishingly constant with the strain that was applied. The resistance gradually decreased during the intervals when a constant strain was maintained, as would be expected with these materials shown in Figure 5.15 (107). A test of 400 cycles with 50% strain was carried out to ensure the robustness of the sensor and examine resistance drift, and also, to look into the changes in the response of sensor with prolonged usage.

In this study done by Xiang et al. (101), the fused fila fabrication 3D printing technique was used to develop very flexible strain sensors made of CNTs, GNP and TPU nanocomposites. The inclusion of GNPs to the array resulted in an increased extensive conducting structure in the generated composite material, lowering the percolation threshold from 1.98 to 1.42 wt.%. A printed CNTs/GNPs (3:1)/TPU sensor demonstrated exceptional sensitivity, with a GF of 136327.4, a wide usable strain range of 250% and a remarkable stability for 3,000 cycles. Because of the synergistic impact of CNTs and GNPs, a wider distribution of CNTs in the TPU matrices was found, and both the electrical and tensile features of the sensor improved significantly. It has also been shown that strain sensors are capable of monitoring speech recognition, physiological processes and limb motions. The study can be presumed to provide an efficient technique for 3D printing of exceptionally efficient flexible strain sensors, with possible applications in wearable computing, health surveillance equipment and human-computer interaction.

5.3 EXISTING CHALLENGES OF THE CURRENT DEVICES

Although there has been significant progress in the research and development of CNT-based flexible sensors, there are still a few challenges that must be resolved in the present scenario. The conjugation of CNTs with other active nanomaterials is an area the researchers are currently addressing. The inclusion of certain nanomaterials, such as metallic nanobeads and nanoplatelets, would increase the active surface area for sensing purposes. The inclusion of metallic nanomaterials can also additionally functionalize the CNTs-based flexible sensors in order to induce characteristics, such as improved carrier mobility, high tunnelling current and electronic band, and diameter-dependent variation in the electron density of states. The hydrophilic nature of the CNTs is another sector that the researchers are constantly trying to optimize over the years. Even though certain surfactants like sodium dodecyl sulfonate (SDS) have been mixed with CNTs to improve their hydrophilic nature, there are certain limitations associated with them. Some of them include increase in toxicity, non-biodegradability, weaker stability and potential hazard to the environment (108). An alternative to these surfactants is the functionalization of CNTs to improve their hydrophilic nature in different solvents. In a similar area, the addition of different functional groups in a single sensor is another sector to work on. The addition of multiple groups plays a significant role in tuning the nature of the surface of the electrodes, and subsequently, providing higher stability during the adsorption of molecules. The optimization of the printing techniques employed to develop CNTs should be done in terms of attachment and integration of CNTs on the polymeric substrates. The transfer process used in certain cases (109, 110) should be avoided, if possible, as it affects the electromechanical properties of the formed electrodes. The uniformity in the roll-to-roll production of the CNTs-based sensors should be further improved to increase the overall production for industrial purposes. To avoid generation of a high amount of electronic waste every year, certain parameters of the sensors like biodegradability, biocompatibility and eco-friendly nature must be addressed. In order to reduce waste, a viable solution to the replacement of the sensors must be stressed; the prototypes should be developed with high robustness and longevity. One of the ways to address this is to induce multifunctional nature of the sensors. The use of CNTs other than SWCNTs and MWCNTs is another interesting area which should be stressed, particularly to form flexible CNTs-based flexible sensors. Since a lot of work has already been done with SWCNTs and MWCNTs over the years, the variation in the number of walls with subsequent variation in their physicochemical nature would be interesting for electrochemical and strain-sensing applications. In respect to applications, the integration of CNT-based sensors with flexible, printed circuit boards (FPCBs) would be cutting-edge in the coming years. The wide application spectrum of CNTs-based sensors can be exploited with the FPCBs to determine and validate the capability of the signal conditioning circuit-embedded CNTs-based sensors. The optimization of the nature of wireless communication protocol to be integrated with CNTs-based flexible sensors is also to be worked upon. This is to capitalize on the band rate and bandwidth of the wireless protocol when operating with multiple sensing prototypes.

5.4 CONCLUSION AND FUTURE PERSPECTIVES

The chapter highlights some of the important manufacturing processes employed in the fabrication and development of CNT-based flexible sensors. It covers the primary aspects on these synthesis processes by exemplifying their utilization to form the electrodes. It highlights the use of certain printing technologies, such as screen printing, inkjet printing, laser ablation and 3D printing, to produce effective flexible sensors. Some of the advantages of these techniques include cost efficiency, quick and easy fabrication process, high roll-to-roll production and high mechanical flexibility of the developed electrodes. It also elucidates some of the challenges faced by the current CNTs-based flexible sensors and their possible remedies. Given the high electromechanical characteristics of CNTs, the market value of the developed and deployed flexible sensors has constantly been increasing. These sensors can be further tested in real-time situations to validate their potential for commercialization for sensing daily activities.

ACKNOWLEDGEMENT

This work was supported by the Free State of Saxony and by the European Union (ESF Plus) by funding the research group "MultiMOD." This study was funded by the German Research Foundation (DFG, Deutsche Forschungsgemeinschaft) as part of Germany's Excellence Strategy – EXC 2050/1 – Project ID 390696704 – Cluster of Excellence "Centre for Tactile Internet with Human-in-the-Loop" (CeTI) of Technische Universität Dresden.

REFERENCES

1. Liang B, Zhang Z, Chen W, Lu D, Yang L, Yang R, et al. Direct patterning of carbon nanotube via stamp contact printing process for stretchable and sensitive sensing devices. Nano-micro Letters. 2019;11:1–11.
2. Zhan Z, Lin R, Tran V-T, An J, Wei Y, Du H, et al. Paper/carbon nanotube-based wearable pressure sensor for physiological signal acquisition and soft robotic skin. ACS Applied Materials & Interfaces. 2017;9(43):37921–8.
3. Horne J, McLoughlin L, Bury E, Koh AS, Wujcik EK. Interfacial phenomena of advanced composite materials toward wearable platforms for biological and environmental monitoring sensors, armor, and soft robotics. Advanced Materials Interfaces. 2020;7(4):1901851.
4. Kim S, Amjadi M, Lee T-I, Jeong Y, Kwon D, Kim MS, et al. Wearable, ultrawide-range, and bending-insensitive pressure sensor based on carbon nanotube network-coated porous elastomer sponges for human interface and healthcare devices. ACS Applied Materials & Interfaces. 2019;11(26):23639–48.

5. Bansal M, Gandhi B, editors. IoT based smart health care system using CNT electrodes (for continuous ECG monitoring). 2017 International Conference on Computing, Communication and Automation (ICCCA); 2017: IEEE.

6. Yamamoto Y, Harada S, Yamamoto D, Honda W, Arie T, Akita S, et al. Printed multifunctional flexible device with an integrated motion sensor for health care monitoring. Science Advances. 2016;2(11):e1601473.

7. Herbert R, Kim J-H, Kim YS, Lee HM, Yeo W-H. Soft material-enabled, flexible hybrid electronics for medicine, healthcare, and human-machine interfaces. Materials. 2018;11(2):187.

8. Gogurla N, Kim S. Self-powered and imperceptible electronic tattoos based on silk protein nanofiber and carbon nanotubes for human–machine interfaces. Advanced Energy Materials. 2021;11(29):2100801.

9. Yiu C, Wong TH, Liu Y, Yao K, Zhao L, Li D, et al. Skin-like strain sensors enabled by elastomer composites for human–machine interfaces. Coatings. 2020;10(8):711.

10. Dey A, Bajpai OP, Sikder AK, Chattopadhyay S, Khan MAS. Recent advances in CNT/graphene based thermoelectric polymer nanocomposite: A proficient move towards waste energy harvesting. Renewable and Sustainable Energy Reviews. 2016;53:653–71.

11. Son M, Park H, Liu L, Choi H, Kim JH, Choi H. Thin-film nanocomposite membrane with CNT positioning in support layer for energy harvesting from saline water. Chemical Engineering Journal. 2016;284:68–77.

12. Das S, Biswal AK, Parida K, Choudhary R, Roy A. Electrical and mechanical behavior of PMN-PT/CNT based polymer composite film for energy harvesting. Applied Surface Science. 2018;428:356–63.

13. Li M, Chen S, Fan B, Wu B, Guo X. Printed flexible strain sensor array for bendable interactive surface. Advanced Functional Materials. 2020;30(34):2003214.

14. Wu Z, Wang Y, Liu X, Lv C, Li Y, Wei D, et al. Carbon-nanomaterial-based flexible batteries for wearable electronics. Advanced Materials. 2019;31(9):1800716.

15. Maddipatla D, Zhang X, Bose A, Masihi S, Panahi M, Palaniappan V, et al., editors. Development of a flexible force sensor using additive print manufacturing process. 2019 IEEE International Conference on Flexible and Printable Sensors and Systems (FLEPS); 2019: IEEE.

16. Kumar B, Castro M, Feller J-F. Controlled conductive junction gap for chitosan–carbon nanotube quantum resistive vapour sensors. Journal of Materials Chemistry. 2012;22(21):10656–64.

17. Muthukumaran P, Sumathi C, Wilson J, Sekar C, Leonardi S, Neri G. Fe2O3/Carbon nanotube-based resistive sensors for the selective ammonia gas sensing. Sensor Letters. 2014;12(1):17–23.

18. Hsiao F-R, Wu I-F, Liao Y-C. Porous CNT/rubber composite for resistive pressure sensor. Journal of the Taiwan Institute of Chemical Engineers. 2019;102:387–93.

19. Han T, Nag A, Afsarimanesh N, Akhter F, Liu H, Sapra S, et al. Gold/polyimide-based resistive strain sensors. Electronics. 2019;8(5):565.

20. Zhao MQ, Ren CE, Ling Z, Lukatskaya MR, Zhang C, Van Aken KL, et al. Flexible MXene/carbon nanotube composite paper with high volumetric capacitance. Advanced Materials. 2015;27(2):339–45.

21. Feng G, Li S, Atchison JS, Presser V, Cummings PT. Molecular insights into carbon nanotube supercapacitors: Capacitance independent of voltage and temperature. The Journal of Physical Chemistry C. 2013;117(18):9178–86.

22. Feng P, Yuan Y, Zhong M, Shao J, Liu X, Xu J, et al. Integrated resistive-capacitive strain sensors based on polymer–nanoparticle composites. ACS Applied Nano Materials. 2020;3(5):4357–66.

23. Nag A, Zia AI, Mukhopadhyay S, Kosel J, editors. Performance enhancement of electronic sensor through mask-less lithography. 2015 9th International Conference on Sensing Technology (ICST); 2015: IEEE.

24. Arani AG, Jamali M, Mosayyebi M, Kolahchi R. Wave propagation in FG-CNT-reinforced piezoelectric composite micro plates using viscoelastic quasi-3D sinusoidal shear deformation theory. Composites Part B: Engineering. 2016;95:209–24.

25. Kim H, Kim Y. High performance flexible piezoelectric pressure sensor based on CNTs-doped 0–3 ceramic-epoxy nanocomposites. Materials & Design. 2018;151:133–40.

26. Zeng W, Deng W, Yang T, Wang S, Sun Y, Zhang J, et al. Gradient CNT/PVDF piezoelectric composite with enhanced force-electric coupling for soccer training. Nano Research. 2023:1–8.

27. Jang D, Yoon H, Nam I, Lee H. Effect of carbonyl iron powder incorporation on the piezoresistive sensing characteristics of CNT-based polymeric sensor. Composite Structures. 2020;244:112260.

28. Lu S, Chen D, Wang X, Shao J, Ma K, Zhang L, et al. Real-time cure behaviour monitoring of polymer composites using a highly flexible and sensitive CNT buckypaper sensor. Composites Science and Technology. 2017;152:181–9.

29. Chen H, Su Z, Song Y, Cheng X, Chen X, Meng B, Song Z, Chen D, Zhang H. Omnidirectional bending and pressure sensor based on stretchable CNT-PU sponge. Advanced Functional Materials. 2017;27(3):1604434.

30. Nag A, Nuthalapati S, Mukhopadhyay SC. Carbon fiber/polymer-based composites for wearable sensors: A review. IEEE Sensors Journal. 2022;22:10235–45.

31. Nag A, Simorangkir RB, Gawade DR, Nuthalapati S, Buckley JL, O'Flynn B, et al. Graphene-based wearable temperature sensors: A review. Materials & Design. 2022:110971.

32. Nag A, Zia AI, Li X, Mukhopadhyay SC, Kosel J. Novel sensing approach for LPG leakage detection: Part I – Operating mechanism and preliminary results. IEEE Sensors journal. 2015;16(4):996–1003.

33. Nag A, Zia AI, Li X, Mukhopadhyay SC, Kosel J. Novel sensing approach for LPG leakage detection – part II: Effects of particle size, composition, and coating layer thickness. IEEE Sensors Journal. 2015;16(4):1088–94.

34. Yao S, Zhu Y. Nanomaterial-enabled stretchable conductors: Strategies, materials and devices. Advanced Materials. 2015;27(9):1480–511.

35. Kim J, Lee M, Shim HJ, Ghaffari R, Cho HR, Son D, et al. Stretchable silicon nanoribbon electronics for skin prosthesis. Nature Communications. 2014;5(1):5747.

36. Beg S, Rizwan M, Sheikh AM, Hasnain MS, Anwer K, Kohli K. Advancement in carbon nanotubes: Basics, biomedical applications and toxicity. Journal of Pharmacy and Pharmacology. 2011;63(2):141–63.

37. Nag A, Menzies B, Mukhopadhyay SC. Performance analysis of flexible printed sensors for robotic arm applications. Sensors and Actuators A: Physical. 2018;276:226–36.

38. Nag A, Mukhopadhyay S, Kosel J, editors. Influence of temperature and humidity on carbon based printed flexible sensors. 2017 Eleventh International Conference on Sensing Technology (ICST); 2017: IEEE.

39. He S, Zhang Y, Gao J, Nag A, Rahaman A. Integration of different graphene nanostructures with PDMS to form wearable sensors. Nanomaterials. 2022;12(6):950.

40. Nag A, Alahi MEE, Mukhopadhyay SC. Recent progress in the fabrication of graphene fibers and their composites for applications of monitoring human activities. Applied Materials Today. 2021;22:100953.

41. Alahi MEE, Nag A, Mukhopadhyay SC, Burkitt L. A temperature-compensated graphene sensor for nitrate monitoring in real-time application. Sensors and Actuators A: Physical. 2018;269:79–90.

42. Gao J, He S, Nag A, Wong JWC. A review of the use of carbon nanotubes and graphene-based sensors for the detection of aflatoxin M1 compounds in milk. Sensors. 2021;21(11):3602.

43. Nag A, Alahi M, Eshrat E, Mukhopadhyay SC, Liu Z. Multi-walled carbon nanotubes-based sensors for strain sensing applications. Sensors. 2021;21(4):1261.
44. Nag A, Mukhopadhyay SC, Kosel J. Printed Flexible Sensors: Springer; 2019.
45. Nag A, Mukhopadhyay SC, Kosel J. Flexible carbon nanotube nanocomposite sensor for multiple physiological parameter monitoring. Sensors and Actuators A: Physical. 2016;251:148–55.
46. Han T, Nag A, Mukhopadhyay SC, Xu Y. Carbon nanotubes and its gas-sensing applications: A review. Sensors and Actuators A: Physical. 2019;291:107–43.
47. Hur J, Park S, Kim JH, Cho JY, Kwon B, Lee JH, et al. Ultrasensitive, transparent, flexible, and ecofriendly NO2 gas sensors enabled by oxidized single-walled carbon nanotube bundles on cellulose with engineered surface roughness. ACS Sustainable Chemistry & Engineering. 2022;10(10):3227–35.
48. Li Y, Ai Q, Mao L, Guo J, Gong T, Lin Y, et al. Hybrid strategy of graphene/carbon nanotube hierarchical networks for highly sensitive, flexible wearable strain sensors. Scientific Reports. 2021;11(1):21006.
49. Nag A, Simorangkir RB, Valentin E, Björninen T, Ukkonen L, Hashmi RM, et al. A transparent strain sensor based on PDMS-embedded conductive fabric for wearable sensing applications. IEEE Access. 2018;6:71020–7.
50. Nag A, Afasrimanesh N, Feng S, Mukhopadhyay SC. Strain induced graphite/PDMS sensors for biomedical applications. Sensors and Actuators A: Physical. 2018;271: 257–69.
51. Nag A, Mukhopadhyay SC, Kosel J. Sensing system for salinity testing using laser-induced graphene sensors. Sensors and Actuators A: Physical. 2017;264:107–16.
52. Nag A, Mukhopadhyay SC. Fabrication and implementation of printed sensors for taste sensing applications. Sensors and Actuators A: Physical. 2018;269:53–61.
53. Nag A, Mukhopadhyay SC, Kosel J. Tactile sensing from laser-ablated metallized PET films. IEEE Sensors Journal. 2016;17(1):7–13.
54. Yaqoob U, Phan D-T, Uddin AI, Chung G-S. Highly flexible room temperature NO2 sensor based on MWCNTs-WO3 nanoparticles hybrid on a PET substrate. Sensors and Actuators B: Chemical. 2015;221:760–8.
55. Pastrana J, Dsouza H, Cao Y, Figueroa J, González I, Vilatela JJ, et al. Electrode effects on flexible and robust polypropylene ferroelectret devices for fully integrated energy harvesters. ACS Applied Materials & Interfaces. 2020;12(20):22815–24.
56. Cao Y, Li W, Sepúlveda N. Performance of self-powered, water-resistant bending sensor using transverse piezoelectric effect of polypropylene ferroelectret polymer. IEEE Sensors Journal. 2019;19(22):10327–35.
57. Hu Q, Nag A, Xu Y, Han T, Zhang L. Use of graphene-based fabric sensors for monitoring human activities. Sensors and Actuators A: Physical. 2021;332:113172.
58. Nag A, Simorangkir RB, Sapra S, Buckley JL, O'Flynn B, Liu Z, et al. Reduced graphene oxide for the development of wearable mechanical energy-harvesters: A review. IEEE Sensors Journal. 2021;21:26415–25.
59. Afsarimanesh N, Nag A, Sarkar S, Sabet GS, Han T, Mukhopadhyay SC. A review on fabrication, characterization and implementation of wearable strain sensors. Sensors and Actuators A: Physical. 2020;315:112355.
60. Manikkavel A, Kumar V, Kim J, Lee DJ, Park SS. Investigation of high temperature vulcanized and room temperature vulcanized silicone rubber based on flexible piezoelectric energy harvesting applications with multi-walled carbon nanotube reinforced composites. Polymer Composites. 2022;43(3):1305–18.
61. He E, Sun Y, Wang X, Chen H, Sun B, Gu B, et al. 3D angle-interlock woven structural wearable triboelectric nanogenerator fabricated with silicone rubber coated graphene oxide/cotton composite yarn. Composites Part B: Engineering. 2020;200:108244.

62. Zhan Y, Hao S, Li Y, Santillo C, Zhang C, Sorrentino L, et al. High sensitivity of multi-sensing materials based on reduced graphene oxide and natural rubber: The synergy between filler segregation and macro-porous morphology. Composites Science and Technology. 2021;205:108689.

63. Lam TN, Lee GS, Kim B, Xuan HD, Kim D, Yoo SI, et al. Microfluidic preparation of highly stretchable natural rubber microfiber containing CNT/PEDOT: PSS hybrid for fabric-sewable wearable strain sensor. Composites Science and Technology. 2021;210:108811.

64. Olivieri F, Rollo G, De Falco F, Avolio R, Bonadies I, Castaldo R, et al. Reduced graphene oxide/polyurethane coatings for wash-durable wearable piezoresistive sensors. Cellulose. 2023:1–20.

65. Chen X, Zhang D, Luan H, Yang C, Yan W, Liu W. Flexible pressure sensors based on molybdenum disulfide/hydroxyethyl cellulose/polyurethane sponge for motion detection and speech recognition using machine learning. ACS Applied Materials & Interfaces. 2022;15(1):2043–53.

66. Ibanez Labiano I, Arslan D, Ozden Yenigun E, Asadi A, Cebeci H, Alomainy A. Screen printing carbon nanotubes textiles antennas for smart wearables. Sensors. 2021;21(14):4934.

67. Jiang L, Hong H, Hu J. Facile thermoplastic polyurethane-based multi-walled carbon nanotube ink for fabrication of screen-printed fabric electrodes of wearable e-textiles with high adhesion and resistance stability under large deformation. Textile Research Journal. 2021;91(21–22):2487–99.

68. Tang X, Wu K, Qi X, Kwon H-j, Wang R, Li Z, et al. Screen printing of silver and carbon nanotube composite inks for flexible and reliable organic integrated devices. ACS Applied Nano Materials. 2022;5(4):4801–11.

69. Hu D, Zhu W, Peng Y, Shen S, Deng Y. Flexible carbon nanotube-enriched silver electrode films with high electrical conductivity and reliability prepared by facile screen printing. Journal of Materials Science & Technology. 2017;33(10):1113–9.

70. Loghin FC, Bobinger M, Rivadeneyra A, Becherer M, Lugli P, editors. Flexible carbon nanotube sensors with screen printed and interdigitated electrodes. 2019 IEEE 19th International Conference on Nanotechnology (IEEE-NANO); 2019: IEEE.

71. Maddipatla D, Narakathu BB, Ali MM, Chlaihawi AA, Atashbar MZ, editors. Development of a novel carbon nanotube based printed and flexible pressure sensor. 2017 IEEE Sensors Applications Symposium (SAS); 2017: IEEE.

72. Pyo S, Lee J-I, Kim M-O, Chung T, Oh Y, Lim S-C, et al. Development of a flexible three-axis tactile sensor based on screen-printed carbon nanotube-polymer composite. Journal of Micromechanics and Microengineering. 2014;24(7):075012.

73. Turkani VS, Maddipatla D, Narakathu BB, Bazuin BJ, Atashbar MZ. A carbon nanotube based NTC thermistor using additive print manufacturing processes. Sensors and Actuators A: Physical. 2018;279:1–9.

74. Wu L, Qian J, Peng J, Wang K, Liu Z, Ma T, et al. Screen-printed flexible temperature sensor based on FG/CNT/PDMS composite with constant TCR. Journal of Materials Science: Materials in Electronics. 2019;30:9593–601.

75. Cao X, Chen H, Gu X, Liu B, Wang W, Cao Y, et al. Screen printing as a scalable and low-cost approach for rigid and flexible thin-film transistors using separated carbon nanotubes. ACS Nano. 2014;8(12):12769–76.

76. Słoma M, Głód MA, Wałpuski B. Printed flexible thermoelectric nanocomposites based on carbon nanotubes and polyaniline. Materials. 2021;14(15):4122.

77. Muhammad A, Hajian R, Yusof NA, Shams N, Abdullah J, Woi PM, et al. A screen printed carbon electrode modified with carbon nanotubes and gold nanoparticles as a sensitive electrochemical sensor for determination of thiamphenicol residue in milk. RSC Advances. 2018;8(5):2714–22.

78. Hernández-Ibáñez N, García-Cruz L, Montiel V, Foster CW, Banks CE, Iniesta J. Electrochemical lactate biosensor based upon chitosan/carbon nanotubes modified screen-printed graphite electrodes for the determination of lactate in embryonic cell cultures. Biosensors and Bioelectronics. 2016;77:1168–74.

79. Khan S, Lorenzelli L, Dahiya RS. Technologies for printing sensors and electronics over large flexible substrates: A review. IEEE Sensors Journal. 2014;15(6):3164–85.

80. Siegel AC, Phillips ST, Dickey MD, Lu N, Suo Z, Whitesides GM. Foldable printed circuit boards on paper substrates. Advanced Functional Materials. 2010;20(1):28–35.

81. Han T, Nag A, Afsarimanesh N, Mukhopadhyay SC, Kundu S, Xu Y. Laser-assisted printed flexible sensors: A review. Sensors. 2019;19(6):1462.

82. Chang-Jian S-K, Ho J-R, Cheng J-WJ. Fabrication of transparent double-walled carbon nanotubes flexible matrix touch panel by laser ablation technique. Optics & Laser Technology. 2011;43(8):1371–6.

83. Chrzanowska J, Hoffman J, Małolepszy A, Mazurkiewicz M, Kowalewski TA, Szymanski Z, et al. Synthesis of carbon nanotubes by the laser ablation method: Effect of laser wavelength. physica status solidi (b). 2015;252(8):1860–7.

84. Huang K, Ning H, Hu N, Liu F, Wu X, Wang S, et al. Ultrasensitive MWCNT/PDMS composite strain sensor fabricated by laser ablation process. Composites Science and Technology. 2020:108105.

85. Lascialfari L, Marsili P, Caporali S, Muniz-Miranda M, Margheri G, Serafini A, et al. Carbon nanotubes/laser ablation gold nanoparticles composites. Thin Solid Films. 2014;569:93–9.

86. Naik SS, Lee SJ, Theerthagiri J, Yu Y, Choi MY. Rapid and highly selective electrochemical sensor based on ZnS/Au-decorated f-multi-walled carbon nanotube nanocomposites produced via pulsed laser technique for detection of toxic nitro compounds. Journal of Hazardous Materials. 2021;418:126269.

87. Araromi OA, Rosset S, Shea HR. High-resolution, large-area fabrication of compliant electrodes via laser ablation for robust, stretchable dielectric elastomer actuators and sensors. ACS Applied Materials & Interfaces. 2015;7(32):18046–53.

88. Tabassum R, Kant R. Laser-ablated core-shell nanostructures of MWCNT@ Ta2O5 as plasmonic framework for implementation of highly sensitive refractive index sensor. Sensors and Actuators A: Physical. 2020;309:112028.

89. Shintake J, Piskarev Y, Jeong SH, Floreano D. Ultrastretchable strain sensors using carbon black-filled elastomer composites and comparison of capacitive versus resistive sensors. Advanced Materials Technologies. 2018;3(3):1700284.

90. Tortorich R, Choi J. Inkjet printing of carbon nanotubes. Nanomaterials. 2013;3:453–68.

91. Chen C-B, Kao H-L, Chang L-C, Cho C-L, Lin Y-C, Huang C-C, et al. Fabrication of inkjet-printed carbon nanotube for enhanced mechanical and strain-sensing performance. ECS Journal of Solid State Science and Technology. 2021;10(12):121001.

92. Alshammari AS, Alenezi MR, Lai K, Silva S. Inkjet printing of polymer functionalized CNT gas sensor with enhanced sensing properties. Materials Letters. 2017;189:299–302.

93. Michelis F, Bodelot L, Bonnassieux Y, Lebental B. Highly reproducible, hysteresis-free, flexible strain sensors by inkjet printing of carbon nanotubes. Carbon. 2015;95:1020–6.

94. George J, Abdelghani A, Bahoumina P, Tantot O, Baillargeat D, Frigui K, et al. CNT-based inkjet-printed RF gas sensor: Modification of substrate properties during the fabrication process. Sensors. 2019;19(8):1768.

95. Kuzubasoglu BA, Sayar E, Bahadir SK. Inkjet-printed CNT/PEDOT: PSS temperature sensor on a textile substrate for wearable intelligent systems. IEEE Sensors Journal. 2021;21(12):13090–7.

96. Kao H-L, Cho C-L, Chang L-C, Chen C-B, Chung W-H, Tsai Y-C. A fully inkjet-printed strain sensor based on carbon nanotubes. Coatings. 2020;10(8):792.

97. Ervasti H, Jarvinen T, Pitkanen O, Bozó É, Hiitola-Keinanen J, Huttunen O-H, et al. Inkjet-deposited single-wall carbon nanotube micropatterns on stretchable PDMS-Ag substrate–electrode structures for piezoresistive strain sensing. ACS Applied Materials & Interfaces. 2021;13(23):27284–94.

98. Akindoyo JO, Ismail NH, Mariatti M. Development of environmentally friendly inkjet printable carbon nanotube-based conductive ink for flexible sensors: Effects of concentration and functionalization. Journal of Materials Science: Materials in Electronics. 2021;32:12648–60.

99. Gaspar C, Olkkonen J, Passoja S, Smolander M. Paper as active layer in inkjet-printed capacitive humidity sensors. Sensors. 2017;17(7):1464.

100. Lee J-Y, An J, Chua CK. Fundamentals and applications of 3D printing for novel materials. Applied Materials Today. 2017;7:120–33.

101. Xiang D, Zhang X, Han Z, Zhang Z, Zhou Z, Harkin-Jones E, et al. 3D printed high-performance flexible strain sensors based on carbon nanotube and graphene nanoplatelet filled polymer composites. Journal of Materials Science. 2020;55:15769–86.

102. Gnanasekaran K, Heijmans T, Van Bennekom S, Woldhuis H, Wijnia S, De With G, et al. 3D printing of CNT-and graphene-based conductive polymer nanocomposites by fused deposition modeling. Applied Materials Today. 2017;9:21–8.

103. Xiao T, Qian C, Yin R, Wang K, Gao Y, Xuan F. 3D printing of flexible strain sensor array based on UV-curable multiwalled carbon nanotube/elastomer composite. Advanced Materials Technologies. 2021;6(1):2000745.

104. Mu Q, Wang L, Dunn CK, Kuang X, Duan F, Zhang Z, et al. Digital light processing 3D printing of conductive complex structures. Additive Manufacturing. 2017;18:74–83.

105. Abshirini M, Charara M, Marashizadeh P, Saha MC, Altan MC, Liu Y. Functional nanocomposites for 3D printing of stretchable and wearable sensors. Applied Nanoscience. 2019;9(8):2071–83.

106. Wan X, Zhang F, Liu Y, Leng J. CNT-based electro-responsive shape memory functionalized 3D printed nanocomposites for liquid sensors. Carbon. 2019;155:77–87.

107. Herren B, Saha MC, Altan MC, Liu Y. Development of ultrastretchable and skin attachable nanocomposites for human motion monitoring via embedded 3D printing. Composites Part B: Engineering. 2020;200:108224.

108. Jun LY, Mubarak N, Yee MJ, Yon LS, Bing CH, Khalid M, et al. An overview of functionalised carbon nanomaterial for organic pollutant removal. Journal of Industrial and Engineering Chemistry. 2018;67:175–86.

109. Chen H, Liu J, Cao W, He H, Li X, Zhang C. 3D printing CO2-activated carbon nanotubes host to promote sulfur loading for high areal capacity lithium-sulfur batteries. Nano Research. 2023;16(6):8281–9.

110. Dore C, Dörling B, Garcia-Pomar JL, Campoy-Quiles M, Mihi A. Hydroxypropyl scellulose adhesives for transfer printing of carbon nanotubes and metallic nanostructures. Small. 2020;16(47):2004795.

6 Carbon Nanotubes/ Polymer-Based Nanocomposite Sensors

*Aniket Chakraborthy, Anindya Nag,
and Mehmet Ercan Altinsoy*

6.1 INTRODUCTION

Flexible and stretchable electronics are expected to be the product of the electronic industry's next evolutionary phase (1, 2). They are expected to play bigger roles in the future age of intelligence and significantly alter our way of living (3). Over the past years, numerous approaches to the design of materials and structures have been explored. For example, newly developed nanostructured materials, including carbon nanotubes and graphene, have been chosen as desirable alternatives to merge with specially designed, micropatterned, stretchable substrates to create flexible, stretchable or both types of sensors (4, 5). Today, there is a growing demand for flexible and stretchable strain and pressure sensors in a variety of applications, including skin-mountable sensors (6–8), human motion detection (9, 10), robotic touch applications (11), prosthetic skin (12), human-machine interface (10–12, 13, 14), structural health monitoring (13, 14), medical devices (15–19) and many others (20). These sensing systems might be used to provide inexpensive, inconspicuous, wearable solutions to continually monitor electrophysiological activity carried out by the human body. They can include a variety of small, physical sensors (temperature, pressure and strain sensors), transmission modules and self-sustaining power supplies (15–19). Piezoresistive materials are those whose electrical resistance depends on internal strain. Examples of flexible and stretchable strain sensors include metallic strain sensors and polymer composite-based strain sensors. Metallic strain sensors are frequently employed and the highest strain that can be measured with them is 5%. Strain sensors also come in various forms based on different kinds of flexible polymeric substrates (20, 21). These substrates affect the overall mechanical flexibility of the prototypes.

Recent research has demonstrated the development of strain sensors based on nanomaterials such as Carbon Nanotubes (CNTs) (22, 23), graphene (24–26), nanoparticles (27, 28) and nanowires (29, 30). Typically, conductive fillers, like Single-Walled Carbon Nanotubes (SWCNTs) (31, 32), Multi-Walled Carbon Nanotubes (MWCNTs) (33, 34), carbon black (35, 36), graphite (37, 38) in a polymer matrix and film composites, are used to make them. For more than 20 years, researchers have been researching CNTs. It is considered that the distinctive properties of CNTs have ushered in a new age in the domain of materials, particularly in the areas of conductive

DOI: 10.1201/9781003376071-6

polymers and CNTs-based nanocomposites (39). Moreover, CNTs are 10–100 times stronger than robust steel, while weighing a small fraction of it (40, 41). Additionally, CNTs exhibit exceptional thermal stability up to 2,800°C in vacuum, electrical conductivity of 10^3 S/cm, thermal conductivity of 1900 W/mK and a very high electric current-carrying capacity, which is about twice as high as that of diamond (42). CNTs are progressively being considered as a viable alternative to traditional smart materials, partly because of their superior electrical characteristics. They have been widely acknowledged as ideal functional nanofillers in polymeric composites to produce multifunctional features because of their high aspect ratio and good mechanical, thermal and electrical properties (43–45).

The objective for designing a strain-sensing method based on a composite structure of thin and compact films is to monitor static and dynamic responses with no modification to the structure. (46, 47). The CNTs-polymer matrix composite has been the subject of intense research, primarily because it is simpler to fabricate than metals and ceramics (48, 49). Polydimethylsiloxane (PDMS) (50, 51), Ecoflex (52, 53), polyethylene terephthalate (PET) (54, 55), Polyimide (PI) (56, 57), Thermoplastic polyurethane (TPU) (58, 59), polypyrrole (PPy) (60, 61), polyaniline (PANI) (62, 63) and poly (3,4-ethylenedioxythiophene) polystyrene sulfonate (PEDOT: PSS) (64, 65) are the most used polymers for the flexible support and substrate for the sensors. PDMS has been a widely used material for microfabrication for decades. PDMS is transparent, has strong thermal stability, stable chemical properties and can be applied to the surface of electronic components very easily (66). Since PDMS preserves its elasticity and extension capability through a wide range of temperatures, it has been suggested for flexible device applications due to the ease and low cost of its production (67–69). TPUs, which have a higher and changeable modulus than the low-modulus silicone-based elastomers used in low-pressure composite-based sensors, can extend the pressure range of elastomer-based composite sensors. (70). PEDOT: PSS possesses adjustable electrical conductivity and work function, great flexibility and stretchability, in addition to optical transparency in the visible range. It is the material that is most frequently used to create the photovoltaics (PVs), displays, transistors and several types of sensing electronics, such as strain, pressure, temperature, humidity and biosensors (71). It is the most successful due to the accessibility to its aqueous solution and outstanding solution processability (72).

6.2   CARBON NANOTUBES-BASED NANOCOMPOSITE SENSORS

A. Nag et al. (34) developed a novel surfactant-induced composite based on CNTs and PDMS. The fabrication of the doped nanocomposite showed that the MWCNTs was initially dispersed in sodium dodecyl sulfonate (SDS) for better electrical and mechanical properties of the nanostructure. The composite was cast on moulds and cured at 80°C, and then, one more layer of PDMS was cast on it as shown in Figure 6.1(a). To obtain the specific impact of the surfactants, various wt.% of SDS were taken into consideration for the individual wt.% of CNTs. The response for the continuous bending cycles was noted. The test was done for 20 minutes, and the strain level was tried to keep almost constant for each cycle. The sensor was tested for tactile application, and the performance was found to be better when compared to other tactile sensors.

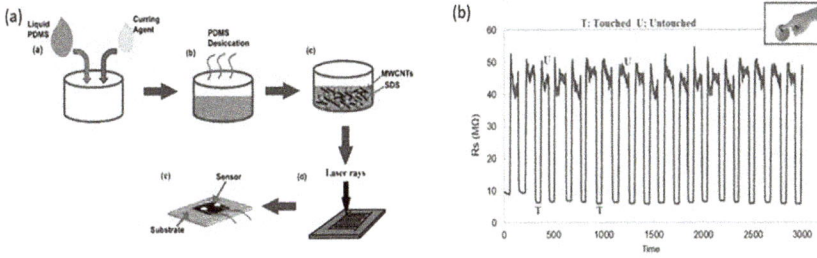

FIGURE 6.1 (a) Schematic diagram of steps involved in the fabrication of MWCNTs/PDMS-based sensor. (b) Response of the sensor to tactile sensing for exerting pressure.

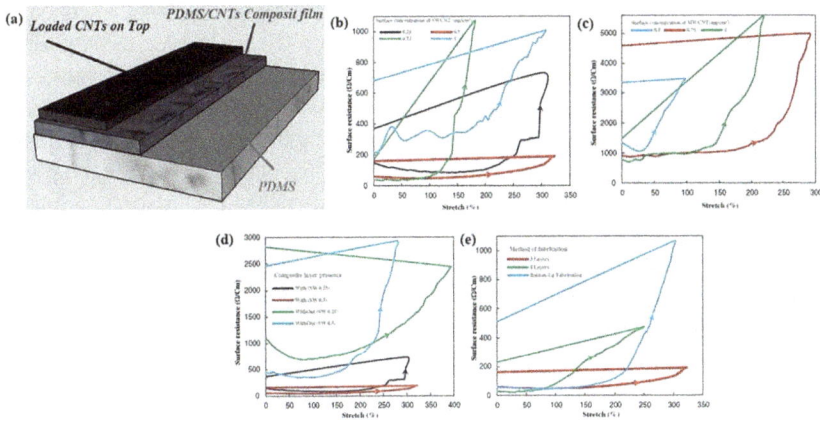

FIGURE 6.2 (a) Schematic representation of structure of the electrodes. (b and c) Surface resistance of SWCNTs and MWCNTs at different concentration. (d and e) Surface resistance change at various thickness of SWCNTs and MWCNTs (73).

From the Figure 6.1(b), it can be observed that the resistance changes as the pressure is applied to the sensor. The prototype has Young's modulus of 2.06 MPa and the response time of 1 millisecond. Although the stress-strain behaviour had 68% fracture point, their response was consistent and repeatable at 40% strain. A. Nilchian et al. (73) made a comparative study for the preparation and characterisation of CNTs based on stretchable electrodes which were aimed to develop electrical biosensors. Here, the sensor structure was composed of three layers, which could improve the electrochemical and mechanical properties. The process of fabrication was comprised of a basic solution-based method, where CNTs were coated with CNTs-PDMS composite on top of a pre cured PDMS layer. The composite structure was made of CNTs and PDMS, which was coated in between the CNTs and the substrate, thus acting a supportive layer. With CNTs on one side and PDMS on the other, CNTs in composite layers make a strong network that enhances the adhesion of electrode on the substrate, as showed in Figure 6.2(a). A variety of electrodes were developed using different amounts and thicknesses of CNTs for the active electrodes.

Figure 6.2(b) and (c) show the change in surface resistance when mechanical stress was applied at different electrodes at various surface concentration.

Similarly, Figure 6.2 (d) and (e) showed the resistance change at the electrode at different thickness of the material. A cyclic, voltametric test was conducted to study the electrochemical activity of the electrodes which were composed of SWCNTs and MWCNTs. The surface resistance and the electrical stability for each of the electrodes were tested while being mechanically, uniaxially stretched using a stretcher machine. The electrodes having surface concentration of 0.5mg/cm^2 and 0.75mg/cm^2 for SWCNTs and MWCNTs, respectively, showed excellent electrochemical redox signals of $K_3[Fe (CN)_6]$ in a phosphate buffer saline solution. S. Azhari et al. (74) reported a chemically modified technique to fabricate CNTs/PDMS composite to study the piezoresistive behaviour of the nanocomposite while applying pressure. A simple screen-printing technique was used to fabricate the nanocomposite where the CNTs were functionalized with nucleophilic group to improve the dispersion rate of the CNTs. Two wt.% of CNTs/PDMS – namely, 10% and 15% nanocomposites – were prepared. The purchased CNTs were physically and chemically conducted to increase the crystallinity and functionality. The fabricated nanocomposite showed that conductance increases with the increase in weight ratio, while electrical resistance decreases at 15 wt. % for the pressure of the nanocomposites. When pressure was applied on the sensor, there was a change in response of the sensor. Loading and unloading cycle was conducted for the resistance of 15 wt.% sensor. The test was continued for 25 cycles with constant load on the sensor, and every time, the sensor was found to retain its original resistance after the load was removed.

F. Akhter et al. (75) developed a low-cost and low-power nitrate sensor for a real-time water-quality monitoring system. The sensor was fabricated via a mould-based method where the electrodes were developed using CNTs and the substrate used in this is PDMS. The sensor was characterized to determine the temperature and the nitrate concentration by using Electrochemical Impedance Spectroscopy (EIS). The fabricated sensor can detect and differentiate the nitrate concentration varying from 0.01 ppm to 30 ppm. The microcontroller-based device is trained to detect temperature and amounts of nitrate in water samples using a machine learning algorithm. Ordinarily, the water temperature varies widely. As a result, as temperature changes, the sensor's performance suffers. The sensor responses were obtained for the nitrate concentration in samples prepared in the laboratory. The sensor has the advantage that it has a wide detection range and provides response in 1 second, which is faster compared to other sensors. The sensing system has negligible cost of maintenance and could be used by anyone with minimal training. The sensor can be reused by washing it with deionised water, which also increases its accuracy. K. Leemets et al. (76) reported a multimodal mechanical sensor for both resistance and capacitance measurements. While it was discovered that the resistance of sensor layers is significantly dependent on humidity, this was overcome by utilising a Wheatstone bridge configuration that removes environmental impacts that affect all the layers of sensor at once. The sensor was able to evaluate the it's extension and curvature simultaneously while distinguishing the signals. The Wheatstone bridge was utilised to quantify sensor bending, while the capacitive component was employed to measure elongation responses. The Wheatstone bridge is made up of four layers, two on either side of the sensor. When the sensor is flexed, the outer side is stretched and the

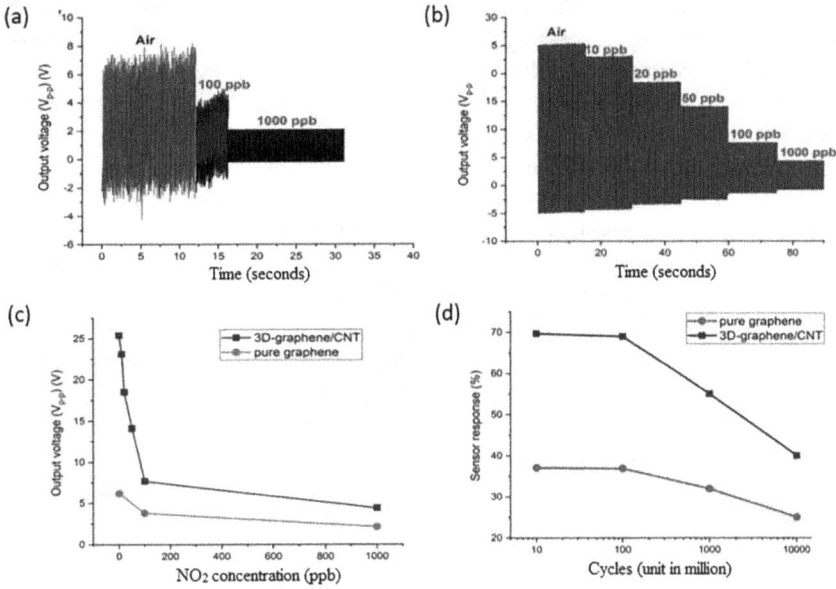

FIGURE 6.3 (a) Open-circuit voltage of pure graphene, (b) open-circuit voltage of 3D graphene/CNTs, (c) output volatge of fabricated sensor for NO_2, (d) linear sensor response for pure graphene and 3D graphene/CNTs (77).

inner side is compressed, resulting in a voltage difference between the detecting signals. It was observed that the conductivity of the sensor increases as the concentration of the CNTs increases, while the contact interface remained unchanged. It was shown that the value of impedance decreased at lower frequency. The comparatively tiny capacitive area and low resistance of the PDMS matrix, when compared to the copper silicone interface, corresponds to the high-frequency transfer of charge that happens between specific MWCNTs in that contact.

Hong et al. (77) developed a self-powered NO_2 gas sensor using graphene and graphene/CNTs composite, where a composite of PDMS/graphene was used as a sensitive layer. The addition of CNTs was to strengthen the graphene to make it mechanically robust to self-power the sensor. Here, surface modification of PDMS could enhance the triboelectric effect which leads the sensor to target for higher sensitivity. Figure 6.3(a) (77) show that, as the concentration of NO_2 gas increases, the triboelectric potential decreases in open circuit mode. The output voltage of the sensor was monitored, while it was exposed to NO_2 gas, at a constant contact force using a servomotor spinning at 250 rpm. Figure 6.3(b) shows that the voltage decreased as the concentration of gas increased up to 1,000 ppb.

The voltage drops much more exceedingly for the 3D graphene/CNTs when the gas concentration increases as compared to pure graphene, as can be seen in Figure 6.3(c). Due to better mechanical stability of the CNTs, the 3D graphene/CNTs showed better sensing performance when compared to pure graphene. In the end, long-term stability of the developed sensor was verified by multiple pressing and releasing process Figure 6.3(d). Lai et al. (78) developed a resistive sensing array that is capable of retaining and erasing the tactile images. The sensor was created by dispersed MWCNTs along with silver nanoparticles in PDMS using the dielectrophoresis (DEP) method. DEP creates an electric field that coordinates the nanoparticles of MWCNTs between the surfaces of the electrodes. It also increases the conductivity of the polymer composites. When a load was applied to the sensing component, some conducting MWCNTs networks were disrupted inside the polymer matrix, increasing the component's resistance. When the load is removed from the element and the conductive networks are restored, the sensor returns to its previous shape. The features of the sensing device were evaluated with different electrode spacing, and the findings revealed that the sensor had high repeatability. Lim et al. (79) developed a pressure sensor that was highly efficient and could be used to detect human motion and monitor health. MWCNTs and PDMS composites were fabricated using a water emulsion state. The intermediate MWCNTs generated showed both the resistive and the capacitive characteristics of the composite, and with higher MWCNTs concentration, had more resistive properties. Moreover, an effective piezo-impedance sensor was exhibited to enhance sensing performance. The sum of the capacitive GF (27) and resistive GF (28) yielded a GF of 55 for the 0.91 wt.% CNTs concentration. This GF value is significantly larger than the same for existing CNTs-based mechanical sensing systems. The GF could also be affected by the pore size of MWCNTs/PDMS composite. A piezo-impedance sensor with an outstanding GF of 55 was revealed in the composite substantially loaded with MWCNTs utilising an enhanced sensing approach based on impedance analysis. This experiment examined the effect on the concentration of MWCNTs on the piezo impedance of the composites.

Lv et al. (80) mentioned that a double, dynamic, covalent, sacrificial system was achieved in the fabrication of CNT-PDMS nanocomposite. The CNTs and PDMS nanocomposites were fabricated by mixing the CNTs with pyrenecarboxaldehyde (PA), which was chosen as a functional molecule. PA can interrelate with CNTs while stirring via π-π interaction. One-pot reaction was performed by combining H2N-PDMS-NH2, 4-aminophenyl disulfide (APDS), 1, 3, 5-triformylbenzene (TFB) and CNTs/PA into a homogeneous slurry. After the solvent had evaporated, nanocomposite elastomers were produced by hot press moulding. In this process, aromatic disulphide bonds and imine bonds acted as sacrificial units and semi crosslinking points, respectively. The unique design renders the composite a high extension, up to 1,420%; high durability, up to 5000 KJ/m³; and an excellent tensile strength, up to 1.10MPa. The mechanical properties were measured by varying chemical configuration of the CNTs concentration. The nanocomposite structure has an outstanding property of self-healing at room temperature. It could be processed multiple times. It was noted that the nanocomposite structure could be customised by adding trifluoroacetic acid, O-ethyl hydroxylamine and benzaldehyde. The previously mentioned characteristics of the nanocomposite could be used for various application. X. Xu et al. (81) proposed an ultra-stable and washable sensor in which CNTs and MXene were embedded into the PDMS matrix. The embedded microstructure and uneven surface of the composite could be comfortable to skin. The CNTs were mixed with sodium dodecyl solution and MXene were mixed with de-ionised water, and both were separately ultrasonicated. A thin film was obtained from the composite. The

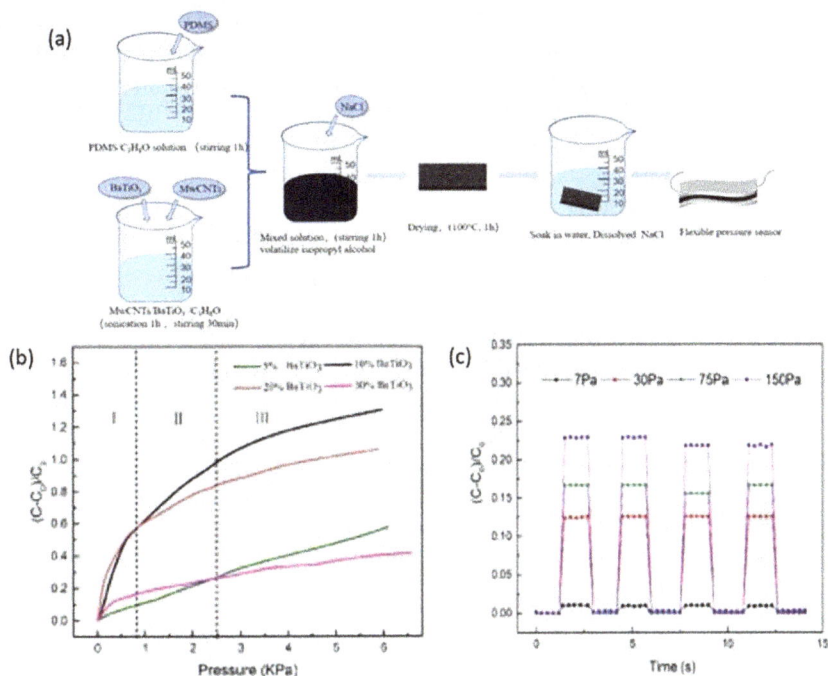

FIGURE 6.4   (a) Fabrication process of the sensor, (b) capacitive curve at different pressures, (c) response of the sensor at different pressures (82).

dry CNTs/MXene composite film was positioned on top of PTFE mould, and PDMS mix was transferred on top of it. After drying, they were cut into strips. The resistance of the CNTs/MXene sensing layer was strain sensitive. This sensor exhibited consistent responses at various frequencies and displayed durability for repeated cycling (around 1,000 cycles). The advantages of the composite strain sensor include higher resistance to temperature variations and a washable nature. The embedded microstructure of the composite sensor could be used for instantaneous supervision of mobility of human joints (fingers, knees and elbows) and a portion of a physiological signal (ECG).

F. Wang et al. (82) developed a flexible, capacitive pressure sensor. A modest method was used to fabricate absorbent PDMS matrix having a nonconducting layer filled with the mixture of MWCNTs and barium titanium oxide ($BaTiO_3$) ceramics. Both MWCNTs and $BaTiO_3$ were separately dispersed in isopropanol solution. The two solutions were then combined to create a composite solution. Later, NaCl was added to the solution while mixing constantly until the isopropanol was entirely volatilised. The solution was transferred on a mould and solidified, and two conductive adhesives were attached to the dielectric layer of the fabricated sensor as shown in Figure 6.4(a). When the fabricated sensor was characterised, the response increased with subsequent increase in amount of the $BaTiO_3$. Since the inclusion of $BaTiO_3$ enhances the polarizing capacity of the connection between the matrix and the filler, the dielectric constant increases. The sensor response figure and pressure-capacitance curves of the dielectric layers at various $BaTiO_3$ concentrations are shown in Figure 6.4(b) as three separate zones. It could be observed that the sensor doped with 10% $BaTiO_3$ exhibited sensitivity of 0.308 K/Pa in Region I, 0.102 K/Pa in 0.8 ~ 2.5 kPa in Region II and 2.5 ~ 6 kPa in Region III. When the concentration of $BaTiO_3$ exceeded 20%, the stiffness of the sensor increased and its compressibility declined, which consequently reduced its sensitivity. Comparable responses of the composite with 20% $BaTiO_3$ at various pressures are shown in Figure 6.4(c). When the pressure was loaded and unloaded, the sensor responded steadily showing excellent recovery.

A highly sensitive and flexible pressure sensor, based on conductive CNTs and PDMS elastomer with a simple production approach employing a polymer wet etching procedure, was described by Y Jung et al. (83). The exposed CNTs on the PDMS surface that has been wet etched may boost the sensitivity of the pressure sensor. The developed sensor could analyse the pressure based on the variations in the contact resistance with micro-pyramid structure. It also obtained high sensitivity at pressure lower than 250 Pa, which corresponds to the sensors being pressed by fingers. The flexible sensor could not only sense the low forces but also showed faster response. The sensor was tested for 10,000 cycles and showed high repeatability. The sensor demonstrated suitability for use as artificial skin on human fingers. It was confirmed that the sensor could respond to gentle pressure. Li Wang et al. (84) developed a wearable device that could measure wrist-pulse pressure and cardiac-electrical activity. The design of the sensing element includes a strain sensor and capacitive electrodes that were tuned using FEM to eliminate disturbance between the physical deformation of the skin and the electric field of the ECG. The sensor is comprised of CNTs-PDMS composite strips and it is covered with PDMS on top and bottom layers. This is done to insulate the noise caused by the electrical cardio-pulse during strain measurements. The polymer substrate would deform when an exterior force is applied to the sensor strip, which alters the internal arrangement of the CNTs and causes a change in resistance. To fix the output of the wrist pulse and ECG sensors, a preconditioning process (regular intervals of extension up to 20% strain) was carried out. The range for the relation between resistance change and pressure was from 0.4 kPa to 14.0 kPa, and the regulated sensitivity was 0.01 Pa^{-1}.

Y. Wang et al. (85) reported a high performance and tuneable microwave shielding with MWCNTs and PDMS composites. Firstly, CNTs were dispersed into the PDMS polymer and were ultrasonicated for several minutes. Then liquid metals (LM) were dispersed into the solution and manually stirred to obtain a uniform assortment. Then, the samples were tailed by curing the composite structure. First, the electrical conductivity of the CNTs-PDMS composites was significantly increased by the inclusion of LM. Furthermore, the electrical conductivity of the CNTs/LM/PDMS composites dropped after the preliminary rise of the LM concentration, and then subsequently, increased with increase in the LM. With the addition of 0.5 vol.% LM, the electrical conductivity of the sensor containing 1.0 vol.% CNTs increased from 0.06 to 0.11s/cm – an improvement of 83.33%. The CNTs/LM/PDMS composites displayed high-performance EMI shielding due to the beneficial interaction between nanoscale CNTs and microscale LM particles. In particular, the composites that contain 2.5 vol.% LM and 3.0 vol.% CNTs particles had an average EMI SE of 42.6 dB.

The level of electrical conductivity and EMI SE might be easily regulated by external stress, due to the deformation that results in the development of cracks and LM particle coalescence in the PDMS matrix. Xu et al. (86) created a flexible sensor for on-site electrochemical analysis. MWCNTs-based flexible sensors and related composite materials are frequently utilised in the development of strain sensors that generate a signal in response to mechanical or physical deformation. Here, MWCNTs/ PDMS electrodes were used instead of the extensively researched strain sensors for electrochemical analysis. The MWCNTs/PDMS flexible electrodes, which were prepared using a quick and easy mould transfer technique, with the ability to be modified into a broad range of sizes, shapes and geometries based on the detection environmental requirements, significantly improved the on-site electrochemical analysis module. In comparison to a two-electrode system, a three-electrode system can provide an electrochemical signal that is relatively more accurate and stable. The MWCNTs/PDMS flexible composite films, which serve as both the working and the opposite electrodes, are fabricated using a conventional moulding transfer technique. The three-electrode analysis module revealed a fast response in a wide linear range of detection, with stability and repeatability. The suggested on-site electrochemical analysis benefits from the simple fabrication process, variable processing capability, low cost of fabrication and excellent conductivity, in association with several biological and medicinal applications.

M. Ren et al. (87) developed a cost-effective, stretchable strain-sensor using CNTs and TPU. Fabrication of CNTs/TPU fibre mats were arranged in the form of a wave microstructure using electrospinning and ultrasonication to boost the sensing range and stability. Excellent conductive paths were aligned outward of the TPU fibres. CNTs/TPU mat sensing characteristics have been investigated in both vertical and parallel directions. The sensing element was aligned CNTs/TPU fibre mats, which used the joint structure of the TU fibres to provide a wide sensing range (0–900%), rapid response time (70 ms) and a low sensing-limit (0.5%). The sensor also had good stability and reproducibility. The sensors have been shown to be capable of tracking real-world applications that include the bending of wrist, squatting, bending of the finger and certain complex actions such as cheek bulging and phonation. These results showed that the sensor could be used for wearable application to monitor human activity. Y. He et al. (88) presented a wearable piezoresistive sensor for a health monitoring system. The microporous CNTs/PU films were fabricated through a simple method. Firstly, the Dimethylformamide (DMF) was mixed with PU into a homogeneous solution and dried. The CNTs were then dispersed in SDS solution to form ink. This porous PU was then dipped in the ink for 1 hr to attach the CNTs uniformly, and later, was dried in an oven at 50°C for 2 hours. The CNTs/PU film was covered with woven lining by hot pressing, and one of them was coated with conductive silver paste to work for interdigitated electrode. As the amount of porous PU on the surface area of the CNTs film increased (corresponding to a rise in PU concentration), its electrical resistance was reduced. From the Figure 6.5(b) and (c), it can be seen that the sensor can differentiate between muscle motions with excellent sensitivity and repeatability when the sensor is attached to the fingers being used for pressing, bending and grasping. Around 50% change in resistance was noticed, along with repeated pressure and relaxation, whereas 40% change was reported during the

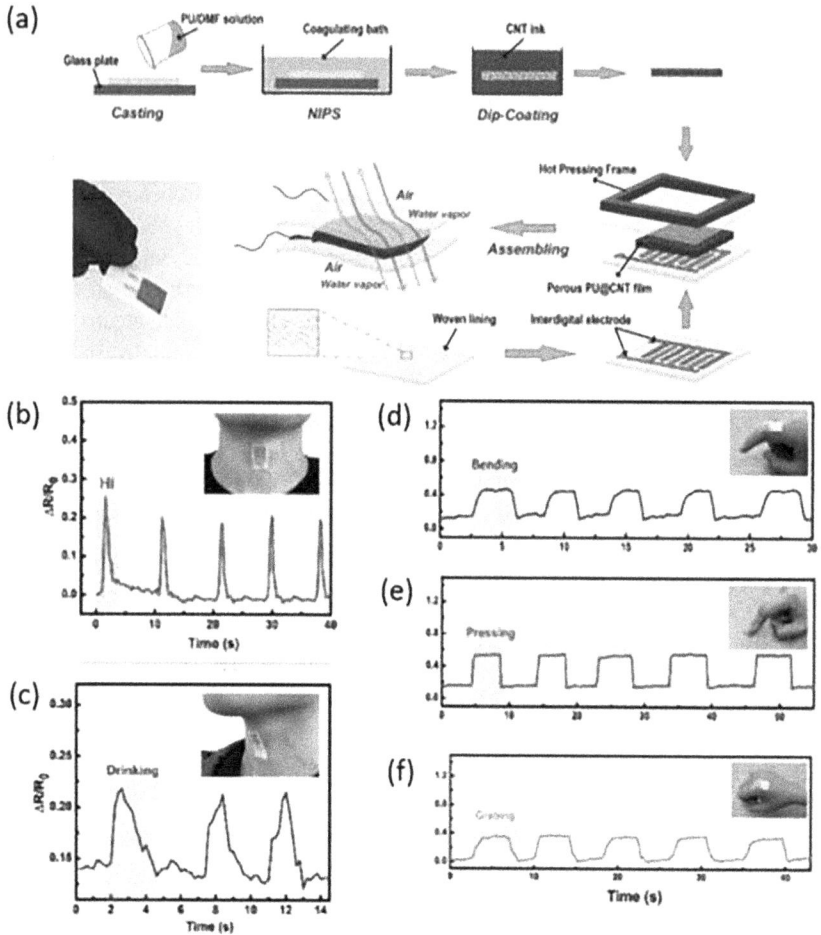

FIGURE 6.5 (a) Fabrication process of the sensor. (b and c) Detection of the muscle movement by the sensor. (d, e, and f) Different movements detected by the sensor when placed on wrist (88).

bending of the sensor and 30% change in resistance was observed while grasping which was notably lower when compared to pressing and bending.

Kim et al. (89) developed 3D printed multiaxial force sensors with the help of CNTs and TPU filaments. A novel method was employed to fabricate 3D structure-based force sensors via fused deposition modelling (FDM) for the nanocomposite filaments. A commercial 3D printer was used to print the sensors in which TPU was used for the structural part, and the CNTs/TPU composite was used for the sensing part of the sensor. Three orthogonal beams were combined to create a 3D cubic cross in a 3D-structure-based sensor. The reason for choosing FDM was its low cost of fabrication, good compatibility and convenience. A 3D cube-like structure was developed where each beam could measure the force applied in a particular direction.

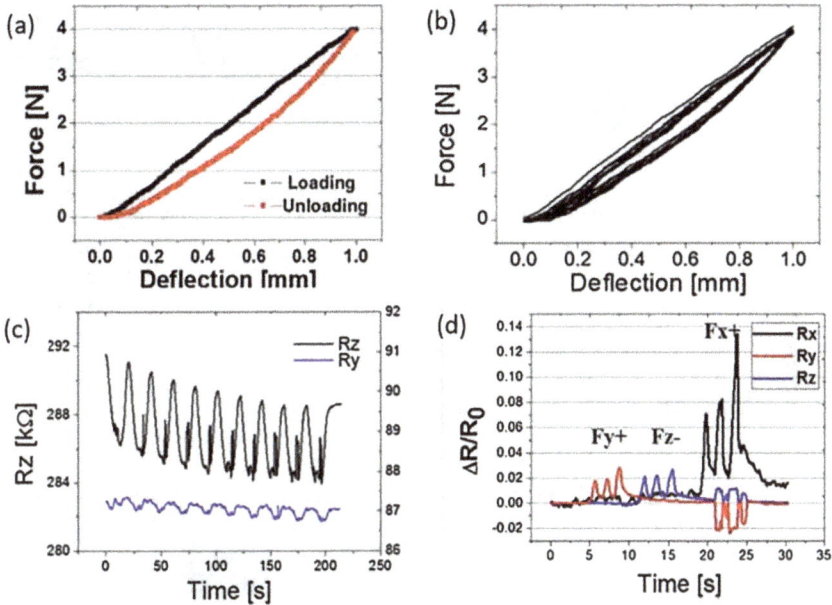

FIGURE 6.6 (a and b) Force vs deflection of the sensor while loading and unloading pressure. (c) Cyclic deflection at Rz and Ry in Z axis. (d) Shows the resistance variation at different axis (89).

There was a linear relation between the applied force and the bending caused by it. The maximum displacement along the z-axis was 1 mm with force of 4 N, whereas along the x and y axes, the forces were 3.46 N and 3.66 N, respectively. A minor hysteresis curve was also observed while loading and unloading because of viscoelastic behaviour of the TPU material. When the load was applied in the z-direction, Rz and Ry were reduced by 2% and 0.2%, respectively, showing that the sensor could self-reliantly perform with minimal crosslink. Figure 6.6(f) demonstrates the resistance variation of Rx, Ry and Rz as it is being measured in real time.

Lee et al. (90) developed a sensor that was coated with silicone rubber, which increases the range of pressure measurement and also improves the response time. The sensor was developed by encasing a layer of TPU within rubberized silicone and then coating it with MWCNTs. The TPU was used as a binding material which provides the interfacial electrical resistance. The sensor can measure pressure as low as 100 Pa and as high as 200 KPa. It is also capable of measuring oscillating pressure above 50 Hz. The sensor exhibited excellent repeatability and robustness for 1,000 cycles, which were operated at a pressure of 360 kPa. The hysteresis was lower when compared with similar sensors due to better adhesion between the material and the electrodes. The sensitivity could be controlled by controlling the amount of silicone. The sensor could be used to measure the complex pressures such as pulse and pressure beneath the heel. Tran et al. (91) showed that quantum resistive sensors (QRS) are the conductive polymer nanocomposite (CPC) where nanofillers are

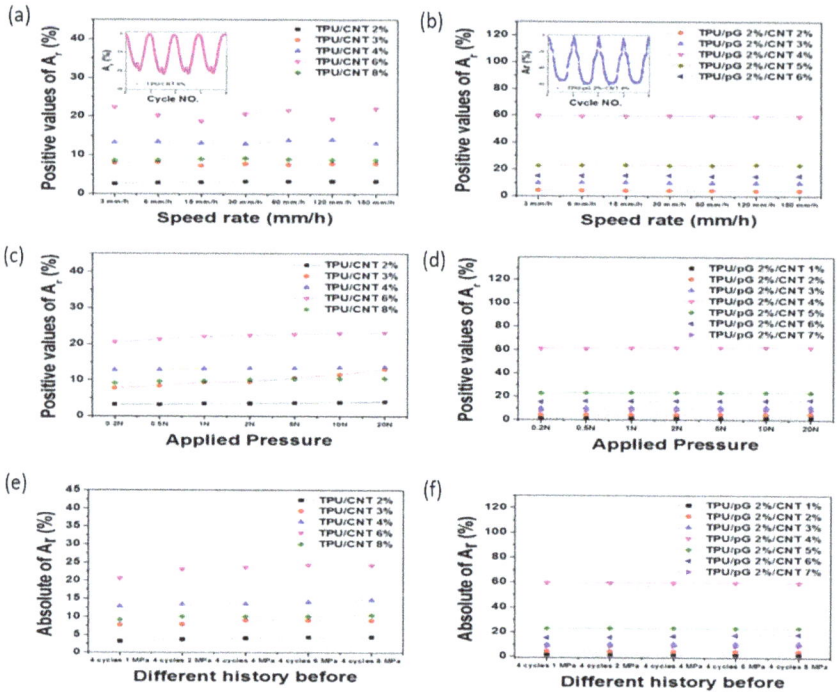

FIGURE 6.7 Effect of solicitation speed on the piezo-resistive response of nanocomposites made of (a) TPU-CNTs and (b) TPU-pG 2%/CNT, effect of various pre-compressions on the nanocomposites, (c) TPU-CNTs and (d) TPU-pG 2%/CNT, piezoresistive responses at different (e) TPU-CNTs and (f) TPU-pG 2%/CNTs (91).

used for percolation for formation of conducting architecture. QRS could be integrated into flexible electronics, smart textiles, robotics, etc. In this research, a series of conductive polymer nanocomposites were dispersed along with CNTs into TPU to form a piezo-resistive sensor, in which 3D spraying was used to create an integrated smart textile. The piezo-resistive responses of TPU-CNTs and TPU-pG 2%/CNTs composite sensors to compression at different speeds, ranging from 3 to 15 mm h1 (0.05e0, 25 mm min1), are shown in Figure 6.7(a) and 7(b). Before performing cyclic compression tests, forces oscillating from 0 to 20 N were applied on to the pQRS in order to examine the impact of precompression load on pressure sensing. Figures 7(c) and (d), respectively, show the findings for TPU-CNTs and TPU-pG 2%/CNTs pQRS at various CNTs filler concentrations. Figures 7(e) and (f) show that this mechanical pre-treatment has essentially no impact on the amplitude of the signals of the sensor.

Wang et al. (92) used the electrospinning process for TPU and MWCNTs to develop a stretchable strain sensor with a wide operating range. TPU was employed as the substrate, which was a fibre membrane with high elasticity and durability. Because of their excellent mechanical strength and electrical conductivity, CNTs were employed as conducting media. Ultrasonication was used to develop very flexible sensors by depositing the CNTs on top of the TPU membrane. Dopamine was used to alter the

TPU membrane, and it revealed a large working range of 710% and outstanding sensitivity with gauge factor of 1,200. The dopamine layer had a washable nature. During the 15,000 cycles of the stretching-releasing test, outstanding durability and good reproducibility were noted. The prototype based on the composite membrane showed exceptional flexibility and sensitivity to human body actions like bending the elbow, bending the fingers and swallowing. Xiang et al. (93) described a 3D printed high elastic strain sensor, developed using CNTs/TPU composite that was printed using FDM and pyrene carboxylic acid (PCA). PCA was adopted to transform the nanotubes non-covalently, without affecting their natural structure. The tensile and electrical characteristics of the nanocomposite were shown to improve with a homogeneous distribution of CNTs. The 3D printed sensor illustrated good properties, having a GF of 117,213 at strain of 250%. It featured a detachable strain range of 0 to 250%, high stability, up to 1,000 loading and unloading cycles, and a wide frequency response range from 0.1 to 1 Hz. The strain-sensing ability was also improved when PCA was implemented. The sensor has the capability to monitor strains at different frequencies. The sensor demonstrated its ability to track human actions, such as joint and finger movements, breathing patterns and voice recognition.

Xie et al. (94) constructed a spirally stacked SWCNT, RGO and TPU composite for strain-sensing applications. There was a connection between the CNTs and rGO as a result of the structure of the composite yarn. The spirally stacked construction was employed to improve the sensor's mechanical characteristics. The advantages of these structures include the prevention of the entrapment of conductive layers from abrasion. They also prevent the formation of a crack, which leads to high sensitivity. The bonding of CNTs with rGO resulted in a homogenous network with good mechanical and electrical characteristics. The sensor exhibited good strength, conductivity, durability, stretchability and sensitivity. These properties allow the sensor to sense a range of human actions, including speaking and knee bending. The yarn form of the strain sensors makes it possible to create electronic textiles, and it was reported to have highly sensitive and better stretchable strain sensors. Zhang et al. (95) showed a highly squeezable WCT fused foam. It was fabricated using modest solution-blending and freeze-drying technique. The dispersion was homogeneous and had good interfacial bonding between cellulose nanocrystal (CNC) and TPU, which improved the compression properties. The entanglement of the CNTs/CNC increased the conductive network and was able to generate outstanding resistance. These networks had 2.5 times higher sensitivity than the ones without CNC. The piezoresistive sensor exhibits high compressibility and a steady piezoresistive sensing output in the 80% compressive strain range due to the porous material and excellent flexibility of WPU. Following the first stabilization step, the piezoresistive sensor demonstrates excellent recovery and repeatability in tests with varying compression stresses and speeds. It also demonstrated outstanding durability and reproducibility over 1,000 compression cycles as well as a fast response time of roughly 30 ms. Conclusively, the piezoresistive sensor was found to detect diverse human movements.

Kuzubasoglu et al. (96) produced a temperature sensor employing CNTs and PEDOT: PSS based inks that were immediately coated on the sticky polyamide oriented taffeta material, using an inkjet printer. CNTs and PEDOT: PSS were evenly disseminated and mixed with triton X-100 before being employed as temperature

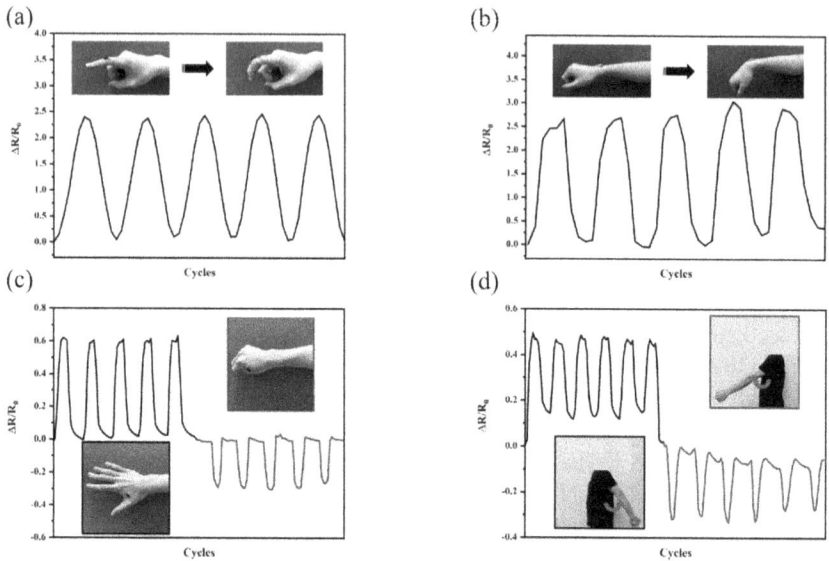

FIGURE 6.8 Application of the sensor at (a) folding fingers, (b) elbow bending, (c) wrist movements, (d) arm stretching (97).

inkjet inks. Three varieties of sensors were developed using the mentioned material composites and were attached to the measuring device with silver paste and silver yarns. For temperature measurement, ranging from room temperature to 50°C, all of the sensors exhibited negative temperature coefficient (NTC) behaviours and sensitivity of 0.15%/°C for CNT, 0.41%/°C for PEDOT:PSS and 0.31%/°C for CNTs/PEDOT:PSS. In comparison to CNTs and PEDOT:PSS printed sensors, CNTs/PEDOT:PSS composite ink printed sensors exhibited higher bending stability and achieved higher sensing repeatability, with a resistance variation of 0.3% up to 1,000 cycles. He et al. (97) developed ultra-stretchable PEDOT:PSS/CNTs composite films, which were fabricated using electrospinning and vacuum filtration. Polycaprolactone (PCL) nanofibres were used as a binder for the composite. The composite film was found to have better stretchability and more than 400% of facture strain. The composite films were used to detect the human motion like finger bending, wrist bending, etc. Figures 6.8(a) and (b) demonstrate how the strain sensor was used for monitoring the human body parts undergoing significant deformations, such as finger flexion and extension and wrist-joint movement. The strain sensors were also attached to the shoulder and palm for the sensing application and recorded some interesting results. The sensor at the shoulder joint could recognise the direction of the arm movement and showed change in resistance signals as shown in Figures 6.8(c) and (d). The results suggest that the composite film could be used for real time human health monitoring systems. The work done by Lam et al. (98) discussed about the wearable strain sensor. Mechanically tough and highly stretchable stormers microfibers consisting of natural rubbers strain sensor were fabricated using PEDOT: PSS and CNTs. Here PEDOT: PSS acts as a bridge in the composite

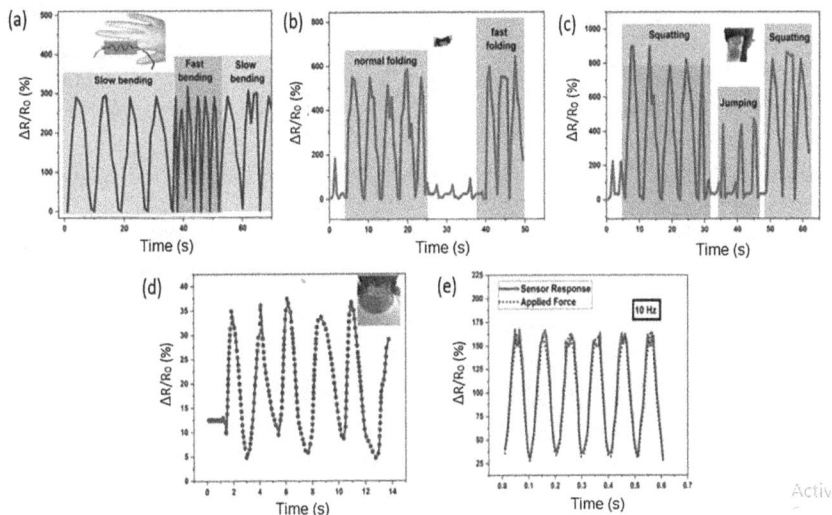

FIGURE 6.9 (a) Shows the movement of fingers at room temperature, (b) shows the sensor being attached to the elbow and the measurements noted for fast and slow folding. (c) The sensor was attached to the knee to check the response of the sensor during jumping and squatting, (d) the sensor was attached to the neck and showed the response of the sensor while drinking water, (e) sensor response when pressure was applied (99).

which increases the conductivity and the linearity of the sensor. The highly stretchable composite sensor has a white sensing range with an elongation of 1,275% and a high linearity of 1,000%. The sensors were formed on a fabric and could measure various motions such as bending, walking and jumping. The sensor showed a fast response of 63 ms and a durability of 200% over 200 repetitive cycles.

Another strain sensor developed by Jabbar et al. (99) was found to be biocompatible and was developed using the same composite. The elastic property of PEDOT:PSS and conductivity of the MWCNTs could change the electrical resistance under the 150% strain. The developed sensor could achieve hysteresis of 1.56% and high linearity of $R^2 = 0.9935$. The sensor had a GF of 89.4 at 150% strain. The fabricated sensor was used for various wearable electronics to detect human motion.

Figure 6.9 (a) shows the movement of fingers at room temperature. The fingers were bent fast and sow to check the change in resistance. Figure 6.9(b) shows that when the sensor was attached to the elbow and the measurements were noted for fast and slow folding. Similarly, the sensor was attached to the knee to check the response of the sensor during jumping and squatting, as shown in Figure 6.9(c). Then the sensor was attached to the neck and showed the response of the sensor while drinking water, as shown in Figure 6.9(d). The sensor responded when the strain of speed of 10 Hz was applied. Mangu et al. (100) developed a gas sensor using PEDOT:PSS and MWCNTs composite which were integrated on a resistive sensor design. The Si/SiO$_2$ were used as a substrate on which the MWCNTs were grown using chemical deposition technique. PEDOT:PSS were dissolved into various solvents for the composite films. The fabrication method renders the composite films gas-specific and offers extensive control over device sensitivity. The commercialisation of sensor technologies based on CNTs has taken a significant step in this direction. The p-type semiconducting characteristics of the composite film have been exhibited by high sensitivity of the MWCNTs-based composite films at lower concentrations of both reducing and oxidising species. At room temperature, the sensor displayed great responsiveness and sensitivity. The sensor was subjected to several chemicals at concentrations of 100 ppm, including NH_3 and NO_2. A PEDOT:PSS/MWCNTs composite sensor measured a sensitivity of 28.1% to 100 ppm of NH_3, whereas a PEDOT:PSS sensor mixed with 0.1 M NaOH measured a sensitivity of 27.8% to 100 ppm of NO_2.

Khalaf et al. (101) fabricated a wearable human-body-temperature sensor using drop-casting technique with the help of PEDOT:PSS as a delicate layer and inkjet printing silver interdigitated electrodes. To examine the effects of changing the number of fingers on sensor sensitivity recital, two designs have been developed. Additionally, PI and Epson glossy paper were also used as the substrates for the sensors. The highest sensitivity of the PI-constructed sensor to temperature was 3.202%/C. Changing the number of fingers, however, did not appear to increase sensitivity. According to the tests, the sensor demonstrated great linearity and repeatability ranging between 28 and 50°C. The flexibility and the repeatability of the response of the sensor were demonstrated, which are crucial to determine a person's body temperature. The sensor showed a response time of 10 s and recovery time of 22 s for every 13°C steps. The sensor showed considerable possibility for applications involving temperature detection, particularly for the human body. Xu et al. (102) had shown a simple and universal method for conducting polymers to achieve high-efficiency, synergistic charge-percolation for neurochemical sensing and neuroelectrical signals. CNTs acting as tie chains generate charge percolation routes that bridge linearly aligned PEDOT grains when CNTs are combined with solution-based post-processing process. The electrochemical kinetics are sped up by CNT-bridged charge percolation, which also catalyses the electrochemical oxidation of vitamin C for selective sensing. The capacity of PEDOT:PSS for charge storage and injection is improved by CNTs. The produced CNTs/PEDOT:PSS fibre was demonstrated to be a biocompatible multifunctional micro sensor, capable of high-performance neuro-sensing in-vivo and mechanically matched with the neural tissues.

Gu et al. (103) developed a transparent conductive film on a PET substrate by a spray-coating method. The CNTs were coated in PEDOT:PSS. Ethyl glycol was used to improve the conductivity of the film. By applying HNO_3 to the films, the transmission was increased, and the resistance of the film was decreased. At a wavelength of 550 nm, TCF sheet resistance as low as 95.15 per square at 87.72% transmission was attained. The highest brightness of the device when the composite was utilised as an anode in an organic light emitting diode was 1,598 cd/cm^2 at 14 volts, and the highest current efficiency was 1.5 cd/A at 13 volts. The sandwich-structured CNTs/PEDOT:PSS/CNT-TCF had high potential when used in OLEDs. Wen et al. (104) showed the effective production of 1D extreme-functioning PEDOT:PSS/TE fibres involving continuous wet spinning and an H_2SO_4 post-treatment. PEDOT:PSS fibres were produced to enhance the proportions of quinoid PEDOT and improve chain packing order, owing to the spatial confinement effect brought on by the unique fibre shape. At room temperature, the optimized PEDOT:PSS fibres had an electrical conductivity of 4,029.5 S/cm and a Seebeck coefficient of 19.2 µV/K, yielding a power factor of 147.8 µW/mK². The PEDOT:PSS fibres also showed a significant breaking strain of 30.5% and a high tensile strength of 389.5 MPa. The PEDOT:PSS fibres outperformed their film counterpart in terms of both TE and mechanical properties, which might primarily be attributed to the existence of a significant amount of quinoid-PEDOT-packing alignment in the 1D formation. The PEDOT:PSS fibres also have special benefits, such as freestanding use and great structural designability.

Although CNTs-polymer composites-based strain sensors provide many benefits and have enormous potential for a range of applications, there are still a number of difficulties that must be overcome. Wearable sensors still need to be improved before they can be effectively used for long-term, continuous and inconspicuous monitoring of human activities. In order to carry out practical uses that accurately identify human activities or signals from the body, it is necessary to further expand the high sensitivity and prolonged operating stability of carbon-based sensor devices as well as address cost concerns. (105). Even though recent studies on wearable technology have shown excellent sensing performances, including high sensitivity, durability and mechanical property, the capacity to differentiate between multiple or mixed stimuli is still limited.

6.3 CONCLUSION

The chapter presents a comprehensive review of flexible CNTs-based nanocomposite strain sensors. The CNTs based composites have various manufacturing techniques that were analysed and discussed. CNTs are extensively used nowadays and have many different uses. The combination of a flexible polymer composite and a carbon nanotube-based pressure sensor demonstrates promising developments in the fields of electronic skin, tactile sensing and other human-machine interfaces. The strain sensors that were developed by CNTs composites have demonstrated exceptional stability and stretchability because of the polymer in the composites. The previously mentioned sensing examples show the viability and efficiency of the conductive phase of the CNTs, thereby opening the door to future healthcare monitoring and intelligent controlling systems.

ACKNOWLEDGEMENT

This work was supported by the Free State of Saxony and by the European Union (ESF Plus) by funding the research group "MultiMOD." This study was funded by the German Research Foundation (DFG, Deutsche Forschungsgemeinschaft) as part of Germany's Excellence Strategy – EXC 2050/1 – Project ID 390696704 – Cluster of Excellence "Centre for Tactile Internet with Human-in-the-Loop" (CeTI) of Technische Universität Dresden.

REFERENCES

1. Huang Q, Zhu Y. Printing conductive nanomaterials for flexible and stretchable electronics: A review of materials, processes, and applications. Advanced Materials Technologies. 2019;4(5):1800546.

2. Nayak L, Mohanty S, Nayak SK, Ramadoss A. A review on inkjet printing of nano-particle inks for flexible electronics. Journal of Materials Chemistry C. 2019;7(29): 8771–95.

3. Han T, Nag A, Afsarimanesh N, Mukhopadhyay SC, Kundu S, Xu Y. Laser-assisted printed flexible sensors: A review. Sensors. 2019;19(6):1462.

4. Nag A, Mukhopadhyay SC, Kosel J. Wearable flexible sensors: A review. IEEE Sensors Journal. 2017;17(13):3949–60.

5. Nag A, Nuthalapati S, Mukhopadhyay SC. Carbon fiber/polymer-based composites for wearable sensors: A Review. IEEE Sensors Journal. 2022;22:10235–45.

6. Sapra S, Chakraborthy A, Nuthalapati S, Nag A, Inglis DW, Mukhopadhyay SC, et al. Printed, wearable e-skin force sensor array. Measurement. 2023;206:112348.

7. He S, Zhang Y, Gao J, Nag A, Rahaman A. Integration of different graphene nanostructures with PDMS to form wearable sensors. Nanomaterials. 2022;12(6):950.

8. Nag A, Simorangkir RB, Gawade DR, Nuthalapati S, Buckley JL, O'Flynn B, et al. Graphene-based wearable temperature sensors: A review. Materials & Design. 2022:110971.

9. Nag A, Mukhopadhyay SC, Kosel J. Flexible carbon nanotube nanocomposite sensor for multiple physiological parameter monitoring. Sensors and Actuators A: Physical. 2016;251:148–55.

10. Nag A, Mukhopadhyay S, Kosel J, editors. Transparent biocompatible sensor patches for touch sensitive prosthetic limbs. 2016 10th International Conference on Sensing Technology (ICST); 2016: IEEE.

11. Nag A, Menzies B, Mukhopadhyay SC. Performance analysis of flexible printed sensors for robotic arm applications. Sensors and Actuators A: Physical. 2018;276:226–36.

12. Chortos A, Liu J, Bao Z. Pursuing prosthetic electronic skin. Nature Materials. 2016;15(9):937–50.

13. Yin R, Wang D, Zhao S, Lou Z, Shen G. Wearable sensors-enabled human–machine interaction systems: From design to application. Advanced Functional Materials. 2021;31(11):2008936.

14. Gogurla N, Kim S. Self-powered and imperceptible electronic tattoos based on silk protein nanofiber and carbon nanotubes for human–machine interfaces. Advanced Energy Materials. 2021;11(29):2100801.

15. Nag A, Alahi M, Eshrat E, Mukhopadhyay SC, Liu Z. Multi-walled carbon nanotubes-based sensors for strain sensing applications. Sensors. 2021;21(4):1261.

16. Afsarimanesh N, Nag A, Sarkar S, Sabet GS, Han T, Mukhopadhyay SC. A review on fabrication, characterization and implementation of wearable strain sensors. Sensors and Actuators A: Physical. 2020;315:112355.

17. Han T, Nag A, Afsarimanesh N, Akhter F, Liu H, Sapra S, et al. Gold/Polyimide-Based Resistive Strain Sensors. Electronics. 2019;8(5):565.

18. Alahi MEE, Nag A, Mukhopadhyay SC, Burkitt L. A temperature-compensated graphene sensor for nitrate monitoring in real-time application. Sensors and Actuators A: Physical. 2018;269:79–90.

19. Nag A, Mukhopadhyay S, Kosel J, editors. Influence of temperature and humidity on carbon based printed flexible sensors. 2017 Eleventh International Conference on Sensing Technology (ICST); 2017: IEEE.

20. Nag A, Simorangkir RB, Valentin E, Björninen T, Ukkonen L, Hashmi RM, et al. A transparent strain sensor based on PDMS-embedded conductive fabric for wearable sensing applications. IEEE Access. 2018;6:71020–7.

21. Nag A, Afasrimanesh N, Feng S, Mukhopadhyay SC. Strain induced graphite/PDMS sensors for biomedical applications. Sensors and Actuators A: Physical. 2018;271:257–69.

22. Han T, Nag A, Mukhopadhyay SC, Xu Y. Carbon nanotubes and its gas-sensing applications: A review. Sensors and Actuators A: Physical. 2019;291:107–43.

23. Gao J, He S, Nag A, Wong JWC. A review of the use of carbon nanotubes and graphene-based sensors for the detection of aflatoxin M1 compounds in milk. Sensors. 2021;21(11):3602.
24. Hu Q, Nag A, Xu Y, Han T, Zhang L. Use of graphene-based fabric sensors for monitoring human activities. Sensors and Actuators A: Physical. 2021;332:113172.
25. Nag A, Simorangkir RB, Sapra S, Buckley JL, O'Flynn B, Liu Z, et al. Reduced graphene oxide for the development of wearable mechanical energy-harvesters: A review. IEEE Sensors Journal. 2021.
26. Nag A, Mitra A, Mukhopadhyay SC. Graphene and its sensor-based applications: A review. Sensors and Actuators A: Physical. 2018;270:177–94.
27. Elahi N, Kamali M, Baghersad MH. Recent biomedical applications of gold nanoparticles: A review. Talanta. 2018;184:537–56.
28. Jeevanandam J, Barhoum A, Chan YS, Dufresne A, Danquah MK. Review on nanoparticles and nanostructured materials: History, sources, toxicity and regulations. Beilstein Journal of Nanotechnology. 2018;9(1):1050–74.
29. Nasr Esfahani M, Alaca BE. A review on size-dependent mechanical properties of nanowires. Advanced Engineering Materials. 2019;21(8):1900192.
30. Liu L, Jiang Y, Jiang J, Zhou J, Xu Z, Li Y. Flexible and transparent silver nanowires integrated with a graphene layer-doping PEDOT: PSS film for detection of hydrogen sulfide. ACS Applied Electronic Materials. 2021;3(10):4579–86.
31. Jena SK, Chakraverty S, Malikan M, Tornabene F. Effects of surface energy and surface residual stresses on vibro-thermal analysis of chiral, zigzag, and armchair types of SWCNTs using refined beam theory. Mechanics Based Design of Structures and Machines. 2020:1–15.
32. Guo X, Huang Y, Zhao Y, Mao L, Gao L, Pan W, et al. Highly stretchable strain sensor based on SWCNTs/CB synergistic conductive network for wearable human-activity monitoring and recognition. Smart Materials and Structures. 2017;26(9):095017.
33. Magar HS, Hassan RY, Abbas MN. Non-enzymatic disposable electrochemical sensors based on CuO/Co3O4@ MWCNTs nanocomposite modified screen-printed electrode for the direct determination of urea. Scientific Reports. 2023;13(1):2034.
34. Nag A, Afsarimanesh N, Nuthalapati S, Altinsoy ME. Novel surfactant-induced MWCNTs/PDMS-based nanocomposites for tactile sensing applications. Materials. 2022;15(13):4504.
35. Xiao Y, Jiang S, Li Y, Zhang W. Screen-printed flexible negative temperature coefficient temperature sensor based on polyvinyl chloride/carbon black composites. Smart Materials and Structures. 2021;30(2):025035.
36. Devaraj H, Schober R, Picard M, Teo MY, Lo C-Y, Gan WC, et al. Highly elastic and flexible multi-layered carbon black/elastomer composite based capacitive sensor arrays for soft robotics. Measurement: Sensors. 2020:100004.
37. Nag A, Feng S, Mukhopadhyay S, Kosel J, Inglis D. 3D printed mould-based graphite/PDMS sensor for low-force applications. Sensors and Actuators A: Physical. 2018;280:525–34.
38. Kurian AS, Mohan VB, Souri H, Leng J, Bhattacharyya D. Multifunctional flexible and stretchable graphite-silicone rubber composites. Journal of Materials Research and Technology. 2020;9(6):15621–30.
39. Kumar S, Gupta TK, Varadarajan K. Strong, stretchable and ultrasensitive MWCNT/ TPU nanocomposites for piezoresistive strain sensing. Composites Part B: Engineering. 2019;177:107285.
40. Sun X, Sun J, Li T, Zheng S, Wang C, Tan W, et al. Flexible tactile electronic skin sensor with 3D force detection based on porous CNTs/PDMS nanocomposites. Nano-Micro Letters. 2019;11(1):1–14.
41. Kim H, Kim Y. High performance flexible piezoelectric pressure sensor based on CNTs-doped 0–3 ceramic-epoxy nanocomposites. Materials & Design. 2018;151:133–40.

42. Nurazzi N, Sabaruddin F, Harussani M, Kamarudin S, Rayung M, Asyraf M, et al. Mechanical performance and applications of CNTs reinforced polymer composites – A review. Nanomaterials. 2021;11(9):2186.
43. Guo Y, Wei X, Gao S, Yue W, Li Y, Shen G. Recent advances in carbon material-based multifunctional sensors and their applications in electronic skin systems. Advanced Functional Materials. 2021;31(40):2104288.
44. Chani MTS, Karimov KS, Asiri AM. Carbon nanotubes and graphene powder based multifunctional pressure, displacement and gradient of temperature sensors. Semiconductors. 2020;54:85–90.
45. Julkapli NM, Bagheri S, Sapuan S. Multifunctionalized carbon nanotubes polymer composites: Properties and applications. In Eco-friendly Polymer Nanocomposites: Springer; 2015. p. 155–214.
46. Son M, Park H, Liu L, Choi H, Kim JH, Choi H. Thin-film nanocomposite membrane with CNT positioning in support layer for energy harvesting from saline water. Chemical Engineering Journal. 2016;284:68–77.
47. Cao X, Chen H, Gu X, Liu B, Wang W, Cao Y, et al. Screen printing as a scalable and low-cost approach for rigid and flexible thin-film transistors using separated carbon nanotubes. ACS Nano. 2014;8(12):12769–76.
48. Li S, Li R, González OG, Chen T, Xiao X. Highly sensitive and flexible piezoresistive sensor based on c-MWCNTs decorated TPU electrospun fibrous network for human motion detection. Composites Science and Technology. 2021;203:108617.
49. Zhou Q, Chen T, Cao S, Xia X, Bi Y, Xiao X. A novel flexible piezoresistive pressure sensor based on PVDF/PVA-CNTs electrospun composite film. Applied Physics A. 2021;127:1–10.
50. Nag A, Alahi MEE, Feng S, Mukhopadhyay SC. IoT-based sensing system for phosphate detection using Graphite/PDMS sensors. Sensors and Actuators A: Physical. 2019;286:43–50.
51. Akhtar I, Chang SH. Stretchable sensor made of MWCNT/ZnO nanohybrid particles in PDMS. Advanced Materials Technologies. 2020;5(9):2000229.
52. Son W, Kim K-B, Lee S, Hyeon G, Hwang K-G, Park W. Ecoflex-passivated graphene–yarn composite for a highly conductive and stretchable strain sensor. Journal of Nanoscience and Nanotechnology. 2019;19(10):6690–5.
53. Zhang S, Wen L, Wang H, Zhu K, Zhang M. Vertical CNT–Ecoflex nanofins for highly linear broad-range-detection wearable strain sensors. Journal of Materials Chemistry C. 2018;6(19):5132–9.
54. Nag A, Mukhopadhyay SC, Kosel J. Tactile sensing from laser-ablated metallized PET films. IEEE Sensors Journal. 2016;17(1):7–13.
55. Emamian S, Narakathu BB, Chlaihawi AA, Bazuin BJ, Atashbar MZ. Screen printing of flexible piezoelectric based device on polyethylene terephthalate (PET) and paper for touch and force sensing applications. Sensors and Actuators A: Physical. 2017;263:639–47.
56. Nag A, Mukhopadhyay SC, Kosel J. Sensing system for salinity testing using laser-induced graphene sensors. Sensors and Actuators A: Physical. 2017;264:107–16.
57. Nag A, Mukhopadhyay SC. Fabrication and implementation of printed sensors for taste sensing applications. Sensors and Actuators A: Physical. 2018;269:53–61.
58. Jiang L, Hong H, Hu J. Facile thermoplastic polyurethane-based multi-walled carbon nanotube ink for fabrication of screen-printed fabric electrodes of wearable e-textiles with high adhesion and resistance stability under large deformation. Textile Research Journal. 2021;91(21–22):2487–99.
59. Stan F, Stanciu N-V, Constantinescu A-M, Fetecau C. 3D Printing of flexible and stretchable parts using multiwall carbon nanotube/polyester-based thermoplastic polyurethane. Journal of Manufacturing Science and Engineering. 2020:1–33.

60. Xiong S, Zhou J, Wu J, Li H, Zhao W, He C, et al. High performance acoustic wave nitrogen dioxide sensor with ultraviolet activated 3D porous architecture of Ag-decorated reduced graphene oxide and polypyrrole aerogel. ACS Applied Materials & Interfaces. 2021;13(35):42094–103.

61. Bai X, Zhang B, Liu M, Hu X, Fang G, Wang S. Molecularly imprinted electrochemical sensor based on polypyrrole/dopamine@ graphene incorporated with surface molecularly imprinted polymers thin film for recognition of olaquindox. Bioelectrochemistry. 2020;132:107398.

62. Zhang X, Wang Y, Fu D, Wang G, Wei H, Ma N. Photo-thermal converting polyaniline/ionic liquid inks for screen printing highly-sensitive flexible uncontacted thermal sensors. European Polymer Journal. 2021;147:110305.

63. Bibi A, Rubio YRM, Xian-Lun L, Sathishkumar N, Chen C-Y, Santiago KS, et al. Detection of hydrogen sulfide using polyaniline incorporated with graphene oxide aerogel. Synthetic Metals. 2021;282:116934.

64. Shen G, Chen B, Liang T, Liu Z, Zhao S, Liu J, et al. Transparent and stretchable strain sensors with improved sensitivity and reliability based on Ag NWs and PEDOT: PSS patterned microstructures. Advanced Electronic Materials. 2020;6(8):1901360.

65. Yu Y, Peng S, Blanloeuil P, Wu S, Wang CH. Wearable temperature sensors with enhanced sensitivity by engineering microcrack morphology in PEDOT: PSS–PDMS sensors. ACS Applied Materials & Interfaces. 2020;12(32):36578–88.

66. Chen J, Zheng J, Gao Q, Zhang J, Zhang J, Omisore OM, et al. Polydimethylsiloxane (PDMS)-based flexible resistive strain sensors for wearable applications. Applied Sciences. 2018;8(3):345.

67. Juárez-Moreno J, Ávila-Ortega A, Oliva A, Avilés F, Cauich-Rodríguez J. Effect of wettability and surface roughness on the adhesion properties of collagen on PDMS films treated by capacitively coupled oxygen plasma. Applied Surface Science. 2015;349:763–73.

68. Chen D, Chen F, Hu X, Zhang H, Yin X, Zhou Y. Thermal stability, mechanical and optical properties of novel addition cured PDMS composites with nano-silica sol and MQ silicone resin. Composites Science and Technology. 2015;117:307–14.

69. Johnston I, McCluskey D, Tan C, Tracey M. Mechanical characterization of bulk Sylgard 184 for microfluidics and microengineering. Journal of Micromechanics and Microengineering. 2014;24(3):035017.

70. Eisape A, Rennoll V, Van Volkenburg T, Xia Z, West JE, Kang SH. Soft CNT-polymer composites for high pressure sensors. Sensors. 2022;22(14):5268.

71. Fan X, Nie W, Tsai H, Wang N, Huang H, Cheng Y, et al. PEDOT: PSS for flexible and stretchable electronics: Modifications, strategies, and applications. Advanced Science. 2019;6(19):1900813.

72. Chen G, Xu W, Zhu D. Recent advances in organic polymer thermoelectric composites. Journal of Materials Chemistry C. 2017;5(18):4350–60.

73. Nilchian A, Li C-Z. Mechanical and electrochemical characterization of CNT/PDMS composited soft and stretchable electrodes fabricated by an efficient solution-based fabrication method. Journal of Electroanalytical Chemistry. 2016;781:166–73.

74. Azhari S, Yousefi AT, Tanaka H, Khajeh A, Kuredemus N, Bigdeli MM, et al. Fabrication of piezoresistive based pressure sensor via purified and functionalized CNTs/PDMS nanocomposite: Toward development of haptic sensors. Sensors and Actuators A: Physical. 2017;266:158–65.

75. Akhter F, Siddiquei H, Alahi MEE, Mukhopadhyay SC. An IoT-enabled portable sensing system with MWCNTs/PDMS sensor for nitrate detection in water. Measurement. 2021;178:109424.

76. Leemets K, Mäeorg U, Aabloo A, Tamm T. Effect of contact material and ambient humidity on the performance of MWCNT/PDMS multimodal deformation sensors. Sensors and Actuators A: Physical. 2018;283:1–8.

77. Hong HS, Ha NH, Thinh DD, Nam NH, Huong NT, Hue NT, et al. Enhanced sensitivity of self-powered NO2 gas sensor to sub-ppb level using triboelectric effect based on surface-modified PDMS and 3D-graphene/CNT network. Nano Energy. 2021;87:106165.
78. Lai Y-T, Chen Y-M, Yang Y-JJ. A novel CNT-PDMS-based tactile sensing array with resistivity retaining and recovering by using dielectrophoresis effect. Journal of Microelectromechanical Systems. 2011;21(1):217–23.
79. Lim SJ, Lim HS, Joo Y, Jeon D-Y. Impact of MWCNT concentration on the piezo-impedance response of porous MWCNT/PDMS composites. Sensors and Actuators A: Physical. 2020;315:112332.
80. Lv C, Wang J, Li Z, Zhao K, Zheng J. Degradable, reprocessable, self-healing PDMS/CNTs nanocomposite elastomers with high stretchability and toughness based on novel dual-dynamic covalent sacrificial system. Composites Part B: Engineering. 2019;177:107270.
81. Xu X, Chen Y, He P, Wang S, Ling K, Liu L, et al. Wearable CNT/Ti3C2T x MXene/PDMS composite strain sensor with enhanced stability for real-time human healthcare monitoring. Nano Research. 2021;14(8):2875–83.
82. Wang F, Tan Y, Peng H, Meng F, Yao X. Investigations on the preparation and properties of high-sensitive BaTiO3/MwCNTs/PDMS flexible capacitive pressure sensor. Materials Letters. 2021;303:130512.
83. Jung Y, Jung KK, Kim DH, Kwak DH, Ahn S, Han JS, et al. Flexible and highly sensitive three-axis pressure sensors based on carbon nanotube/polydimethylsiloxane composite pyramid arrays. Sensors and Actuators A: Physical. 2021;331:113034.
84. Wang L, Dou W, Chen J, Lu K, Zhang F, Abdulaziz M, et al. A CNT-PDMS wearable device for simultaneous measurement of wrist pulse pressure and cardiac electrical activity. Materials Science and Engineering: C. 2020;117:111345.
85. Wang Y, Gao Y-N, Yue T-N, Chen X-D, Wang M. Achieving high-performance and tunable microwave shielding in multi-walled carbon nanotubes/polydimethylsiloxane composites containing liquid metals. Applied Surface Science. 2021;563:150255.
86. Xu J, Su W, Li Z, Liu W, Liu S, Ding X. A modularized and flexible sensor based on MWCNT/PDMS composite film for on-site electrochemical analysis. Journal of Electroanalytical Chemistry. 2017;806:68–74.
87. Ren M, Zhou Y, Wang Y, Zheng G, Dai K, Liu C, et al. Highly stretchable and durable strain sensor based on carbon nanotubes decorated thermoplastic polyurethane fibrous network with aligned wave-like structure. Chemical Engineering Journal. 2019;360:762–77.
88. He Y, Zhao L, Zhang J, Liu L, Liu H, Liu L. A breathable, sensitive and wearable piezoresistive sensor based on hierarchical micro-porous PU@ CNT films for long-term health monitoring. Composites Science and Technology. 2020;200:108419.
89. Kim K, Park J, Suh J-h, Kim M, Jeong Y, Park I. 3D printing of multiaxial force sensors using carbon nanotube (CNT)/thermoplastic polyurethane (TPU) filaments. Sensors and Actuators A: Physical. 2017;263:493–500.
90. Lee J, Kim J, Shin Y, Jung I. Ultra-robust wide-range pressure sensor with fast response based on polyurethane foam doubly coated with conformal silicone rubber and CNT/TPU nanocomposites islands. Composites Part B: Engineering. 2019;177:107364.
91. Tran M, Tung T, Sachan A, Losic D, Castro M, Feller J-F. 3D sprayed polyurethane functionalized graphene/carbon nanotubes hybrid architectures to enhance the piezo-resistive response of quantum resistive pressure sensors. Carbon. 2020;168:564–79.
92. Wang Y, Li W, Li C, Zhou B, Zhou Y, Jiang L, et al. Fabrication of ultra-high working range strain sensor using carboxyl CNTs coated electrospun TPU assisted with dopamine. Applied Surface Science. 2021;566:150705.
93. Xiang D, Zhang X, Li Y, Harkin-Jones E, Zheng Y, Wang L, et al. Enhanced performance of 3D printed highly elastic strain sensors of carbon nanotube/thermoplastic

polyurethane nanocomposites via non-covalent interactions. Composites Part B: Engineering. 2019;176:107250.

94. Xie X, Huang H, Zhu J, Yu J, Wang Y, Hu Z. A spirally layered carbon nanotube-graphene/polyurethane composite yarn for highly sensitive and stretchable strain sensor. Composites Part A: Applied Science and Manufacturing. 2020;135:105932.

95. Zhang S, Sun K, Liu H, Chen X, Zheng Y, Shi X, et al. Enhanced piezoresistive performance of conductive WPU/CNT composite foam through incorporating brittle cellulose nanocrystal. Chemical Engineering Journal. 2020;387:124045.

96. Kuzubasoglu BA, Sayar E, Bahadir SK. Inkjet-printed CNT/PEDOT: PSS temperature sensor on a textile substrate for wearable intelligent systems. IEEE Sensors Journal. 2021;21(12):13090–7.

97. He X, Shi J, Hao Y, Wang L, Qin X, Yu J. PEDOT: PSS/CNT composites based ultra-stretchable thermoelectrics and their application as strain sensors. Composites Communications. 2021;27:100822.

98. Lam TN, Lee GS, Kim B, Xuan HD, Kim D, Yoo SI, et al. Microfluidic preparation of highly stretchable natural rubber microfiber containing CNT/PEDOT: PSS hybrid for fabric-sewable wearable strain sensor. Composites Science and Technology. 2021;210:108811.

99. Jabbar F, Soomro AM, Lee J-w, Ali M, Kim YS, Lee S-h, et al. Robust fluidic biocompatible strain sensor based on PEDOT: PSS/CNT composite for human-wearable and high-end robotic applications. Sensors & Materials. 2020;32.

100. Mangu R, Rajaputra S, Singh VP. MWCNT–polymer composites as highly sensitive and selective room temperature gas sensors. Nanotechnology. 2011;22(21):215502.

101. Khalaf AM, Ramírez JL, Mohamed SA, Issa HH. Highly sensitive interdigitated thermistor based on PEDOT: PSS for human body temperature monitoring. Flexible and Printed Electronics. 2022;7(4):045012.

102. Xu T, Ji W, Zhang Y, Wang X, Gao N, Mao L, et al. Synergistic charge percolation in conducting polymers enables high-performance in vivo sensing of neurochemical and neuroelectrical signals. Angewandte Chemie. 2022;134(41):e202204344.

103. Gu Z-Z, Tian Y, Geng H-Z, Rhen DS, Ethiraj AS, Zhang X, et al. Highly conductive sandwich-structured CNT/PEDOT: PSS/CNT transparent conductive films for OLED electrodes. Applied Nanoscience. 2019;9:1971–9.

104. Wen N, Fan Z, Yang S, Zhao Y, Cong T, Xu S, et al. Highly conductive, ultra-flexible and continuously processable PEDOT: PSS fibers with high thermoelectric properties for wearable energy harvesting. Nano Energy. 2020;78:105361.

105. Matos MA, Pinho ST, Tagarielli VL. Application of machine learning to predict the multiaxial strain-sensing response of CNT-polymer composites. Carbon. 2019;146:265–75.

7 Carbon Nanotubes as Versatile Elements for Multiple Sensory Analysis in Biological Applications

Alisha Mary Manoj, Merlin R. Charlotte,
Mahalakshmi, Suresh Nuthalapati, Anindya Nag,
and Leema Rose Viannie

7.1 INTRODUCTION

A sensor is a device that helps in the detection of changes in physical quantities like temperature, pressure, force, etc., and converts it into a measurable electrical signal (1). A transducer may be defined as a device that helps to convert one form of energy to the other. The sensors technology is now widely used in various applications ranging from automotive (2), space sciences (3), defence, environmental, food quality monitoring, and medical diagnostics. The sensors which utilise biological reactions to detect and quantify a specific analyte are called biosensors (4).

Owing to their amazing electrical and mechanical properties, CNTs have attracted increasing interest since their discovery in 1991 for use in nanoscale electronic devices and chemical and biological sensors (5). CNTs are pseudo-one-dimensional allotropes of carbon that may be distinguished by their seamless cylinder structures made of one or more layers of carbon atoms with either open or closed ends. Depending on their architectures, CNTs exhibit metal or semiconductor behaviour (6, 7). CNTs have attracted interest in the healthcare sector during the past few years as one of the potential candidates for medical devices because of their many distinguishing characteristics, including outstanding mechanical strength (8), enhanced surface area, excellent electrical conductivity (9), stable behaviour in aqueous and nonaqueous solutions, and high thermal conductivity (10). Biosensors can be used to identify analytes that mix biological components with physical detectors. The excellent biocompatibility of carbon nanotubes makes them exceptional candidates for applications in biosensing. The major components of a sensor include a sensor for target recognition and a signal-transduction mechanism, converting the recognition event into a measurable signal. The target recognition elements mainly include antibodies, DNA sequences, aptamers, molecular imprints, etc., the binding of which

DOI: 10.1201/9781003376071-7

results in change in the physical or chemical properties at the sensor surface. The transduction is sometimes achieved by labelling the sensor with enzymes or other chemical agents that convert the response signal into a measurable form. In most cases, the signal change changes in resistance, impedance, capacitance, or changes in optical signals at the sensor and analyte interface.

Biosensors are being used in a variety of industries, including the food industry, clinical inspection, biomedicine, and environmental assessment. They are also being added to emerging, sophisticated technologies, including nanotechnology, optoelectronics, microfabrication, and molecular biology. Traditional analytical procedures that need a trained expert, such as high-performance liquid chromatography, mass spectroscopy, nuclear magnetic-resonance spectroscopy, and immunological approaches, are too expensive. By using electrical biosensors, these problems can be avoided. Additionally, electrical biosensors have the benefits of being inexpensive, quick to analyse, and field ready. These merits turn biosensor research into a modern hotspot for science and technology (11).

In this chapter, we'll go over several ways to make CNTs from various carbon sources and their uses in medicine as well as many kinds of biosensors based on electrode configuration and how to categorise them according to various transduction mechanisms.

7.2 CLASSIFICATIONS OF SENSORS

7.2.1 BROAD CLASSIFICATION OF SENSORS BASED ON THE TYPE OF ANALYTES

Sensors are classified into a broad range, depending on the physical quantity to be measured. The physical quantity may be temperature, pressure, force, or any analyte of interest, which may be a chemical substance or biological unit. Based on the type of signal the sensors can be broadly divided into the following types.

(i) *Physical sensors:* These kinds of sensors measure physical quantities like force, pressure, temperature, flow rate, etc. With the help of advancements in MEMS technology and flexible and wearable electronics, they are largely used in the medical field for monitoring movements and as motion sensors in various on-body applications.

(ii) *Chemical sensors:* Chemical sensors generate an analytically useful signal on interaction with a chemical compound. They are mostly used in food and drug analysis, drug release, volatile organic compounds, and other chemical agents which are secreted from the human body.

(iii) *Biological sensors:* These sensors are used to directly monitor biological processes such as interactions with whole cells, antigen/antibody interactions, enzymes, and hormones with the analyte of interest.

7.2.1.1 Factors Affecting Transducers Choice

The transducer is the major element in the biosensor. The transducer for any sensing is selected based mainly on the type of physical quantity to be measured, which, in turn, depends on the location and environment in which the sensor is being operated. The energy source, physical contact, etc., are the other factors that affect the sensor transduction. In certain biosensors, the measurement of the signal is based on the direct changes in physical properties at the sensor due to changes in the physical conditions. Some of the most common types of sensors include pressure sensors, strain sensors, etc. For example, in a pressure sensor, the pressure change at the sensing element produces a measurable electrical signal. In certain sensors, the sensing is based on the changes that occur during physical interactions between the sensing surface and the analyte medium.

7.2.1.2 Components of Biological Sensors

In most biological sensors, the sensor is aided by a biological element such as an enzyme or antibody which converts the event of interaction into a useful form. Major components of a typical biosensor are as shown in Figure 7.1 and is discussed in detail in the following list.

1) *Analyte:* The analyte is the chemical or biological constituents that are to detected by the sensor.
2) *Bioreceptor:* A biological molecule or biological element that recognises the analyte. Most used receptors include enzymes, aptamers, DNA or RNA, and antibodies. The interaction between bioreceptor and analyte is known as biorecognition.
3) *Transducer:* The function of the transducer is to convert the signal from biorecognition into a measurable electrical form through the process called signalisation. The signal generated by any transducer is directly proportional to the analyte-bioreceptor interactions or the analyte concentration in the detection environment.
4) *Electronics*: The electronics integrated into the sensing platform processes the signal and convert it into a readable, understandable form and are sent to the display unit.

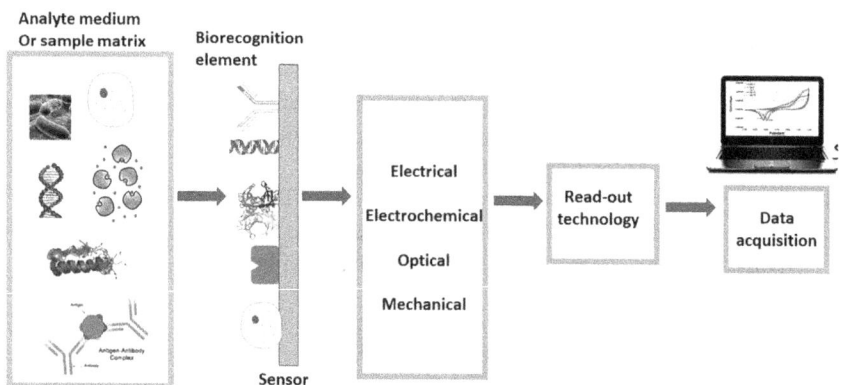

FIGURE 7.1   Components of a sensor.

5) ***Display unit:*** Consisting of the user interpretation system, the display unit displays a numerical, tabular, or graphical output that can help in medical diagnostics.

7.2.2 Common Electrode Designs Used in Biomedical Sensing

An electrode is employed to establish electrical connectivity with the non-metallic elements of a circuit. Different fields such as electrochemistry, semiconductors, and medical devices frequently employ electrodes. The following discussion examines various designs and setups of electrodes (12).

7.2.2.1 Three-Electrode System

As depicted in Figure 7.2, the three-electrode system is comprised of a reference, counter, and working electrode.

7.2.2.1.1 Reference Electrode (RE)

The purpose of reference electrode is to serve as a standard for calculating and altering the potential of the working electrode, without any current transmission. It should possess stable electrochemical potential at low density of current. The iR drop (iRU) between the working and reference electrodes is small, since RE conducts very little current. As a result, the potential of the reference is considerably stable with the three-electrode configuration, and the iR decrease across the solution is compensated for. This results in better control over the potential of working electrode. The Saturated Calomel Electrode and the Ag/AgCl Electrode are the commonly utilised reference electrodes in laboratories (12).

7.2.2.1.2 Counter Electrode (CE)

The use of the counter electrode is necessary to achieve a uniform current flow by counterbalancing the current produced at the WE. This is achieved by the counter electrode, which sometimes fluctuates to remarkably high-voltage levels. Typically, platinum wire is the most employed counter electrode.

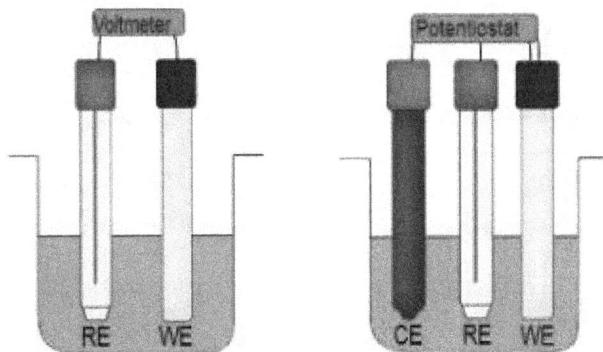

FIGURE 7.2 Schematization of a three-electrode system (1).

7.2.2.1.3   Working Electrode (WE)

The most significant element of an electrochemical cell is the working electrode. The transfer of electron occurs at the junction between the WE and the solution. WE are often a disc of glassy carbon (GC). In specialised applications where a different electrode-specific reactivity is sought, metallic discs (Au, Ni, Cu, Ag, Pt), boron doped diamond, or edge plane pyrolytic graphite discs can also be utilised as substitutes for glassy carbon.

7.2.2.2   Two-Electrode System

One of the electrodes in a two-electrode cell configuration (Figure 7.3(a)) has CE and RE shorted to it. It measures the potential over the entire cell. Both the electrolyte itself and the CE/electrolyte interaction contribute to this. Since fine tuning of the interfacial voltage across the WE interface is not required often, the two-electrode arrangement can be employed whenever the behaviour of the entire cell is being studied. Batteries, fuel cells, solar panels, and other energy storage or conversion technologies are frequently employed in conjunction with this system. It is also utilised for electrochemical impedance measurements at high frequencies (> 100 kHz) or investigations of the ultrafast dynamics of electrode processes (13).

The two-electrode design consists of interdigitated electrodes (IDEs), as illustrated in Figure 7.3(b), which are composed of two individually addressable comb-like electrode architectures. Due to their affordability, ease of manufacturing process, and fast response, IDEs are popular among transducers and are frequently used in technical applications such as biological and chemical sensors. In different sensing technologies, such as surface acoustic-wave (SAW) sensors, chemical sensors, and modern MEMS biosensors, IDEs are tuned. IDEs, when employed in biosensing, can operate in faradaic or non-faradaic modes. The difference between faradaic and non-faradaic biosensors must be clearly understood. In contrast to non-faradaic processes, which involve current flow resulting from capacitive nature, faradaic

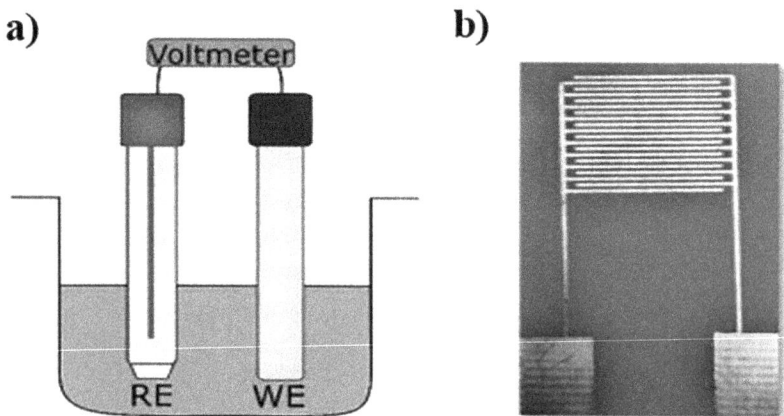

FIGURE 7.3   (a) Schematization of a two-electrode system (1), (b) interdigitated electrode design (14).

processes involve current flow because of electronic transfer. The electrochemical impedance spectroscopy (EIS), which helps to measure double layer capacitance and electron transfer resistance within a frequency range, is frequently the foundation of biosensors that operate in faradaic mode. The non-faradaic mode, on the other hand, relies on variations in capacitance between interdigitated electrodes to signal molecule binding events at the electrode surface (2).

7.2.2.3 Transistors

Due to its potential for downsizing, parallel sensing, quick response time, and smooth integration with electronic manufacturing processes, field-effect transistor (FET)-based sensing has drawn a lot of attention among the numerous potentiometric approaches. A gate, drain, and source-fabricated active device requires a FET as a key electronic component to function. The FET-based biosensor design is depicted in Figure 7.4. The gate is the component that connects the source and drain electrodes, which are the starting and ending points of the charge carrier (4).

A usual FET system seizes the objectives by restricting the components on the conduits that are connected to the source and drain terminals. To a third electrode, a bias potential is applied and regulated. An electrical measuring system takes note of the channel conductance, which changes because of the targets being detected, and processes it further. FETs come in two different varieties: n-type with electrons serving as the majority carriers and p-type with holes. The conductance will be reduced because of the electrons being depleted, if negatively charged objects are detected. In contrast, binding with positive charges causes a p-type FET system to lose conductance because there are fewer charge carriers (holes), while collecting negative charges causes the conductance to increase because of hole build-up (3, 16).

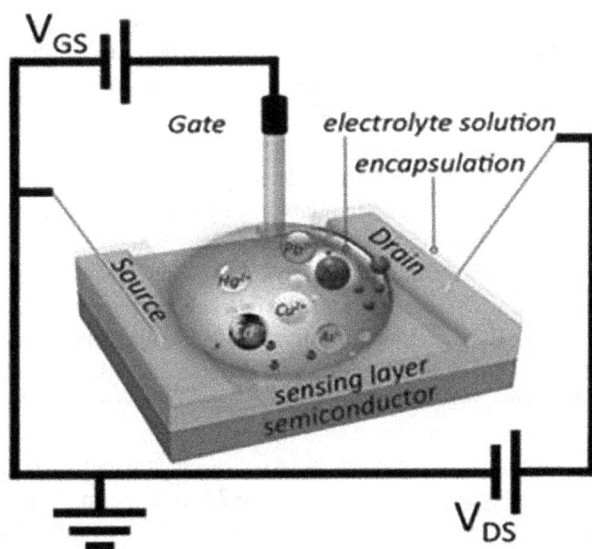

FIGURE 7.4   Design of FET Biosensor (15).

7.2.2.4 Micro-Electromechanical Systems (MEMS)

Microminiature sensors and actuators are used in micro-electromechanical systems. The advantages of MEMS technology are its small size, light weight, great performance, simplicity in mass production, and low cost. By leveraging IC fabrication technology, micromachining is a technique for creating microsystems or MEMS. To create three-dimensional, mechanical devices like cantilevers, diaphragms, and channels, MEMS are micromachined by selectively etching Si, glass, and deposited thin-films. Three subcategories of micromachining technologies exist: bulk, surface, and LIGA techniques. The MEMS sensors utilise various methods of measurement, such as mechanical, thermal, and biological, in order to gather information about the surroundings (17). The most prevalent form of MEMS transducers is those that convert one signal type into another. These transducers come in two versions: the cantilever and the string, with the latter being double-clamped and beam-like. The energy transduction of MEMS relies on transmitting the data from the sensing unit to a controller, which then utilises a control formula to instruct the actuation unit. In recent times, MEMS have been extensively researched within the biomedical field, particularly in the development of drug synthesis and delivery as well as applications in microsurgery. MEMS benefits from their compact size, ease of integration, low power-consumption, high resonance-frequency, potential for electronic-circuit integration, and high level of accuracy, sensitivity, and throughput give them vast potential (7).

7.2.2.5 Flexible Electrode Designs

In recent years, the scientific community has paid a lot of attention to flexible sensing devices. Flexible sensors are used in a variety of industries, including robotics, security, industrial automation, medical, and human-machine interface (18). Non-invasive health-monitoring gadgets, in particular, are anticipated to be crucial in enhancing patient quality of life and lowering the costs of clinical and biomedical diagnostic procedures (19). The biomedical sensors are mostly non-meddling, and analysing the composition helps in identify a disease and its symptoms, deficiency, or abundance of components such as skin, blood, saliva, bp, urine, tear, neural activity, etc. Numerous methods of sensing these things have been created in recent years. Using transduction principles, these sensors can detect physical changes and convert them into electrical signals, which can be validated using known input and output values.

A sensor's base, or substrate, is where its electrical connection and sensing capabilities are built. The most crucial step in making any sensor flexible is to make sure that the substrate is flexible. Various flexible substrate types have been created flexible intentionally, whereas others are inherently flexible (paper, some polymers, etc.). Metal foils often cost a lot of money and possess a rough surface, but they can withstand very high temperatures. Making biosensors affordable is a key component of their development, allowing for onsite use and accessibility to those for whom medical facilities are not readily available. As a result, it is believed that paper and plastic are the best materials, as shown in Figure 7.5, for creating biosensors (8).

One of the most crucial components of any sensor is the sensing substance. These are incorporated into the flexible template, and electrodes are used to create electrical connections. The signal transducer examines recognition events to determine a sensor's sensitivity, which is achieved by translating the recognition event into a

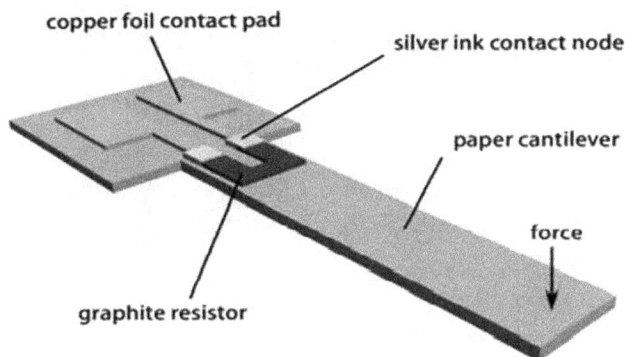

FIGURE 7.5 Schematic for a paper-based sensor for the detection of force (8).

signal using optical, electrochemical, pressure, thermal, etc., or by transforming the signal into an electrical signal. The physical input to the sensing element can be transformed in a variety of ways to produce usable electrical signals. Pressure and temperature can be detected and expressed using variations in electrical properties such capacitance, resistance, triboelectricity, and piezoelectricity.

Numerous operating theories have been investigated for sensing and creating biological devices. The two distinct building parts that make up flexible physical sensors are typically the flexible substrate and the sensing material, which can be solid or liquid. Flexible and wearable sensors' basic sensing mechanisms are based on the mechanical deformations that the sensing devices must withstand, such as bending, pressing, stretching, and twisting. These flexible sensors are used for various purpose in healthcare and bioengineering.

7.2.3 CNTs-Based Biosensors Providing Multiple Transduction Mechanisms towards Biomedical Sensing Applications

Sensors generally convert nonelectrical effects into electrical signals. Different transformative steps are required to convert the generated signal into an electrical output. The following section includes discussion on various physical stimulus that can be directly converted to electrical signal. A brief review of these principles is described.

7.2.3.1 Electrical Transducers

7.2.3.1.1 Triboelectric Mechanism

Triboelectric sensors work on the basic principle of charge separation resulting from movement of objects, friction, air turbulence, etc., and hence, are a result of mechanical charge redistribution. The mechanism of a triboelectric sensor is as shown in Figure 7.6. A triboelectric nanogenerator using PDMS-CNT has been developed for harvesting human motion-energy. The schematic of a triboelectric nanogenerator comprising PDMS-TBCNT nanostructures on the surface is shown in Figure 7.6(b) (20). A vertical contact-separation triboelectric nanogenerator-based autonomous biosensing device was created to solely detect Gram-positive bacteria in solutions.

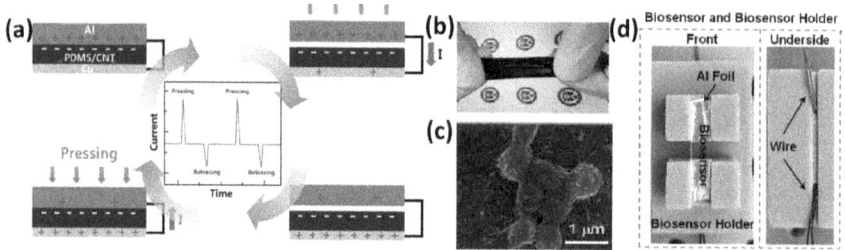

FIGURE 7.6 (a) Schematic of proposed flexible triboelectric nanogenerator using PDMS-CNT composite matrix, (b) stretchable PDMS–CNT film (9), (c) the CNT-Arg solution's molecular processes of *S. aureus* adhesion and capture, *S. aureus* cells, CNT-Arg, and backdrop were distinguished by their different hues of yellow, blue, and purple, and (d) TENG and a biosensor holder powering the self-powered biosensing device (10).

Vancomycin was employed to recognise and capture Gram-positive bacterial cells. Multi-walled carbon nanotubes functionalised with guanidine were used for the purpose of signal amplification for its improved conductivity. The adhesion of bacteria as shown in Figure 7.6(c), resulted in a variation in the output voltage. This stable self-powered biosensing system, when applied for the detection of *S. aureus*, showed a low limit of detection (2×10^3 CFU mL) with high selectivity.

7.2.3.1.1.1 Resistive Sensors The resistive sensors make use of the changes in resistance with differences in the concentration of analyte species interacting with the sensing electrode. Kahng et al. made use of a SWCNT functionalised with polyclonal antibodies for the detection of antigens from *Mycobacterium tuberculosis* (MTB). This low-cost, straightforward TB screening uses an immuno-resistive sensor that was created on a plastic film. The detection limit was determined to be 10 CFU/mL for the target spiked sputum samples in the presence of magnetic nanoparticles. The prototype is as shown in Figure 7.7(d) (21). In the past few years, there have been significant developments in the field of flexible and stretchable sensor designs. The mechanism of working for these flexible resistive sensors is as shown in Figure 7.7. These have satisfied several high demands, like applications of prosthetics, wearables, and health monitoring owing to their customizability, easy integration, and low-cost attainment. The bioinspired hydrogel containing CNTs for resistive sensing was developed by Wei et al. (Figure 7.7(f)). The sensor was developed as an integrated sensing platform for strain detection and reacts immediately to pressure changes in the human body brought on by motions like breathing, finger flexion, and knee flexion. Figures 7.7(g) and 7.7(h), respectively, demonstrate the variations in capacitance and resistance (21).

7.2.3.2   Optical Transducers

For optical biosensing single-walled carbon nanotubes (SWCNTs) have gained a lot of attention recently. They are useful for extremely selective biosensing because of their optoelectronic features, which are sensitive to the environment. Semiconducting SWCNTs, with a wavelength in the near-infrared (NIR, 900–1600 nm), have a

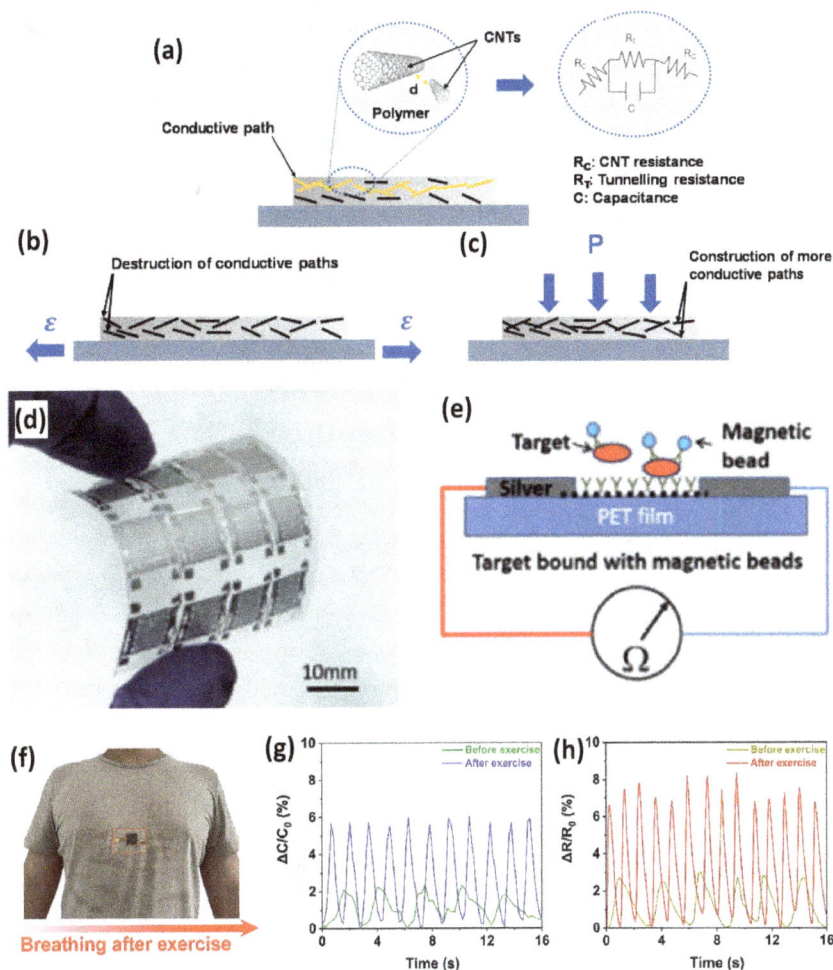

FIGURE 7.7 (a) The polymer/CNT sensor's resistive sensing technique without a load, (b) under strain, (c) under pressure (11), (d) SWCNT-based sensor on a flexible PET film, (e) the resistive transduction is achieved by the binding of target with the functionalized antibody (21), (f) strain sensor placed to the chest that measures the breathing process in real-time, (g) capacitance, (h) changes in resistance values before and after exercise (22).

band-gap, which results in outstanding fluorescence (23). The mechanism of working is as shown in Figure 7.8 (a). To track the level of H_2O_2 in wounds, optical microfibrous material with encapsulated SWCNT-based wearable sensors was developed by Safaee et al. (Figure 7.8 (b)). The sensor operated on the premise that the (8,7)/(9,4)-SWCNT chiralities' ratiometric signals, which varied in their responses to H_2O_2, would be used. Since the fluorescence signal was independent of the excitation source location and exposure time, commercial wound dressings could be detected using a wireless readout. For at least 21 days, these microfibers contained the SWCNTs without causing structural alterations (17).

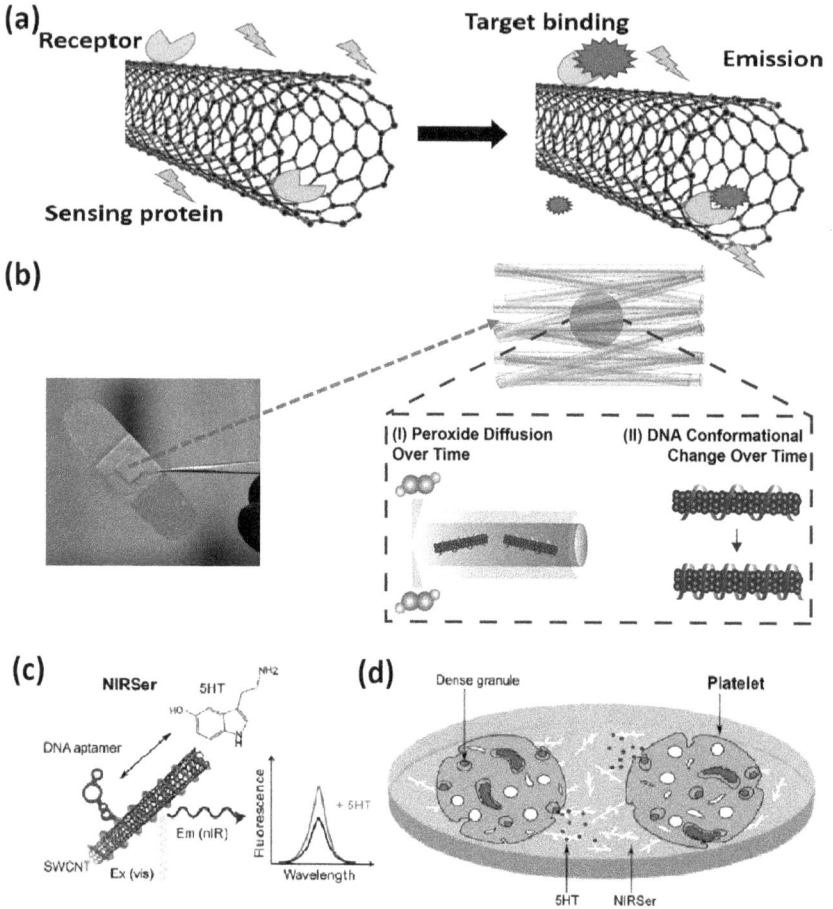

FIGURE 7.8 **(a)** Mechanism of working of CNTs based optical sensors, **(b)** optical fibrous CNTs integrated on a commercial wound bandage and factors affecting the peroxide calibration curve's temporal dependence (17). The nIR fluorescent serotonin nanosensor NIRSer was designed. **(c)** The shape of the aptamer changes when serotonin (5HT) binds to the NIRSer, boosting the fluorescence of the SWCNT. When serotonin (5HT) attaches to the NIRSer, the aptamer's conformation changes, increasing the SWCNT's fluorescence. **(d)** Diagram showing platelets sticking to an NIRSer-coated surface. Serotonin is released by activated platelets from dense granules, and the sensors' images show the change in spatiotemporal concentration (24).

Since the majority of serotonin is kept in blood platelets rather than the human brain, Dinarvand et al. were able to record the release of serotonin from human blood platelets in real time. Figure 7.8(c) and (d) show that the serotonin sensor (NIRSer), which was made of a serotonin-binding aptamer on a SWCNT, showed an up to 80% increase in fluorescence emission in response to serotonin. By placing the sensors underneath and surrounding serotonin-releasing cells, high-resolution images of serotonin release patterns from single cells were captured (24).

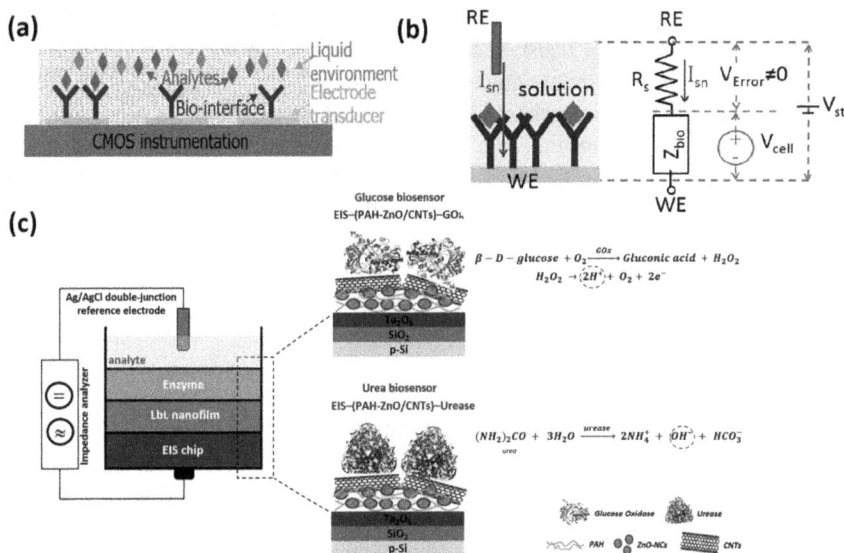

FIGURE 7.9 (a) Components of an electrochemical biosensor with a generalised design. (b) The simple, two-electrode electrochemical system's equivalent circuit model (12). (c) A capacitive EIS (p-Si-SiO2-Ta$_2$O$_5$) sensor modified with PAH-ZnO/CNTs LbL film was investigated for the detection of glucose [EIS-(PAH-ZnO/CNTs)-GOx] and urea [EIS-(PAH-ZnO/CNTs)-urease] are illustrated schematically (25).

7.2.3.3 Electrochemical Transducers

Analyte, a bio-recognition element, a transducer, and instrumentation are the four functional parts that make up an electrochemical biosensor, as illustrated in Figure 7.9. When an analyte and a bio-recognition element bind together, the flow of ions produced by those binding processes is converted into an electrical current or voltage by an electrode acting as the conventional transducer for an electrochemical biosensor. Instrumentation is frequently made up of the electrical circuitry that records, enhances, and aggregates biorecognition signals from the transducer. Figure 7.9(a). The instrumentation component is frequently the electrical circuitry that detects, amplifies, and accumulates biorecognition signals from the transducer (23). The electrode that conducts electrochemical reactions of interest is the working electrode (WE). The electrode that provides a steady and well-known potential is known as the reference electrode (RE) (Figure 7.9(b)). The simplest electrode configuration consists of these two electrodes. A known stimulus voltage, V_{st}, is placed between RE and WE in such two-electrode setups. The ideal voltage at the electrode-electrolyte interface of the electrochemical cell, V_{cell}, should be equal to V_{st} throughout the whole electrochemical reaction. The ion flow channel to RE through the electrolyte solution is represented by R_s, an inherent characteristic of the electrochemical cell, in addition to the electrode-electrolyte contact, as shown in Figure 7.3(a), which is represented by impedance Z_{bio}. The RE supplies the sensing current I_{sn}, resulting in V_{cell} always less than V_{st} because of the voltage drop, V_{Error}, across R_s. I = current, R = resistance, and this decrease in voltage is referred to as the "IR drop." In order to detect glucose and urea, two essential analytes for clinical diagnosis in healthcare, researchers like Morais et al. investigated the effects of using LbL PAH-ZnO/CNTs nanofilms as a matrix for enzymatic biosensors on capacitive EIS chips (Figure 7.9(c)). Immobilizing enzymes, such as glucose oxidase and urease for the detection of urea, allowed for the specialised sensing.

7.3 MOST COMMON BIOMEDICAL SENSING APPLICATIONS USING CNTS

7.3.1 Pressure Sensing

Pressure monitoring devices have gained considerable interest for their application in wearable electronics, electronic skins, and cell phones. These sensors employ different transduction methods, including resistive, capacitive, and piezoelectric to categorise them (11). Modifying the spacing between two parallel plate electrodes is the basic method for implementing capacitive pressure sensors. They typically have great linearity and transparency, but to reduce electromagnetic interference, they must be carefully insulated (EMI) (26). Utilising a composite made using polydimethylsiloxane, low-polarity liquid crystal (TS029), and multi-walled carbon nanotubes, Pan et al. created a resistive-pressure sensor (MWCNT-LPLC-PDMS). The LPLC reduces the modulus of the PDMS elastomer and aids in the creation of conductive pathways in the active layer by forming polymer-dispersed liquid crystal (PDLC) droplets. The performance of the piezo-resistivity sensor was tested at pressures ranging from 0 to 80 kilopascals. Figure 7.10(a) indicates that when pressure was applied, the sensor's resistance decreased. This observation may be associated with the resistance at the contact between the droplets of TS029 and the MWCNT

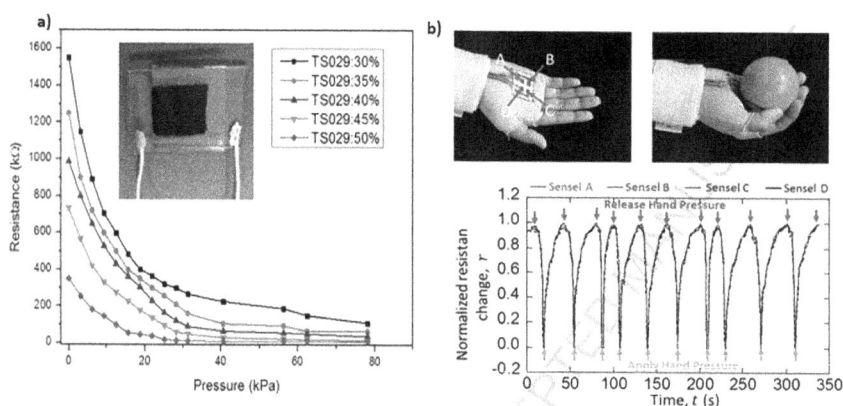

FIGURE 7.10 (a) Resistance of sensor with different concentrations of TS029 in PDMS in the pressure range of 0–80 kPa (26). (b) Prototype sensors with sensels A, B, C and D integrated into a sports glove and normalized resistance change (r) of sensor (20).

particles along with a shrinkage in the conductive path. When 30 wt% TS029 was mixed with PDMS, the sensitivity was computed as 0.63 kPa/1 and 0.057 kPa/1 within the pressure ranges of 0–15 kPa and 15–80 kPa, respectively. The sensitivity for the two pressure ranges increased by 8.49 times and 2.3 times, correspondingly, while the Young's modulus decreased with increasing concentration from 30 wt% to 50 wt%. This pressure sensor has potential applications in wearable technology and e-skin (26).

Zuruzi et al. conducted research on the capacity of MWCNT and PDMS nanocomposite foams to sense pressure. They developed sensing devices by inserting copper between MWCNT and PDMS composites. A 2 x 2 sensor array was constructed using 10 wt% MWCNT and integrated into a commercial sports glove. A 43-year-old man was instructed to hold a rubber ball with the sports glove and squeeze it repeatedly for 11 cycles. Figure 7.10(b) displays a graphical representation of a normalised change in resistance over time. The ball was initially pressed, resulting in a normalised resistance change of approximately 0.96. At 10 seconds, the pressure was released, and the resistance gradually increased to a normalised resistance change of 0. When pressure was reapplied to the ball approximately nine seconds later, the resistance began to decrease, and finally, reached a value of 0.99. The cycle restarted when pressure was released at 41 seconds. This process was repeated 11 times, and the highest normalised resistance change recorded was 0.981 with a standard deviation of 0.013 (20).

Yang et al. fabricated a capacitive-based pressure sensor that is highly sensitive and flexible by using a PVDF nanofibre membrane with CNTs. The sensor consists of a 3D composite nanofibre membrane that sits between two ITO-PET sheets. Four devices were created by adding carbon nanotubes to various solutions of PVDF in weight ratios of 0.03, 0.05, 0.1, and 0.2 wt%. During the experiment, a force gauge applied pressure to the sensor, while an impedance analyser recorded the capacitance (27).

The graph in Figure 7.11(a) displays the capacitance fluctuations of the sensor at different pressure levels and weight percentages of CNT. When pressure was applied to the sensor, the composite nanofibre dielectric layer consisting of CNT

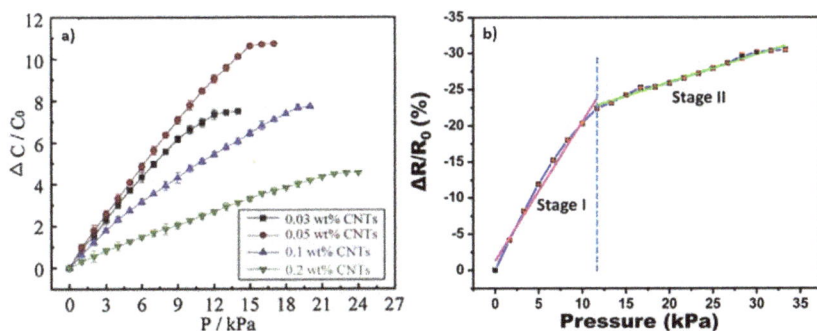

FIGURE 7.11   (a) The relative change in capacitance of the sensor for different weight ratio CNTs inclusion under pressure applied (28). (b) Responsiveness of the sensor at different applied pressures (29).

and PVDF was compressed, resulting in a drop in the space between electrodes and CNTs, leading to an instant rise in capacitance. On exertion of pressure below 1.2 kPa, the sensitivity of the 0.05 wt% CNT composite nanofibre layer was the highest, with an average magnitude of 0.99/kPa, but it declined to 0.63/kPa when the pressure was increased to 15.0 kPa. The sensor has multiple applications in areas such as soft robotics, electronic skin, and pressure measurement (27). Wen-Yin et al. created a delicate, wearable, piezoresistive sensor that depended on the silver nanoparticle (AgNP) and multi-walled carbon nanotube (CNT) nanocomposite films and polydimethylsiloxane. Copper electrodes were sandwiched between a layer of PDMS and AgNP/MWCNT nanocomposite. To evaluate the device's pressure sensitivity, resistance variations were measured at different pressure levels, as displayed in Figure 7.11(b). The sensor exhibits two linear ranges with two distinct slopes. The sensitivity was around 0.02 kPa and 0.004 kPa in low-pressure (11.67 kPa) and high-pressure (11.67–33.3 kPa) ranges, respectively. The wearable sensors were evaluated for their cycle stability at loading and unloading pressures of 6.67 kPa and 16.67 kPa, and their response was demonstrated to be repeatable and reliable, with no significant hysteresis. This sensor has potential uses in wearable sensing devices (30).

7.3.2 STRAIN SENSING

There is a high demand for flexible strain sensors in various technological areas such as biomedicine, healthcare equipment, soft robotics, interactive games, virtual reality, and industrial applications. These sensors are valuable due to their ability to stretch and their softness, which enables them to convert mechanical deformations into electrical signals such as changes in resistance or capacitance (11). Demidenko et al. successfully developed resistive sensors using a sandwich construction of Ecoflex-CNT-Ecoflex and laser structuring to create a sturdy, conductive CNT network over a silicone matrix. They also designed an electronic device that utilises the ATXMEGA8E5-AU microcontroller for processing and reading sensor signals (31). The strain sensors showed linear behaviour when subjected to elongation, as demonstrated by the relative change in resistance in Figure 7.12(a) and (b).

According to research on the strain sensors' response times, the resistance fully recovers and reaches its previous value in 4 seconds after a deformation of 50%. Some of the existing bonds between nanotubes were distorted and not reestablished throughout the cyclic loading-unloading cycles, while new bonds emerged. Hysteresis was therefore seen in the dependences of R/R0 on at the beginning of the cycle, but after many load-unload cycles, hysteresis is reduced because of the creation of a steady equilibrium between "lost" and "restored" bonds between nanotubes. The sensor was used to measure the motion of the finger joints. The software created allows for the tracking of finger joint movements in real-time, as shown in Figure 7.12(c), by receiving variation in resistance of the strain sensor caused by bending of the finger joint via an electrical device and sending this information to the computer (31). Using an aerodynamically focused nanomaterials (AFN) printing technique, Min et al. created and assessed AgNPs/MWCNTs nanocomposites strain sensors.

To provide mechanical and electrical insulation, a flexible polymer cover layer was included in the construction of a printed sensor that measures 3.5 mm in width and 7 mm in length, with 0.5 mm between each line pattern (32). An in-situ

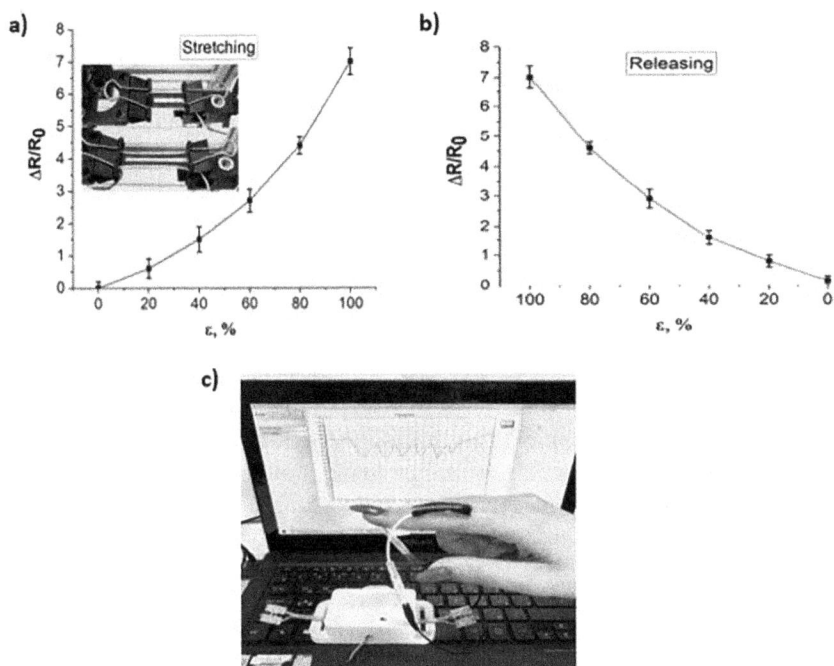

FIGURE 7.12   Resistance dependent elongation (**a**) during elongation, and the insert shows the sensor's stretching process, (**b**) releasing, (**c**) real-time measurement of finger joint movement (31).

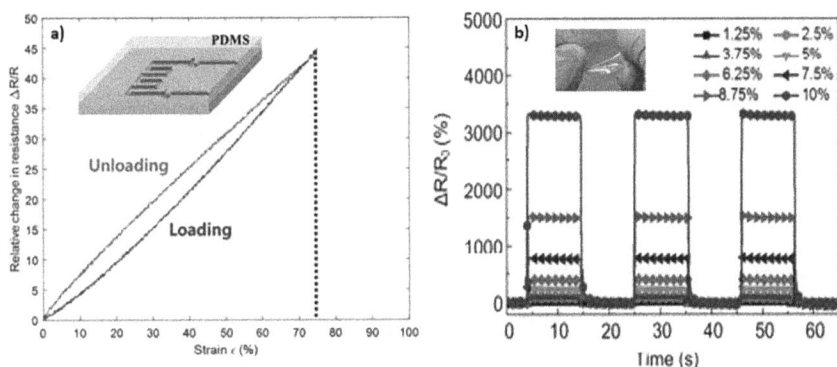

FIGURE 7.13   (**a**) Test for measuring the sensitivity and stretchability of the developed nanocomposites sensors (33). (**b**) The responses of $\Delta R/R0$ to typical strains (34).

configuration was created to determine the sensor's sensitivity and stretchability by using a multi-axis stage to conduct a stretching test, with changes in resistance detected using LabVIEW 2015 and NI USB 6009 modules. Figure 7.13(a) illustrates the basic performance of the sensor, which has a maximum 74% strain limit. The use of a 1,000-cycle test revealed the stability of the printed sensor, with only a slight variation in peak resistance occurring after 10–20 stabilisation runs. When subjected to elongation tests, most resistance variances behave similarly. As a result, printed nanocomposite sensors have the potential to revolutionise numerous engineering fields, such as wearable technology and artificial electronic skins (32).

Nie et al. developed novel strain sensors that have superior sensitivity and are both transparent and flexible by embedding multi-walled carbon nanotubes (MWCNTs) into microtrenches of PDMS sheets. To measure the strains, the researchers utilised a linear motor's test platform that was equipped with the manufactured strain sensors. The sensors were coated in silver paste on both ends and connected with conductive tapes to a source meter that supplied the voltage and recorded the current flow. From Figure 7.13(b), the sensors displayed an increase in normalised resistance (R/R0) as the strain increased, with a value of 22% observed under a small strain of 1.25%. When the strain reached approximately 10%, the R/R0 increased considerably to 3,300%. The rise in resistance occurred due to the formation of microcracks in the MWCNTs within the microtrenches of the PDMS film; these microcracks grew to match the magnitude of the strain when the sensor was subjected to external stresses, leading to an increase in the electrical resistance of the sensor. The gauge factors of the sensors ranged from 16 to 330 as the strains increased from 1.25% to 10%. Applications for these strain sensors could include wearable technology and medical diagnosis (34).

7.3.3 GAS SENSING

Sarkar and team made a SWNT-TPP hybrid nanostructure using electrochemical polymerisation. They examined its ability to sense VOCs such as acetone. Results are shown in Figure 7.14(a) as a plot of steady-state, normalised resistance change vs applied charge density during electro-polymerisation.

At all levels of acetone vapour, the device's reaction initially increased as charge density rose, reaching its peak at 19.56 mC/cm^2 before falling. This pattern may be explained by the fact that the initial availability of porphyrin molecules for interaction/binding with the analyte increased, followed by a decrease in the conductivity of the polymer coating due to the decreased conductivity of the porphyrins, which hindered the charge transfer process. As illustrated in Figure 7.14(b), the sensor showed a quick rise in resistance after each exposure to acetone concentrations ranging from 50 ppm to about 230,000 ppm, followed by a delayed and incomplete recovery. Additionally, it demonstrated excellent stability over a 180-day period (38). Swager's team has demonstrated a reliable chemiresistive sensor that can detect cyclohexanone, a target analyte for explosives, with high specificity. The trifunctional selector increases the overall resilience of SWCNT-based chemiresistors by covalently functionalising SWCNTs via cofacial "-" contacts and hydrogen bonding with cyclohexanone. Their sensors responded to 10 ppm of cyclohexanone in less than 30 seconds in a reversible and repeatable manner, with an average theoretical limit of detection (LOD) of 5 ppm (39).

MWCNTs with amino functionality were placed on interdigitated electrodes in a gas sensor made by Xie et al. With 5.55% and 18.19% amino group-MWCNTs, respectively, MWCNTs were functionalised. On the basis of raw MWCNTs/Nafion

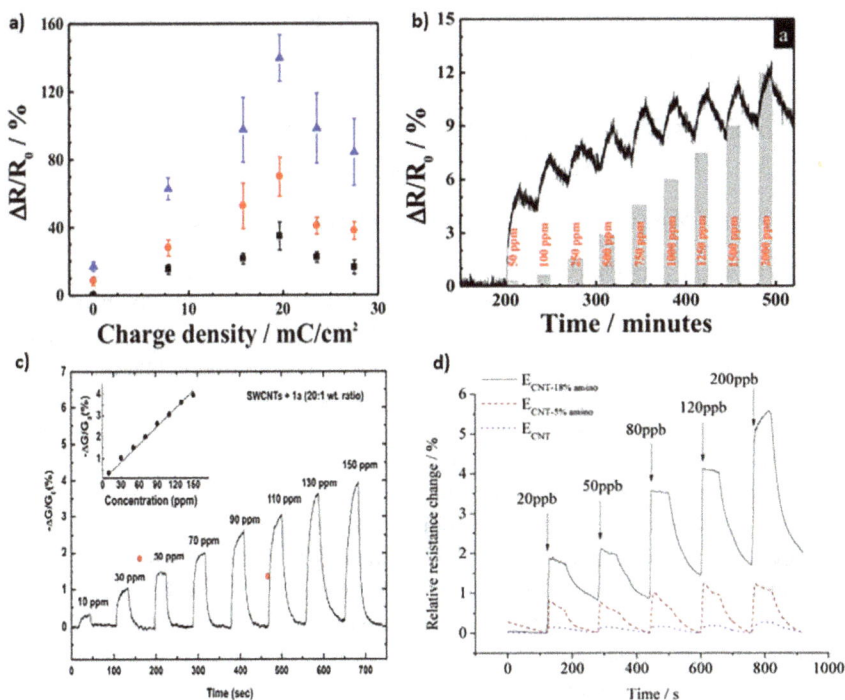

FIGURE 7.14 (a) Tetraphenyl porphyrin (TPP) electropolymerisation at constant potential (CP) (at 2 V) sensor responses to acetone vapour exposure at 12.5 (black square), 25 (red circle), and 50% (blue triangle) of saturation vs charge density applied. (b) Real-time responses of SWNTs and poly(TPP) hybrids to acetone fumes ranging in concentration from 50 to 2000 ppm (35). (c) A trifunctional selector 1a-based SWCNT-based sensor's normalised conductive change (-G/G0 (%)) in response to cyclohexanone at various doses. The inset figure displays the size of the normalised conductivity change in relation to cyclohexanone concentration (36). (d) Typical relative resistance shifts of the ECNT, ECNT-5%amino, and ECNT-18%amino sensors in the presence of formaldehyde gas at concentrations between 20 ppb and 200 ppb (37).

film, MWNTs-N-1/Nafion film, and MWNTs-N-2/Nafion film, three distinct gas sensors were created. (37). In Figure 7.14(d), the relative resistance variations of three sensors to formaldehyde gas at various concentrations are depicted. The only factor affecting the resistance of an ECNT sensor is physical adsorption, in which the van der Waals interaction between the molecules of formaldehyde and MWCNTs causes a change in the density of charge carriers. However, the responses of the resistance change for the aminogroup-modified sensors ECNT-18%amino and ECNT-5%amino are due to both chemical and physical absorption of the formaldehyde gas molecules. The sensors ECNT-18%amino, ECNT-5%amino, and ECNT had relative resistance changes of 1.73%, 0.56%, and 0.13%, respectively. When compared to ECNT-5%amino and ECNT, ECNT-18%amino's relative resistance changes are roughly 2.4 and 13 times greater, respectively. While exhibiting a quicker response than the sensor ECNT, the ECNT-18%amino and ECNT-5%amino display a slower rate of desorption. This presents intriguing prospects for their use as formaldehyde low-concentration sensor applications (37).

7.4   RECENT ADVANCES IN MODERN PLATFORMS FOR SMART DIAGNOSTIC SYSTEMS

7.4.1   Point-of-Care Diagnostics Using Paper-Based Sensing Platforms

The development of paper-based sensing platforms is of widespread interest these days. The advantages of these systems include low cost, disposability, ease of disposal of biological contaminates without much processing, etc. These systems are widely used for rapid diagnostics in healthcare as well as environmental applications. One of the most used applications is the point-of-care diagnostics in healthcare systems. Most of the paper-based systems use inkjet or screen-printing for the electrode fabrication on the desired substrate due to ability for rapid fabrication on a large scale.

These include wide range of applications ranging from pH sensors for sweat, wound monitoring, and biofluids. A carbon nanotube-based pH sensing platform has been developed for paper-based microfluidics. It consisted of CNT ink patterned on a filter paper substrate and was used to detect pH over a range from 5 to 9 (40).

A fully inkjet-printed electrochemical sensing-platform consisting of carbon nanotube was constructed on a paper. The sensor was also aided with a hydrophobic barrier for efficient sensing in specific area. Further, a sensor was used in dopamine (DA) detection, with limits as low as 10 µM (41).

Harbi et.al created a choline detection sensing platform in metabolite-present plasma and urine. The sensing studies in spiked human plasma gave precise and accurate results for measurement of catechol to the point of concentration of 1×10^{-7} M. This turned out to be a good option for detecting volatile organic chemicals (42).

Shen et.al developed a chemiresistive biosensor using SWCNT for point-of-care detection of human serum albumin (HSA). The ink for electrode fabrication was achieved by using pyrene carboxylic acid and CNT stacking. The sensor surface was functionalised by specific antibodies for selectivity. The creation of the hydrophilic channels for efficient fluid transmission was made possible by combining wax printing and vacuum filtering. A limit of detection as low as 1 pM was achieved (43). Further, other than electrical transduction, sensing platforms based on chemical mechanisms were also developed. A rotatable, paper-based, analytical instrument with multi-step electrochemiluminescence immunoassays was created by Sun et al. The platform was fabricated by the assembling of paper discs in a specific fashion. The carbon nanotube ink so formed was added and dried on the immunozone. The system showed an excellent analytical performance in a linear range of 0.1–100 ng mL^{-1} and 0.1–50 ng mL^{-1} with detection limits down to 0.07 ng mL^{-1} and 0.07 ng mL^{-1}, respectively, for CEA and PSA (44).

7.4.2   Self-Powered Designs

Wearable, self-powered biosensors are tools that run without an internal battery or external power source and employ a biorecognition detector component to transmit sensing data. It may be able to detect physiological data in different bodily fluids

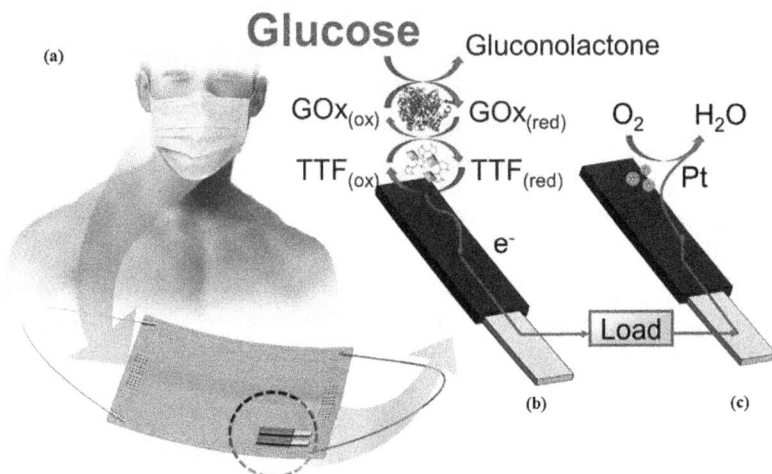

FIGURE 7.15 (a) The idea of a bioelectronic mask that generates electricity and a self-powered signal from perspiration that contains glucose. (b–c) The functional parts of the mask-based bioelectronic device and the redox reactions that relevantly takes place at the (b) the bioanode and (c) cathode (46).

that can be obtained with little to no invasiveness, such as heart rate and human activity. This is because these gadgets are portable, can adhere to skin, and can withstand mechanical distortion. By using Bluetooth or near-field communication (NFC) modules, it ensures that the system collects data on human health and transmits the information in real time to mobile devices. For a Self-Powered Biosensor for cholesterol analysis, high surface-area carbon-cloth electrodes were designed (45). Figure 7.15 depicts a self-powered device based on a mask for bioelectronics. The system's open-circuit voltage (OCV) was 0.37 V with a maximum power density of 14. It runs with modest sample volumes (less than 100 L) $\mu W \ cm^{-1}$ (46).

CNTs exhibit fascinating properties for their applications as biomedical sensors. Apart from their unique sensing capabilities, these nanostructures possess the ability to cross tissue barriers and can result in acute inflammations in the user. The presence of amorphous or metallic forms exhibits more toxicity than the other forms. Hence, the choice of the type of nanotube and functionalisation plays a major role, when used for sensing applications. CNTs are also potential candidates for medical diagnostic imaging and ultrasonographic applications. 1,3-dipolar cycloaddition of azomethineylides (ox-MWCNT-NH3 $^+$) functionalised CNTs was used by Delogu et al. (47). The sonographic studies in heart and liver showed signal response comparable to commercial agents. Recently, PhotoAcoustic (PA) properties of pristine CNTs has also been studied (48). The site-specific action for CNTs was improved by functionalisation with Indo Cyanine Green (ICG) and Arg-Gly-Asp (arginylglyc-ylaspartic acid–RGD) peptides to PEGylated CNTs. This makes the nanotubes to target the integrins at the tumour vasculature (48, 49). An ultraselective nanoparticle consisting for targeting Ly-6C monocytes and macrophages using PA has been developed by Gifani et al. (50). The single-walled carbon nanotubes are selectively and abundantly taken up by the monocytes and shows 6 fold greater signal compared to controls (50).

7.4.3 FLEXIBLE, PRINTABLE, AND WEARABLE DESIGNS FOR BIO-INTEGRATION OF SENSORS USING IoT

The development of thin-film transistors and integrated circuits has recently drawn a lot of attention to a flexible platform. This is accomplished using the chemical vapour deposition (CVD) of CNTs and subsequent techniques, including gas-phase filtering. The nanotube had Y-shaped connections connecting lengthy (10 m) nanotubes (Figure 7.16(a)) (18). The mobility of the transistors developed was found to be 35 cm2 V^{-1} s^{-1} and an on/off ratio of 6×10^6 (18). Recently, plastic-based sensors have been developed that interface with skin, and further, integrate with a circuit board for signal processing and real-time assessment of the physiological state of an individual through perspiration analysis (Figure 7.16(b)) (51). CNTs are often mixed with other polymers and find many healthcare monitoring applications. Cellulose and CNT-based thin films has been developed as E-skin and used for personal healthcare monitoring (Figure 7.16(c)) (52). A temperature sensor was fabricated by Zhao using CNT-GO on a flexible PET substrate. The results of the research demonstrated that temperature had an impact on the sensor's resistance. A thermally induced change in the resistance was used for the temperature detection; hence, a negative temperature coefficient was measured (α). The typical sensitivity was calculated to be 60×10^{-3}/°C. The little degradation, and hence, excellent repeatability was shown by the CNT-based sensor (Figure 7.16(d) and (e)) (53). CNT-based thin-film transistors have also been developed and are found to have efficient bio-integration capabilities (Figure 7.16(f)) (54). An electronic skin with a combination of CNTs for mechanically flexible sensors was developed by Wang et al.; an active matrix of OLED system showed the magnitude of applied pressure and finds applications in medical/healthcare-monitoring devices (Figure 7.16(g)) (55). For real-time health monitoring applications, a disposable cyclic voltammetry device (CV) tag was printed on a plastic film by combining a wireless power transmitter, a triangle wave generator, a power transmitter electrochemical cell, and signage. Thin film transistors (cnTFTs) based on single-walled carbon nanotube networks served as the fundamental component of the system (Figure 7.16(h)) (56). Core-shell-structured CNT has also attracted the attention of researchers for consistent and reliable humidity sensing. A CNT@CPM has been recently used for highly sensitive humidity sensor. The sensor has been used for sensing experiments in smart clothing, such as masks that might do real-time, multi-respiratory monitoring through autonomous ventilation systems (Figure 7.16(i), (j) and (k)) (57).

7.5 RECENT DEVELOPMENTS IN THE EARLY-STAGE CHRONIC DISEASE DIAGNOSTICS

Enzymatic (biomolecular and chemical agents-chronic disease biomarkers, hormonal sensors, infection biomarkers detection [Covid 19], vital signs monitoring, continuous physical signals monitoring) nano-sensors have a vital role in recent

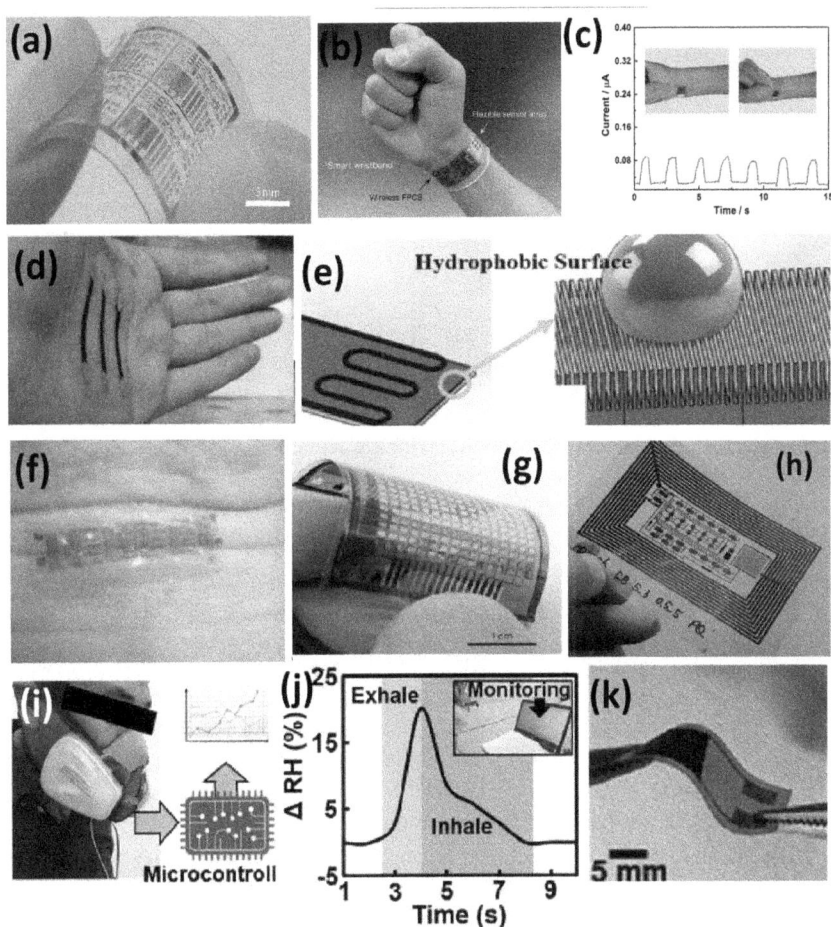

FIGURE 7.16 (a) A picture of a transparent PEN substrate with integrated circuits and TFTs made of carbon nanotubes (18). (b) The multiplexed sweat sensor array and a wireless, flexible, printed circuit-board FPCB are fully incorporated into the wearable sensor known as FISA for multiplexed perspiration analysis (51). (c) An RC-CNT composite film at wrist joint bending (52). (d) A CNT-GO trace-bearing adhesive PET tape on the palm. (e) Sensor that measures temperature on highly hydrophobic surfaces (53). (f) Biointegration capability of carbon nanotube-based thin-film transistors and integrated circuits (54). (g) A completely constructed interactive e-skin with a backplane made of carbon nanotube TFT active pixels that measures 3 by 3.5 cm² in size (55). (h) A fully printed wireless cyclic voltammetry tag involving carbon nanotube network-based thin film transistors (cnTFTs) (56). (i) A mask with micro-controlled flexible humidity sensor (CNT@CPM-3). (j) Variations in response with breathing cycles. (k) Developed sensor (57).

applications of molecular recognition in applications like biomolecular and chemical agents in acute or chronic diseases, infections, or other physiological conditions. The applications of CNT-based sensors for some of the most prevalent illnesses and physical disorders are covered in the section that follows.

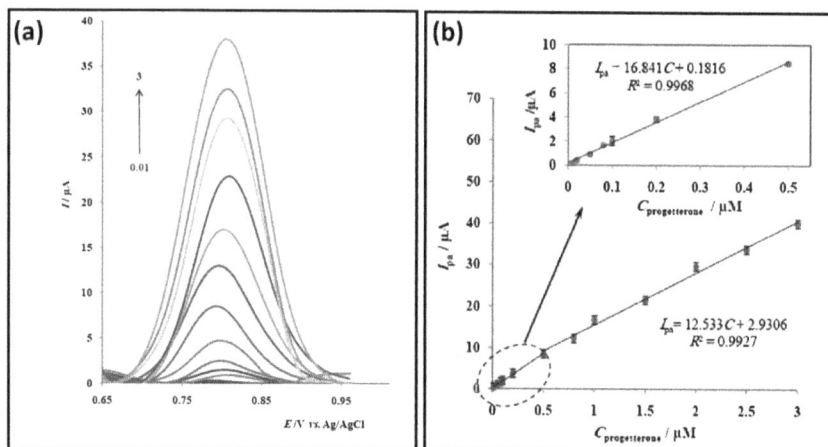

FIGURE 7.17 (a) DPVs obtained at the Fe3O4@GQD/f–MWCNTs/GCE for different concentrations of progesterone. (b) Calibration curve for the sensor (59).

7.5.1 CNT-Based Sensors in Women's Health Monitoring

Estrogens such as 17-βestradiol (E2) and 17-alpha-ethynyl estradiol (EE2), whose unbalanced levels might signify substantial endocrine disruption, are among the hormones. A thionine conjugated MWCNT was coated with Au NPs to produce an aptasensor for the electrochemical detection of 17-beta-estradiol (E2). For this, a potentiostatic insertion approach was used to assemble the E2 aptmaer on the surface of the sensing platform. Using differential pulse voltammetry, real-time female sample sensing characteristics were examined (DPV). The sensor showed a limit of detection as low as 1.5 pM (58). Arvand et al. created an electrochemical sensing platform for the precise detection of progesterone (P4) using graphene quantum dots (GQDs) and MWCNTs functionalised with Fe3O4 nanoparticles. The electrocatalytic characteristics of a modified electrode towards the oxidation of P4 served as the foundation for the biorecognition (Figure 7.17 (a) and (b)). The sensor response was found to vary linearly over a linear range of 0.01–0.5 and 0.5–3.0 μM. The detection limit and sensitivity were 2.18 nM and 16.84 μAμM^{-1} (59)

7.5.2 CNT-Based Sensors in Detection of Various Cancer Biomarkers

Cancer is a group of diseases involving abnormal growth of cells in certain parts of the body, further invading healthy cells and spreading to other parts of the body. These diseases have a higher chance of being cured when diagnosed at an early stage. Abnormal levels of biomarkers in blood or body fluids are often an indicator of underlying conditions. Some of the major biomarkers involved in the most common types of cancer include prostate cancer (57). Among the biomarkers in prostate cancer, the main biomarker identified is prostate-specific antigen (PSA). For the ultrasensitive detection of PSA, a label-free voltametric immunosensor has recently been

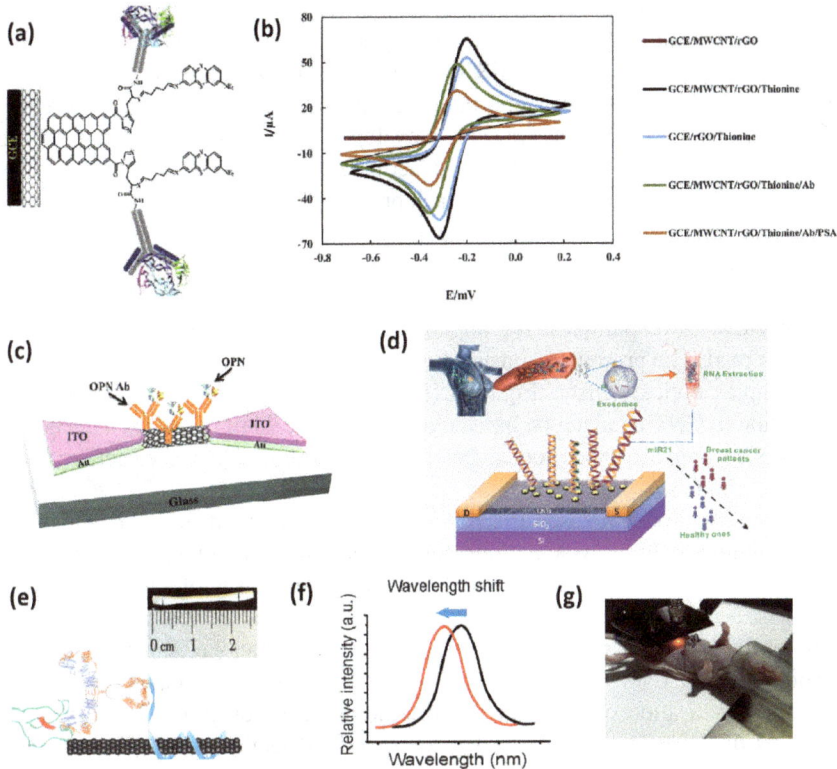

FIGURE 7.18 CNT-based sensors used for the detection of various cancer biomarkers. **(a)** GCE modified with functionalised SWCNT and antibody for the specific detection of prostate specific antigen. **(b)** Change in electrochemical current with each surface modification step (63). **(c)** Between the source and drain electrodes of a transparent SWCNT-based immunosensor for OPN detection, OPN antibodies were placed on the SWCNT surface (64). **(d)** Ultrasensitive detection of exosomal miR21 using the DNA-functionalised CNT FET biosensor (60). **(e)** An Ab-DNA-SWCNT complex synthesised using a MWCO membrane. **(f)** PL excitation/emission plot of the Ab-DNA-SWCNT sensor. **(g)** Photograph of data acquisition using a probe-based system from NIR emission from the implanted sensor in mice (61).

created. The sensing platform consisted of a L-histidine-functionalised, reduced graphene oxide/MWCNT nanocomposite (Figure 7.18(a)). (Figure 7.18(b)) In human saliva samples, the immunosensor was able to detect PSA with a limit of detection (LOD) of 2.8 fg mL-1 (58). Another most common biomarker used for the detection of prostate cancer is osteopontin (OPN), a phosphoprotein that is used by cancer cells. A. Sharma et al. developed a label-free, transparent immunosensor based on SWCNT for OPN detection (Figure 7.18(c)). The sensors were characterised in human serum and showed an LOD of 0.3 pg/mL (59). Another major cancer biomarker is Galectin-3, a β-galactoside-binding protein; the levels of this protein can be directly correlated to the potential of malignancy. Park et al. developed a D-(+)-galactose-conjugated SWCNT for detecting galectin-3 (60).

Breast cancer: The exosomal miRNAs derived from tumours have important functions in the initial stages of cancer progression that results in the onset of progression. A label-free field effect transistor (FET) developed using semiconducting-CNT film was achieved by Li et al. Floating gate-structure achieved using Y_2O_3 and assembled Au nanoparticles acted as linkers to attach probe DNA (Figure 7.18(d)). The current change associated with the attachment of target mRNA was quantified with an LOD of 0.87 aM (60).

Ovarian cancer: Because significant cancer antigens can't always be detected in early-stage blood samples. An implantable sensor prototype was developed by Williams et.al. The prototype consisted of an antibody-functionalised carbon nanotube complex optical sensor (Figure 7.18(e)). The sensor response is based on the modulation of CNT bandgap on biorecognition (Figure 7.18(f) and (g)). This in-vivo, non-invasive optical sensor for cancer biomarkers proved to be effective for the orthotopic models of diseases (61).

Lung cancer: Several gases and volatile organic compounds have been identified as the biomarkers in early-stage lung cancer diagnosis. The quantification of volatile organic compounds (VOCs) has emerged as an easy way of early-stage cancer diagnosis. Some of the most common used sensors include electronic nose for the detection of polar vapours like methanol, ethanol, 2-butanone, and propanol, etc., and non-polar vapours such as benzene, isoprene, etc. (16). Most cases required specific functionalisation of CNT for the specific detection of gases. An Rh drugged CNT was developed by Wan et al. for the specific detection and differentiation of C_6H_6 and C_6H_7N (62).

7.5.3 CNT-BASED SENSORS FOR DETECTION OF ENVIRONMENTAL TOXINS

Dihydorxybenzene isomers are environmental contaminants, the exposure of humans to which increases due to pollution. Aragon et al. achieved the accurate detection and isolation of these environmental pollutants using an electrochemical sensing platform based on reduced graphene oxide (ErGO), carboxylated carbon nanotubes (cMWCNT), and gold nanoparticles (AuNPs). The careful monitoring of the Differential Pulse Voltammetry led to the detection and differentiation of signals from each of these contaminants based on their electrocatalytic oxidation peaks. The sensor system proved to be a simultaneous and precise detection on complex polluted environmental samples as well (65).

Tertrabromobisphenol A is a contaminant which is present in many aspects of daily life, as in plastics, textiles, and other electronic products as well; hence, it is of significant environmental concern. TPPBA was recently detected extensively in human bodies and is found to be a mediator for various types of cancers in human body. Recently, an electrochemical sensor using Fe_2O_3 hybridised MWCNT was constructed to determine TPPBA in real samples. The modification of sensor with TOAB enhanced the signal transduction at the MWCNT surface and led to an effective detection limit of 0.72 nM (66).

7.6 SYNTHESIS AND FABRICATION METHODS FOR CNT-BASED BIOSENSORS

7.6.1 COMMON SYNTHESIS METHODS FOR CNTs

There are many techniques for building carbon-nanotube structures, most of them involving gas phase activities. The three most popular processes for making CNTs are:

(1) Carbon electric arc-discharge technique.
(2) Laser-ablation technique.
(3) Chemical vapour deposition technique.

7.6.1.1 Carbon Electric-Arc Discharge Technique

Electric current between two electrodes passing through an ionised gas column is known as an arc. Two high-purity graphite electrodes – the anode and cathode – are kept close together in the arc discharge process while being surrounded by helium. The process of creating plasma involves supplying a potential difference between two electrodes and creating gas to conduct electricity. It reveals a variety of unwanted impurities, like metallic particles, various graphitised carbon compounds, and amorphous carbon, which prevents their prospective usage. Without the use of any centrifugation procedure, the effectiveness of impurity removal with this method is around 66–75% (67).

A purification of materials produced by electric-arc discharge technique is shown in Figure 7.19. Single-wall nanotubes (SWNTs) are made by using a variety of catalyst forerunners, and they are expanded in an arc discharge using a sophisticated anode

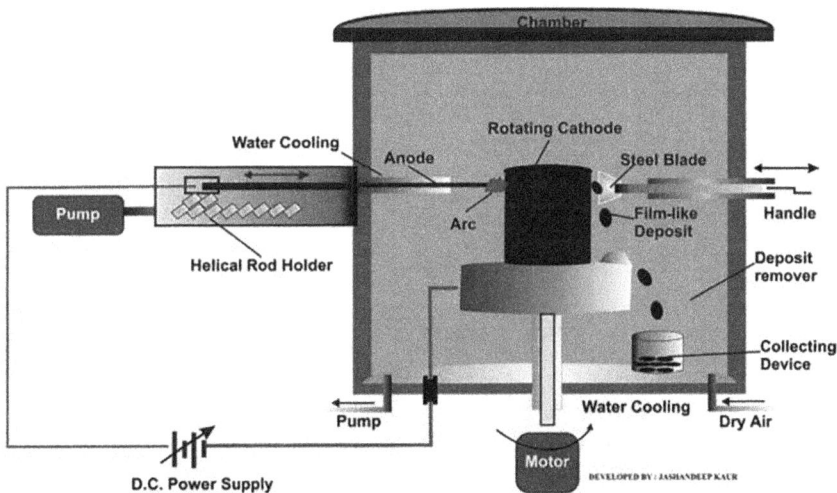

FIGURE 7.19 Schematic diagram of arc discharge method of synthesis (5).

consisting of metal and a graphite (68). But in the multi-wall nanotubes (MWNTs), synthesis can be produced without the inclusion of catalyst precursor.

7.6.1.2   Laser Ablation Technique

Laser ablation or photoablation (also called laser blasting). To distinguish vapour deposition from "laser evaporation," which involves evaporating material and heating it while it is in a thermodynamically stable state, the term "laser ablation" is used to emphasise the nonequilibrium vapour conditions formed at the surface by a potent laser pulse. The graphite target is ablated with a pulsed YAG laser at 1,200 °C in a furnace to create nanotubes. This approach aids in the fabrication of different nanomaterials such as semiconducting quantum dots, carbon nanotubes, nanowires, and core-shell nanoparticles. This procedure has a tremendous benefit in producing high-yield and relatively low-quality nanoparticles, mainly in the quantum range of sizes (10 nm), since the metallic atoms involved have a propensity to evaporate from the tube's end (68).

7.6.1.3   Chemical Vapour Deposition Technique

Chemical vapour deposition (CVD), a convenient and efficient method for producing CNTs at lower temperatures and atmospheric pressure, is shown in Figure 7.20. CVD is a flexible process, so it promises to harness many petroleum products in every state (solid, liquid, or gas) and it permits CNTs synthesis in a different range of forms like thin films, aligned nanotubes, powder nanotubes, and coiled nanotubes or the preferred structures of nanotubes on predetermined locations of such a shaped material (69).

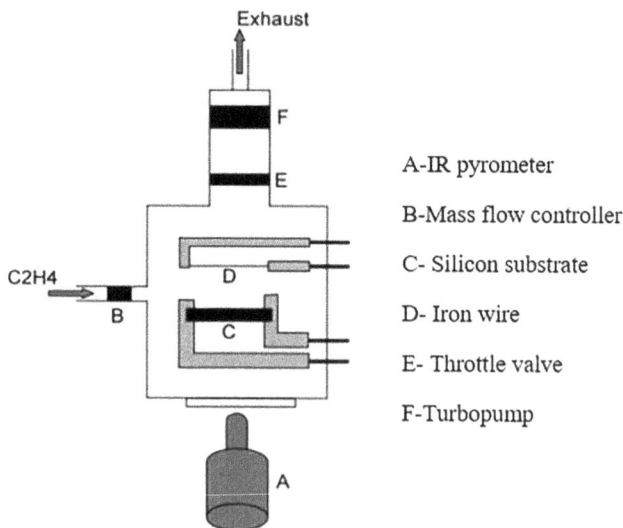

FIGURE 7.20   Schematic diagram of chemical vapour deposition method of synthesis (70). (a) IR pyrometer; (b) Mass flow controller; (c) Silicon substrate; (d) Iron wire.

There are several varieties of CVD methods; notably,

- Plasma enhanced (PE) oxygen-assisted CVD.
- Microwave plasma-enhanced (MPECVD).
- Water-assisted CVD.
- Catalytic chemical vapour deposition (CCVD).
- Hot filament (HF-CVD).
- Radiofrequency plasma-enhanced CVD (RF-PECVD).

Catalysts are still needed, even though CVD operates at lower pressure and temperature than laser ablation and electric arc-discharge techniques. It is increasingly used because of its inexpensive price and capacity to produce huge amounts of CNTs that are quite clean (69). The CVD equipment for synthesizing single-walled carbon nanotubes (SWNTs) is shown in a schematics (70).

7.6.2 FABRICATION TECHNIQUES

7.6.2.1 Screen Printing

Foldable, screen-printed electrodes exhibiting excellent performance and high resolution should concurrently meet the following criteria:

(i) Highly reliable and with excellent printability for micrometer-level printing.
(ii) To print into porous electrode structures and maximize electrode performance, materials must (a) be devoid of non-functional compounds; (b) be able to support deformation after printing; and (c) have excellent mechanical compliance.
(iii) Excellent voltage conductivity for quick electron-charge transfer without rigorous post-treatments.

In this study, we created an interdigitated, flexible energy storage system using an electrode made of banana peel biomass and activated carbon. The surface area of these microelectrodes is 62.03 m²/g. The supercapacitor was created by screen printing a conducting silver ink interdigitated current collector, which concurrently offered all the necessary features. It displays a high capacitance of 33.18 mF/cm² at a scan rate of 1 mV/s. Additionally, the created supercapacitor had outstanding cyclic stability, and even after 5,000 cycles, a capacitance retention of nearly 90% was noted (71).

7.6.2.2 Brush Painting

The mass manufacture of organic devices depends on a roll-to-roll system; brush painting is a very cheap, easy-to-handle, and quick solution-processing approach. It was reported that brush painting methods were used to create highly efficient organic solar cells. These high-performing organic solar cells benefited from the effective brush paintings of excellent-quality conducting electrodes, dielectric layers, and semi-conductive layers (72).

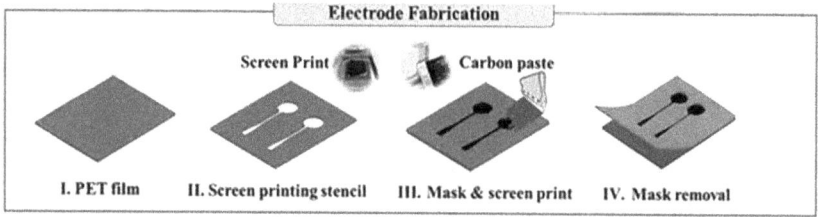

FIGURE 7.21   Shows screen-printing electrode fabrication procedures (74).

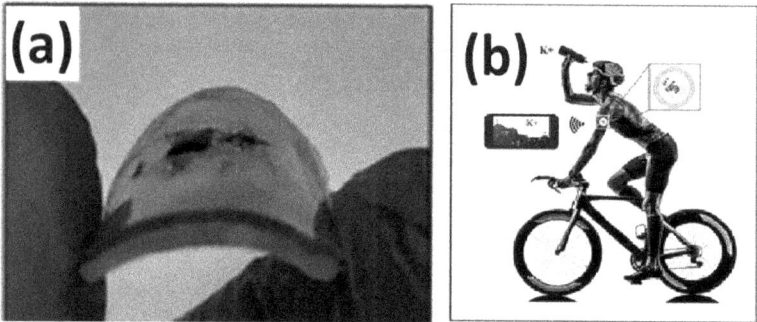

FIGURE 7.22   (a) Indicates the encapsulation of flexible circuits with PDMS. (b) Illustrates the health practitioners may arrange tailored water and electrolyte replenishment by measuring athletes' sweat electrolytes (74).

7.6.2.3   Mask Painting for Flexible Wearable Sensors

The most important phase in making high components for many devices is proper masking. The symmetric micro-supercapacitors (MSC) made by the mask-painting approach on customisable, pore-size, scalable, N-doped, hierarchical, porous carbon showed a high energy density of 5.07 W h cm^{-2} at a power density of 0.25 mW cm^{-2}, and it has a high areal capacitance of up to 36.5 mF cm^{-2} at a current density of 0.25 mA cm^{-2} as well as remarkable cycling stability (98.2% after 10,000 cycles) (73).

Carbon nanotube-based wearable biomarker biosensors have been developed to identify several important human health related species, such as insulin, cytochrome, amino acids, glucose, nicotinamide adenine dinucleotide (NADH), and others. In contrast to laboratory testing, it offers distinct advantages such as lighter, less expensive, real-time response, conformable, lesser volume samples, continuous monitoring, real-time, more sensitivity in low biomarker concentrations, and more flexibility. In this study, Shipeng examined human sweat to measure the potassium ion concentration (K^+), and the electrode preparation is illustrated in Figure 7.21 (74).

As shown in the schematic design of the system in Figure 7.22(a), for on-site bioanalysis monitoring in human perspiration, a wearable wireless microfluidics patch with a skin interface is utilised. This patch is combined with electrochemical sensing technology and does not need batteries. As demonstrated in Figure 7.22(b), the circuits' polydimethylsiloxane (PDMS) encapsulation shields the Near Field Communication (NFC) electronics from perspiration and other external ambient (74). One is on the top of the other self-assembled glucose biosensor made of glucose oxidase (GOx) and multi-walled nanotubes (MWNTs) on the polymer substrate. Flexible biosensors have a huge potential in adjustable biosensor devices whose linear response ranges about 0.02–2.2 mM and poor detectability limit is 10 µm (75). For new wearable/flexible electronic applications, activated carbon made from biomass (banana peel) was developed to have a flexible, high-energy storage capacity. This flexible device outperforms other flexible wearable devices – 60 in terms of energy storage capacity during several cycles of mechanical bending and repeated electrical cycling tests (around 5,000 cycles) (71).

7.6.3 ROLE OF BIO-DERIVED CNTs IN SENSING APPLICATIONS

7.6.3.1 Chitosan-Derived CNTs

Chitin is a fibrous material that is mostly found on the tough outer shells of crustaceans and in certain fungi's cell walls, from which chitosan is made. Chitosan is a recyclable material that has been utilised to create antibacterial food packaging films. It is allowed for use in wound dressings due to its ability to gel. A chitosan-bovine serum albumin (Chi-BSA) cryogel was modified with glucose oxidase, ferrocene (Fc), and MWCNTs to create a reliable glucose biosensor. The MWCNTs/Chi-BSA-Fc/GOD biosensor confirmed excellent operating reliability (RSD = 3.6%), a wide linear range from 0.010 to 30 mM, and a poor Michaelis-Menten constant after more than 350 injections (1.5mM). Figure 7.23 shows how well glucose biosensors respond when using the modified electrode and its great sensitivity. The revised electrode's measurement of the blood plasma's content of glucose was in perfect agreement with the

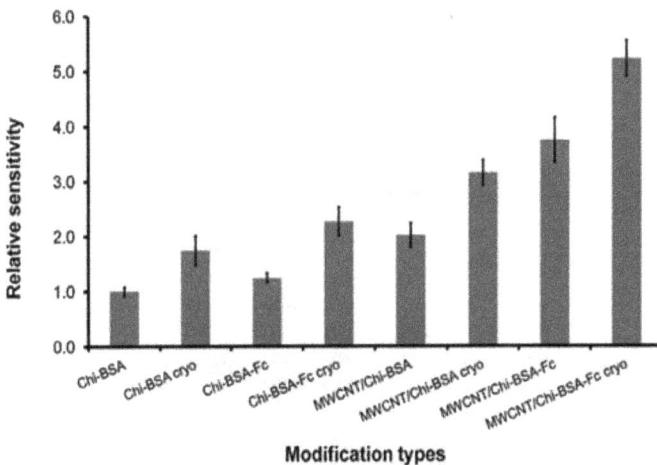

FIGURE 7.23 Displays a comparison of the relative sensitivity of several GOD-modified electrodes for the detection of glucose (76).

hospital's standard hexokinase-spectrophotometric method (P > 0.05) (76). Glucose dehydrogenase (GDH) and chitosan hydro-bonded multi-walled carbon nanotubes (GDH/CS-MWCNT) nano-bio composites were absorbed onto screen-printed carbon electrodes to make a robust glucose sensor that could be detected using the direct electron transfer technique. The sensor's response increased across the glucose level range of 0 to 5.5 mM. Finally, throughout the course of 10 days, the short-term glucose stability was assessed. The GDH enzymatic activity throughout this period was maintained at 80% for 6 days before declining to 50% of its original activities for the last 4 days (77).

7.6.3.2 Carbon Black Derived CNTs

Acetylene black, also known as carbon black (CB), is a well-known by product of incomplete oil combustion that is used as a pigment and as a tyre-reinforcing material in the rubber and paint industries, respectively. Depending on the way the materials are provided – they may be offered as polymeric dispersions, solid materials, or thin films – CB-based electrochemical sensing technologies have been reported to demonstrate an efficient functionalisation, quick charge transfer rates, large surface area, and ultimately, like carbon nanotubes (CNTs) and graphene. Figure 7.24 displays the CNTs that were generated from screen-printed carbon black. There is great potential for highly flexible piezoresistive detectors to be used in the monitoring of human health. The MWCNTs@CB sensor exhibits outstanding findings because of its microporous shape and high adsorption of both conductive MWCNT and CB. It has a low detection limit of 20 Pa, a reaction time of 15 milliseconds, a sensitivity of 48.26 kPa1, a working range of 12.5 kPa–20 kPa, and exceptional stability (> 250,000 loading sequences). With this MWCNTs@CB pressure sensor with flexibility, physiologic movements may be detected (78).

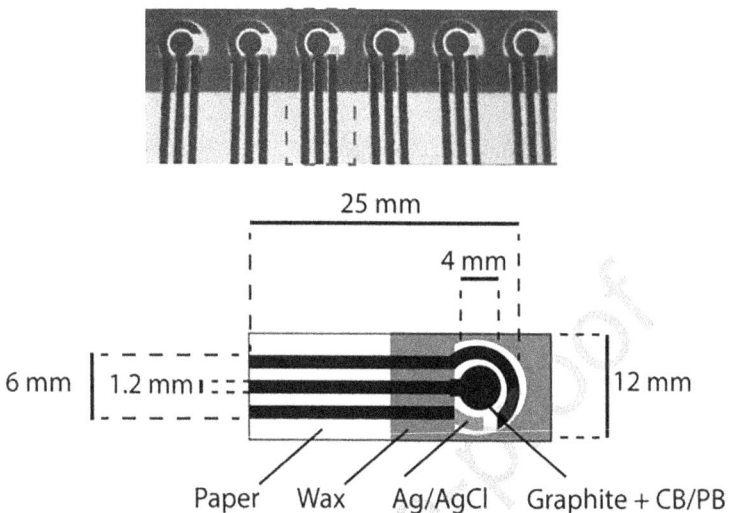

FIGURE 7.24 Exhibits the size and composition of the paper-based screen-printed electrodes (79).

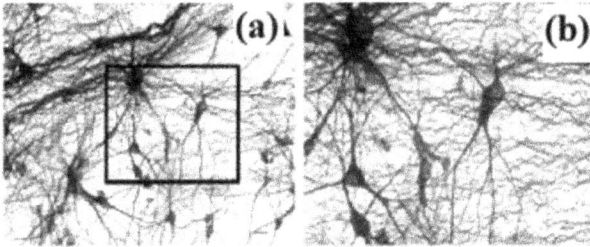

FIGURE 7.25 Hippocampal cultures containing astrocytes and neurons grown on CNT or glass were immunocytochemically analysed. 8-day in vitro cultures (**a**) and (**b**): hippocampal neurons grown on CNT are represented by MAP-2 positive cells, which are abundant and evenly distributed (80).

7.6.3.3 Biomass-Derived CNTs

As CNT is used increasingly often as a basic material for biological applications, its effect on live cells is drawing more and more interest. CNT made from biomass serves as an excellent surface for cellular development and enhances the transmission of brain signals (80). Figure 7.25 displays hippocampus cultures with astrocytes and neurons cultured on glass or CNT. The area in the inset in (a) is magnified more in (b), which displays calibrations of 20 μm in (a) and 10 μm in (b). Additionally, research into the coupling model between hippocampus cells and SWCNT has confirmed that SWCNT may directly trigger brain circuit function, indicating SWCNT as a potential substance. Horseradish peroxidase (HRP)-doped polypyrrole (PPy) had a greater potential than PPy doped with HRP-modified chloride. Due to the very open reticular architecture of the nanocomposite film, the findings reveal increased electron transfer and better accessibility of the enzyme-active sites for the substrate at 0.3 V. In conclusion, the integration of SWCNT-protein as a dopant led to an increase in the loading of HRP in the film, which, in turn, led to an improvement in the effectiveness of protein immobilisation by adsorption on the SWCNTs over the traditional technique of trapping in the film, as shown by the results (81).

7.7 CONCLUSIONS

This chapter described the general aspects of CNTs and their application as biomedical sensors. CNTs constitute a versatile nanostructure that combines the potential to serve for therapeutic and diagnostic applications. Recently, they have found their potential to serve as biosensors of detection and quantification of various biomarkers through different transduction mechanisms. These versatile elements provide characteristics of reliability, high sensitivity, and inexpensive microfabrication for cost effectiveness. The ease of fabrication and integration onto flexible and wearable systems with efficient conductive properties makes them exceptional candidates to be used for early-stage diagnosis of various health conditions. Most importantly, the possibilities of multifunctionalities through surface modification of CNTs helps in the ease of fabrication of the design of biosensors for simultaneous detection of several biomarkers on a single platform. This constitutes a challenging objective to be achieved in the medical field; however, the recent advancements in surface treatments, surface coatings, and additive incorporation of chemical functional groups enable the integration of multiple biosensing moieties. Often, surface functionalisation through chemical and biological moeities has been shown to improve the biocompatibility, thereby making them efficient for on-body measurements and studies in physiological fluids.

REFERENCES

1. Vivaldi F, Salvo P, Poma N, Bonini A, Biagini D, Del Noce L, et al. Recent advances in optical, electrochemical and field effect pH sensors. Chemosensors. 2021;9(2):1–17.
2. Mazlan NS, Ramli MM, Abdullah MMAB, Halin DSC, Isa SSM, Talip LFA, et al. *Interdigitated electrodes as impedance and capacitance biosensors: A review.* AIP Conference Proceedings. 2017. Available from: https://ui.adsabs.harvard.edu/abs/2017AIPC.1885b0276M/abstract
3. Sung D, Koo J. A review of BioFET's basic principles and materials for biomedical applications. Biomed Eng Lett [Internet]. 2021;11(2):85–96. Available from: https://doi.org/10.1007/s13534-021-00187-8
4. Allen BL, Kichambare PD, Star A. Carbon nanotube field-effect-transistor-based biosensors. Adv Mater. 2007;19(11):1439–51.
5. Kaur J, Gill GS, Jeet K. Applications of carbon nanotubes in drug delivery: A comprehensive review [Internet]. *Characterization and biology of nanomaterials for drug delivery: Nanoscience and nanotechnology in drug delivery.* Elsevier Inc. 2018;113–135. Available from: http://dx.doi.org/10.1016/B978-0-12-814031-4.00005-2
6. Vilela D, Romeo A, Sánchez S. Flexible sensors for biomedical technology. Lab Chip. 2016;16(3):402–8.
7. Chircov C, Grumezescu AM. Microelectromechanical systems (MEMS) for biomedical applications. Micromachines. 2022;13.
8. Ren TL, Tian H, Xie D, Yang Y. Flexible graphite-on-paper piezoresistive sensors. Sensors (Switzerland). 2012;12(5):6685–94.
9. Kim MK, Kim MS, Kwon HB, Jo SE, Kim YJ. Wearable triboelectric nanogenerator using a plasma-etched PDMS-CNT composite for a physical activity sensor. RSC Adv. 2017;7(76):48368–73.
10. Wang C, Wang P, Chen J, Zhu L, Zhang D, Wan Y, et al. Self-powered biosensing system driven by triboelectric nanogenerator for specific detection of Gram-positive bacteria. Nano Energy [Internet]. 2022;93(December 2021):106828. Available from: https://doi.org/10.1016/j.nanoen.2021.106828
11. Kanoun O, Bouhamed A, Ramalingame R, Bautista-Quijano JR, Rajendran D, Al-Hamry A. Review on conductive polymer/CNTs nanocomposites based flexible and stretchable strain and pressure sensors. Sensors. 2021;21(2):1–29.
12. Li H, Liu X, Li L, Mu X, Genov R, Mason AJ. CMOS electrochemical instrumentation for biosensor microsystems: A review. Sensors (Switzerland). 2017;17.
13. Scheinberg DA, Villa CH, Escorcia F, Mcdevitt MR. Carbon nanotubes. In *Drug delivery in oncology: From basic research to cancer therapy.* Vol. 2. Wiley. 2011;1163–85.
14. Dudala S, Srikanth S, Dubey SK, Javed A, Goel S. Rapid inkjet-printed miniaturized interdigitated electrodes for electrochemical sensing of nitrite and taste stimuli. Micromachines. 2021;12(9):1–13.
15. Falina S, Syamsul M, Rhaffor NA, Sal Hamid S, Mohamed Zain KA, Abd Manaf A, et al. Ten years progress of electrical detection of heavy metal ions (Hmis) using various field-effect transistor (fet) nanosensors: A review. Biosensors. 2021;11(12).
16. Vu C, Chen W-Y. Field-Effect transistor biosensors for biomedical. Sensors. 2019;19(19):22.

17. Safaee MM, Gravely M, Roxbury D. A wearable optical microfibrous biomaterial with encapsulated nanosensors enables wireless monitoring of oxidative stress. Adv Funct Mater. 2021;31(13):1–14.
18. Sun DM, Timmermans MY, Tian Y, Nasibulin AG, Kauppinen EI, Kishimoto S, et al. Flexible high-performance carbon nanotube integrated circuits. Nat Nanotechnol. 2011;6(3):156–61.
19. Sheikhpour M, Naghinejad M, Kasaeian A, Lohrasbi A, Shahraeini SS, Zomorodbakhsh S. The applications of carbon nanotubes in the diagnosis and treatment of lung cancer: A critical review. Int J Nanomedicine. 2020;15:7063–78.
20. Zuruzi AS, Haffiz TM, Affidah D, Amirul A, Norfatriah A, Nurmawati MH. Towards wearable pressure sensors using multiwall carbon nanotube/polydimethylsiloxane nanocomposite foams. Mater Des [Internet]. 2017;132:449–58. Available from: http://dx.doi.org/10.1016/j.matdes.2017.06.059
21. Kahng SJ, Soelberg SD, Fondjo F, Kim JH, Furlong CE, Chung JH. Carbon nanotube-based thin-film resistive sensor for point-of-care screening of tuberculosis. Biomed Microdevices. 2020;22(3).
22. Wei J, Xie J, Zhang P, Zou Z, Ping H, Wang W, et al. Bioinspired 3D printable, self-healable, and stretchable hydrogels with multiple conductivities for skin-like wearable strain sensors. ACS Appl Mater Interfaces. 2021;13(2):2952–60.
23. Ackermann J, Metternich JT, Herbertz S, Kruss S. Biosensing with fluorescent carbon nanotubes. Angew Chemie – Int Ed. 2022;61(18).
24. Dinarvand M, Neubert E, Meyer D, Selvaggio G, Mann FA, Erpenbeck L, et al. Near-infrared imaging of serotonin release from cells with fluorescent nanosensors. Nano Lett. 2019;19(9):6604–11.
25. Morais PV, Gomes VF, Silva ACA, Dantas NO, Schöning MJ, Siqueira JR. Nanofilm of ZnO nanocrystals/carbon nanotubes as biocompatible layer for enzymatic biosensors in capacitive field-effect devices. J Mater Sci. 2017;52(20):12314–25.
26. Pan J, Liu S, Yang Y, Lu J. A highly sensitive resistive pressure sensor based on a carbon nanotube-liquid crystal-PDMS composite. Nanomaterials. 2018;8(6).
27. Yang X, Wang Y, Qing X. Sensors and actuators A: Physical A flexible capacitive sensor based on the electrospun PVDF nanofiber membrane with carbon nanotubes. Sensors Actuators A Phys [Internet]. 2019;299:111579. Available from: https://doi.org/10.1016/j.sna.2019.111579
28. Yang X, Wang Y, Qing X. A flexible capacitive sensor based on the electrospun PVDF nanofiber membrane with carbon nanotubes. Sensors Actuators, A Phys [Internet]. 2019;299:111579. Available from: https://doi.org/10.1016/j.sna.2019.111579
29. Ko WY, Huang LT, Lin KJ. Green technique solvent-free fabrication of silver nanoparticle–carbon nanotube flexible films for wearable sensors. Sensors Actuators, A Phys [Internet]. 2021;317:112437. Available from: https://doi.org/10.1016/j.sna.2020.112437
30. Anzar N, Hasan R, Tyagi M, Yadav N, Narang J. Carbon nanotube – a review on synthesis, properties and plethora of applications in the field of biomedical science. Sensors Int. 2020;1(February).
31. Demidenko NA, Kuksin A V, Molodykh V V, Pyankov ES, Ichkitidze LP, Zaborova VA, et al. Flexible strain-sensitive silicone-CNT sensor for human motion detection. Bioengineering. 2022;9(1).
32. Min SH, Lee GY, Ahn SH. Direct printing of highly sensitive, stretchable, and durable strain sensor based on silver nanoparticles/multi-walled carbon nanotubes composites. Compos Part B Eng [Internet]. 2019;161(December 2018):395–401. Available from: https://doi.org/10.1016/j.compositesb.2018.12.107
33. Min SH, Lee GY, Ahn SH. Direct printing of highly sensitive, stretchable, and durable strain sensor based on silver nanoparticles/multi-walled carbon nanotubes composites. Compos Part B Eng [Internet]. 2019;161(November 2018):395–401. Available from: https://doi.org/10.1016/j.compositesb.2018.12.107

34. Nie B, Li X, Shao J, Li X, Tian H, Wang D, et al. Flexible and transparent strain sensors with embedded multiwalled carbon nanotubes meshes. ACS Appl Mater Interfaces. 2017;9(46):40681–9.

35. Sarkar T, Srinives S, Sarkar S, Haddon RC, Mulchandani A. Single-walled carbon nanotube-poly(porphyrin) hybrid for volatile organic compounds detection. J Phys Chem C. 2014;118(3):1602–10.

36. Terasawa N. Based on single-walled carbon nanotubes †. RSC Adv [Internet]. 2016;7(Il):2443–9. Available from: http://dx.doi.org/10.1039/C6RA24925F

37. Xie H, Sheng C, Chen X, Wang X, Li Z, Zhou J. Multi-wall carbon nanotube gas sensors modified with amino-group to detect low concentration of formaldehyde. Sensors Actuators, B Chem [Internet]. 2012;168:34–8. Available from: http://dx.doi.org/10.1016/j.snb.2011.12.112

38. Tung TT, Tripathi KM, Kim T, Krebsz M, Pasinszki T, Losic D. Carbon nanomaterial sensors for cancer and disease diagnosis. In *Carbon nanomaterials for bioimaging, bioanalysis, and therapy*. Wiley. 2018;167–202.

39. Frazier KM, Swager TM. Robust cyclohexanone selective chemiresistors based on single-walled carbon nanotubes. Anal Chem. 2013;85(15):7154–8.

40. Lei KF, Lee KF, Yang SI. Fabrication of carbon nanotube-based pH sensor for paper-based microfluidics. Microelectron Eng. 2012;100:1–5.

41. da Costa TH, Song E, Tortorich RP, Choi J-W. A paper-based electrochemical sensor using inkjet-printed carbon nanotube electrodes. ECS J Solid State Sci Technol. 2015;4(10):S3044–7.

42. Al-Harbi EA, Abdelrahman MH, El-Kosasy AM. Ecofriendly long life nanocomposite sensors for determination of carbachol in presence of choline: Application in ophthalmic solutions and biological fluids. Sensors (Switzerland). 2019;19(10).

43. Shen Y, Tran TT, Modha S, Tsutsui H, Mulchandani A. A paper-based chemiresistive biosensor employing single-walled carbon nanotubes for low-cost, point-of-care detection. Biosens Bioelectron [Internet]. 2019;130:367–73. Available from: https://doi.org/10.1016/j.bios.2018.09.041

44. Sun X, Li B, Tian C, Yu F, Zhou N, Zhan Y, et al. Rotational paper-based electrochemiluminescence immunodevices for sensitive and multiplexed detection of cancer biomarkers. Anal Chim Acta [Internet]. 2018;1007:33–9. Available from: https://doi.org/10.1016/j.aca.2017.12.005

45. Sekretaryova AN, Beni V, Eriksson M, Karyakin AA, Turner APF, Vagin MY. Cholesterol self-powered biosensor. Anal Chem. 2014;86(19):9540–7.

46. Jeerapan I, Sangsudcha W, Phokhonwong P. Wearable energy devices on mask-based printed electrodes for self-powered glucose biosensors. Sens Bio-Sensing Res [Internet]. 2022;38:100525. Available from: https://doi.org/10.1016/j.sbsr.2022.100525

47. Speranza G. Carbon nanomaterials: Synthesis, functionalization and sensing applications. Nanomaterials. 2021;11(4).

48. De La Zerda A, Bodapati S, Teed R, May SY, Tabakman SM, Liu Z, et al. Family of enhanced photoacoustic imaging agents for high-sensitivity and multiplexing studies in living mice. ACS Nano. 2012;6(6):4694–701.

49. De La Zerda A, Zavaleta C, Keren S, Vaithilingam S, Bodapati S, Liu Z, et al. Carbon nanotubes as photoacoustic molecular imaging agents in living mice. Nat Nanotechnol. 2008;3(9):557–62.

50. Gifani M, Eddins DJ, Kosuge H, Zhang Y, Paluri SLA, Larson T, et al. Ultraselective carbon nanotubes for photoacoustic imaging of inflamed atherosclerotic plaques. Adv Funct Mater. 2021;31(37):1–8.

51. Gao W, Emaminejad S, Nyein HYY, Challa S, Chen K, Peck A, et al. Fully integrated wearable sensor arrays for multiplexed in situ perspiration analysis. Nature [Internet]. 2016;529(7587):509–14. Available from: http://dx.doi.org/10.1038/nature16521

52. Xie Y, Xu H, He X, Hu Y, Zhu E, Gao Y, et al. Flexible electronic skin sensor based on regenerated cellulose/carbon nanotube composite films. Cellulose [Internet]. 2020;27(17):10199–211. Available from: https://doi.org/10.1007/s10570-020-03496-w

53. Zhao B, Sivasankar VS, Dasgupta A, Das S. Ultrathin and ultrasensitive printed carbon nanotube-based temperature sensors capable of repeated uses on surfaces of widely varying curvatures and wettabilities. ACS Appl Mater Interfaces. 2021;13(8):10257–70.

54. Xiang L, Zhang H, Dong G, Zhong D, Han J, Liang X, et al. Low-power carbon nanotube-based integrated circuits that can be transferred to biological surfaces. Nat Electron [Internet]. 2018;1(4):237–45. Available from: http://dx.doi.org/10.1038/s41928-018-0056-6

55. Wang C, Hwang D, Yu Z, Takei K, Park J, Chen T, et al. User-interactive electronic skin for instantaneous pressure visualization. Nat Mater [Internet]. 2013;12(10):899–904. Available from: http://dx.doi.org/10.1038/nmat3711

56. Jung Y, Park H, Park JA, Noh J, Choi Y, Jung M, et al. Fully printed flexible and disposable wireless cyclic voltammetry tag. Sci Rep. 2015;5:8105.

57. Kim HS, Kang JH, Hwang JY, Shin US. Wearable CNTs-based humidity sensors with high sensitivity and flexibility for real-time multiple respiratory monitoring. Nano Converg [Internet]. 2022;9(1). Available from: https://doi.org/10.1186/s40580-022-00326-6

58. Liu X, Deng K, Wang H, Li C, Zhang S, Huang H. Aptamer based ratiometric electrochemical sensing of 17β-estradiol using an electrode modified with gold nanoparticles, thionine, and multiwalled carbon nanotubes. Microchim Acta. 2019;186(6):2–9.

59. Arvand M, Hemmati S. Magnetic nanoparticles embedded with graphene quantum dots and multiwalled carbon nanotubes as a sensing platform for electrochemical detection of progesterone. Sensors Actuators, B Chem [Internet]. 2017;238:346–56. Available from: http://dx.doi.org/10.1016/j.snb.2016.07.066

60. Li T, Liang Y, Li J, Yu Y, Xiao MM, Ni W, et al. Carbon nanotube field-effect transistor biosensor for ultrasensitive and label-free detection of breast cancer exosomal miRNA21. Anal Chem. 2021;93(46):15501–7.

61. Williams RM, Lee C, Galassi TV., Harvey JD, Leicher R, Sirenko M, et al. Noninvasive ovarian cancer biomarker detection via an optical nanosensor implant. Sci Adv. 2018;4(4).

62. Wan Q, Xu Y, Chen X, Xiao H. Exhaled gas detection by a novel Rh-doped CNT biosensor for prediagnosis of lung cancer: A DFT study. Mol Phys [Internet]. 2018;116(17):2205–12. Available from: https://doi.org/00268976.2018.1467057

63. Farzin L, Sadjadi S, Shamsipur M, Sheibani S. An immunosensing device based on inhibition of mediator's faradaic process for early diagnosis of prostate cancer using bifunctional nanoplatform reinforced by carbon nanotube. J Pharm Biomed Anal [Internet]. 2019;172:259–67. Available from: https://doi.org/10.1016/j.jpba.2019.05.008

64. Sharma A, Hong S, Singh R, Jang J. Single-walled carbon nanotube based transparent immunosensor for detection of a prostate cancer biomarker osteopontin. Anal Chim Acta [Internet]. 2015;869:68–73. Available from: http://dx.doi.org/10.1016/j.aca.2015.02.010

65. Domínguez-Aragón A, Dominguez RB, Zaragoza-Contreras EA. Simultaneous detection of dihydroxybenzene isomers using electrochemically reduced graphene oxide-carboxylated carbon nanotubes/gold nanoparticles nanocomposite. Biosensors. 2021;11(9).

66. Zhou F, Wang Y, Wu W, Jing T, Mei S, Zhou Y. Synergetic signal amplification of multi-walled carbon nanotubes-Fe3O4 hybrid and trimethyloctadecylammonium bromide as a highly sensitive detection platform for tetrabromobisphenol A. Sci Rep. 2016;6(September):1–12.

67. Ribeiro H, Schnitzler MC, da Silva WM, Santos AP. Purification of carbon nanotubes produced by the electric arc-discharge method. Surfaces and Interfaces. 2021;26(April).

68. Yadav MD, Dasgupta K, Patwardhan AW, Joshi JB. High performance fibers from carbon nanotubes: Synthesis, characterization, and applications in composites – a review. Ind Eng Chem Res. 2017;56(44):12407–37.

69. Patole SP, Alegaonkar PS, Lee HC, Yoo JB. Optimization of water assisted chemical vapor deposition parameters for super growth of carbon nanotubes. Carbon NY. 2008;46(14):1987–93.
70. Liao H, Hafner JH. Low-temperature single-wall carbon nanotube synthesis by thermal chemical vapor deposition. J Phys Chem B. 2004;108(22):6941–3.
71. Singh A, Kumar S, Goswami P, Ghosh K, Agarwal AK, Jassal M, et al. *Interdigitated flexible supercapacitor using activated carbon synthesized from biomass for wearable energy storage Metal recovery using bioelectrochemical system: An update of current progress view project my current research involves the development of n.* 2019;(December). Available from: www.researchgate.net/publication/331561759
72. Qi Z, Zhang F, Di CA, Wang J, Zhu D. All-brush-painted top-gate organic thin-film transistors. J Mater Chem C. 2013;1(18):3072–7.
73. Tiwari SK, Sahoo S, Wang N, Huczko A. Graphene research and their outputs: Status and prospect. J Sci Adv Mater Devices [Internet]. 2020;5(1):10–29. Available from: https://doi.org/10.1016/j.jsamd.2020.01.006
74. Zhang S, Zahed MA, Sharifuzzaman M, Yoon S, Hui X, Chandra Barman S, et al. A wearable battery-free wireless and skin-interfaced microfluidics integrated electrochemical sensing patch for on-site biomarkers monitoring in human perspiration. Biosens Bioelectron [Internet]. 2021;175(November 2020):112844. Available from: https://doi.org/10.1016/j.bios.2020.112844
75. Yan XB, Chen XJ, Tay BK, Khor KA. Transparent and flexible glucose biosensor via layer-by-layer assembly of multi-wall carbon nanotubes and glucose oxidase. Electrochem Commun. 2007;9(6):1269–75.
76. Fatoni A, Numnuam A, Kanatharana P, Limbut W, Thammakhet C, Thavarungkul P. A highly stable oxygen-independent glucose biosensor based on a chitosan-albumin cryogel incorporated with carbon nanotubes and ferrocene. Sensors Actuators, B Chem [Internet]. 2013;185:725–34. Available from: http://dx.doi.org/10.1016/j.snb.2013.05.056
77. Jeon WY, Kim HS, Jang HW, Lee YS, Shin US, Kim HH, et al. A stable glucose sensor with direct electron transfer, based on glucose dehydrogenase and chitosan hydro bonded multi-walled carbon nanotubes. Biochem Eng J [Internet]. 2022;187(March):108589. Available from: https://doi.org/10.1016/j.bej.2022.108589
78. Liu C, Tan Q, Deng Y, Ye P, Kong L, Ma X, et al. Highly sensitive and stable 3D flexible pressure sensor based on carbon black and multi-walled carbon nanotubes prepared by hydrothermal method. Compos Commun. 2022;32(March).
79. Ferreira LMC, Silva PS, Augusto KKL, Gomes-Júnior PC, Farra SOD, Silva TA, et al. Using nanostructured carbon black-based electrochemical (bio)sensors for pharmaceutical and biomedical analyses: A comprehensive review. J Pharm Biomed Anal [Internet]. 2022;221:115032. Available from: https://doi.org/10.1016/j.jpba.2022.115032
80. Lovat V, Pantarotto D, Lagostena L, Cacciari B, Grandolfo M, Righi M, et al. Carbon nanotube substrates boost neuronal electrical signaling. Nano Lett. 2005;5(6):1107–10.
81. Mazzatenta A, Giugliano M, Campidelli S, Gambazzi L, Businaro L, Markram H, et al. Interfacing neurons with carbon nanotubes: Electrical signal transfer and synaptic stimulation in cultured brain circuits. J Neurosci. 2007;27(26):6931–6.

8 Carbon Nanotubes
Synthesis, Properties, Sensing Mechanisms, and Applications in Sensor Technologies

Md Eshrat E. Alahi and Fahmida Wazed Tina

8.1 INTRODUCTION

Carbon is an essential element due to its unique bonding characteristics. With an atomic number of six, carbon has six electrons that can occupy the $1s^2$, $2s^2$, and $2p^2$ atomic orbitals. Four valence electrons can undergo hybridization in the sp, sp^2, or sp^3 forms. Carbon can form a diverse range of structures, both in bulk and at the nanoscale. Depending on the hybridization, these structures exhibit different properties, leading to a wide range of carbon allotropes. The most commonly known allotropes of carbon are graphite, soft and conductive (sp^2 hybridization), and diamond, hard and insulating (sp^3 hybridization) (1–3).

Additionally, newer allotropes such as graphene, fullerene or buckyball, and carbon nanotubes (CNTs) have been discovered, offering exciting opportunities for scientific research and demonstrating exceptional properties with various applications. Carbon is a unique element known for its ability to form numerous allotropes, ranging from zero-dimensional (0D) to three-dimensional (3D) structures. Carbon exhibits remarkable manifestations, such as quantum dots and nanoclusters within zero-dimensional systems. On the other hand, one-dimensional carbon structures encompass nanofibres, nanowires, nanotubes, and nanorods. This distinctive feature further underscores carbon's versatility and extensive possibilities across various dimensional scales (4–7).

CNTs are a highly promising and extensively investigated carbon allotrope characterized by sp^2 hybridization (8). CNTs display an extraordinary range of structures and possess diverse structure-property relationships while maintaining a simple chemical composition and atomic bonding configuration. The exploration of CNTs began soon after the successful laboratory creation of fullerene in 1991 by Sumio Iijima. Iijima inadvertently stumbled upon CNTs while examining the soot produced by an arc discharge device. Subsequently, in 1993, both NEC and IBM reported the existence of CNTs comprising a single layer of graphene. Since their momentous discovery, CNTs have considerably contributed to various scientific disciplines, including physics, chemistry, mathematical modelling, and material sciences (9–15).

DOI: 10.1201/9781003376071-8

The properties exhibited by CNTs stem from the underlying characteristics of graphene, which features a regular sp^2-bonded atomic-scale honeycomb pattern. This honeycomb arrangement is the fundamental structure for other carbon materials, such as fullerenes and CNTs. Conceptually, the structure of a CNT can be envisioned as a rolled-up graphene sheet forming a tube. CNTs have garnered extensive research attention among all carbon allotropes due to their distinctive hollow structure, imparting extraordinary mechanical, thermal and electrical properties (8, 15, 16).

CNTs possess many exceptional attributes that make them highly desirable for various applications. Remarkable characteristics distinguish this material, including its low density, unparalleled tensile strength, and stiffness surpassing any metal. Moreover, it exhibits exceptional thermal conductivity, surpassing that of diamond, while boasting high electrical conductivity, remarkable flexibility, and robust thermal and chemical stability (17, 18). These inherent characteristics render CNTs suitable for numerous practical applications. Notably, they function as one-dimensional ballistic conductors, enabling the transmission of electrons over significant distances. Examining specific heat and thermal conductivity at low temperatures can provide direct evidence of the one-dimensional quantization of the phonon band structure. Consequently, considerable research has been conducted to investigate the structure, properties, potential applications, and associated challenges of CNTs (19–21).

While numerous literature reviews (8, 22–24) have explored various facets of CNTs, including their classification, synthesis, and properties, limited attention has been devoted to their potential in heat transfer applications. A scarcity of reports comprehensively covers the classification, synthesis, properties, and holistic applications of CNTs. Hence, the central focus of this book chapter is to bridge this gap by offering a comprehensive analysis of the methods utilized in the fabrication of CNTs and the factors that impact their properties. Additionally, this review aims to provide a broad perspective on the wide range of applications of CNTs, with a specific emphasis on their potential in sensor-based technology. By harnessing the exceptional properties inherent to CNTs, this study seeks to elucidate their contribution to enhancing heat transfer processes (12, 25, 26).

The chapter will encompass an in-depth examination of the various techniques employed in synthesizing CNTs, including both established and emerging methods. Furthermore, the factors affecting the structural and chemical properties of CNTs will be critically analyzed, with a particular emphasis on their impact on heat transfer characteristics. The comprehensive overview will encompass both fundamental aspects and recent advancements in the field, thus offering valuable insights into the current state of CNTs research. Moreover, the chapter will delve into the vast array of applications where CNTs have demonstrated their potential for heat transfer enhancement. This includes, but is not limited to, thermal interface materials, heat exchangers, thermal management systems, and energy storage devices. By highlighting the unique features of CNTs, such as their high thermal conductivity, low density, and exceptional mechanical strength, this review aims to showcase their suitability and efficacy in addressing the challenges encountered in various heat transfer applications.

8.2 TYPE AND STRUCTURE

CNTs are a unique allotrope of carbon, characterized by sp^2 hybridization. They can be visualized as rolled-up sheets of graphene, a planar hexagonal lattice of carbon atoms (27–30). Typically, CNTs possess a high aspect-ratio, with nanometer-scale diameters and micrometre-scale lengths. They can be classified into different types based on various characteristics. Various types of CNTs exist, differing in their features and applications. CNTs can be categorized based on length, resulting in either long or short CNTs. Furthermore, they can be classified as single-walled, double-walled, or multi-walled CNTs, depending on the number of concentric cylindrical layers present in their nanostructure (refer to Figure 8.1). CNTs also come in different configurations, either open or closed. Available types have cylindrical tube shapes with open ends, while closed types possess capped ends (refer to Figure 8.2). Additionally, spiral structures are another variation observed in CNTs. Every kind of CNT entails distinct production costs and finds specific applications (12, 31, 32).

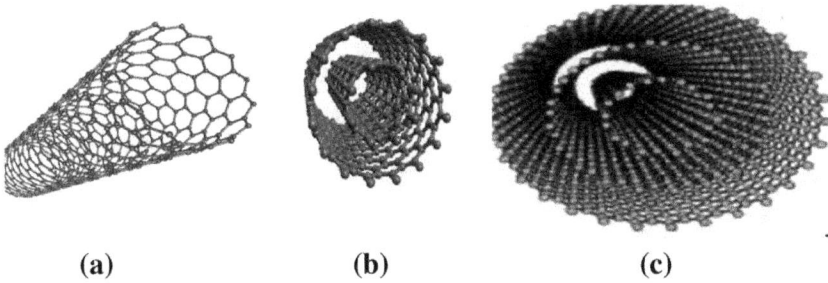

(a) **(b)** **(c)**

FIGURE 8.1 CNTs can be classified based on the number of walls. The types include: a) Single-Walled Nanotubes (SWNTs): These CNTs consist of a single cylindrical wall resembling a hollow tube. b) Double-Walled Nanotubes (DWNTs): DWNTs comprise two concentric cylindrical walls, forming a tube within a tube structure. c) Multi-Walled Nanotubes (MWNTs): MWNTs contain multiple concentric cylindrical walls, with each wall encapsulating the previous one, similar to a series of tubes nested inside each other (33).

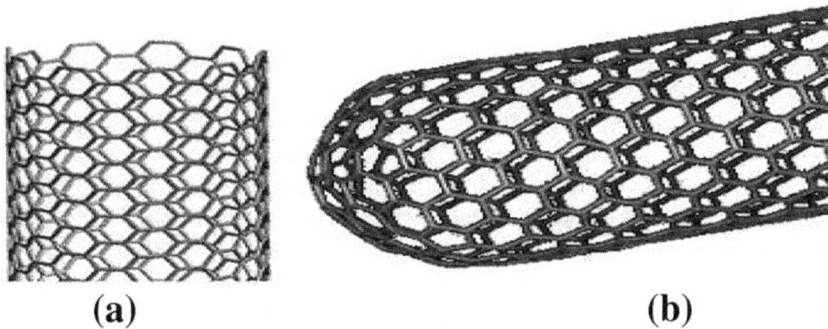

(a) **(b)**

FIGURE 8.2 CNTs can be categorized based on the configuration of their ends. The two main types are: (a) Open-Ended Nanotubes and (b) Close-Ended Nanotubes (33).

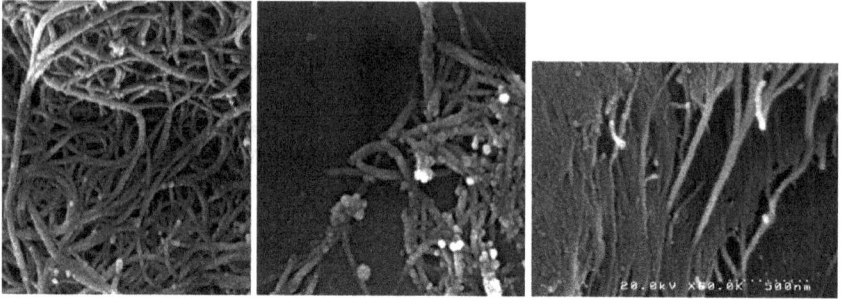

FIGURE 8.3   SEM image of functionalization of SWCNT (39)

CNTs can be further classified based on their crystallographic configurations, categorized as zigzag, armchair, and chiral. The electrical properties of a CNTs are influenced by its chiral vector, represented by a pair of indices (n, m). These indices describe the number of unit vectors along two specific directions within the honeycomb crystal lattice of graphene. In essence, the chiral vector determines the structural arrangement of the carbon atoms in the nanotube, and thus, plays a crucial role in its electrical behaviour – two parameters, m and n, determine the degree of twist in a nanotube. When m = 0, the nanotubes are called zigzag nanotubes, represented by indices like (1, 0), (2, 0), (5, 0), and so on. On the other hand, if n = m, the nanotubes are referred to as armchair nanotubes, characterized by indices such as (1, 1), (2, 2), (3, 3), and so forth. Nanotubes that do not satisfy these conditions are classified as chiral nanotubes, with indices like (4, 2), (3, 2), (4, 1), and others. Armchair and zigzag nanotubes exhibit the highest degree of symmetry (9, 12, 29, 34, 35). Single-walled carbon nanotubes (SWCNTs or SWNTs) demonstrate varying electrical properties, depending on their chirality, and they can be classified as metallic or semiconducting. The band gaps of SWCNTs range from 0.4 to 2 eV. Metallic behaviour is observed only in armchair-type SWNTs, whereas all other SWNTs are semiconducting (9, 11, 12, 34–37) (refer to Figure 8.3).

Multi-walled carbon nanotubes (MWCNTs or MWNTs) are cylindrical structures consisting of concentric layers of single-walled carbon nanotubes with varying diameters. Due to spatial limitations, it is physically impossible for tubes of different diameters to fit perfectly within one another, resulting in a gap between the layers. This arrangement, known as the Russian Doll model, describes a CNT that contains concentric nanotubes of different diameters. In contrast, the Parchment model illustrates a single graphene sheet spirally wrapped multiple times. The multilayer structure of MWNTs provides several advantages. It protects the inner CNT from chemical interactions. It enhances tensile strength, properties that are either absent or only partially present in SWNTs (33–35, 38). Consequently, MWNTs are widely used in many nanotube applications due to their ease of large-scale production and cost-effectiveness compared to SWNTs. However, the structure of MWNTs is more complex and diverse, leading to a lesser understanding than single-walled nanotubes (33) (refer to Figure 8.4).

To achieve perfect crystallinity in CNTs, it is crucial for the hexagonal, aromatic structure of carbon atoms along the tube's graphene sheet to remain consistent

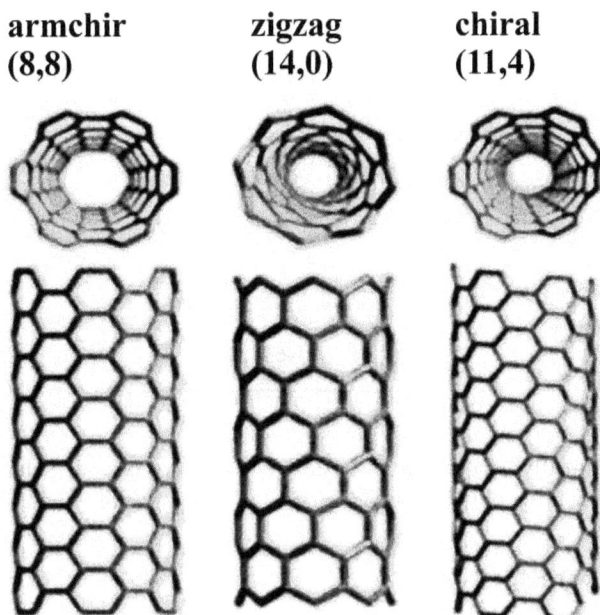

**armchir
(8,8)** **zigzag
(14,0)** **chiral
(11,4)**

FIGURE 8.4 The structure of a multi-walled carbon nanotube consists of three layers of hexagonal lattice sheets with varying chirality (40).

(29). However, current synthesis techniques face significant challenges in producing defect-free walls and caps in crystalline single-walled nanotubes and multi-walled nanotubes. This makes it difficult to characterize them accurately. Although high-resolution transmission electron microscopy (HRTEM) imaging is often used to assess the crystallinity of CNTs, it remains a demanding task (34). One prevalent defect observed in MWNTs is the presence of bamboo structures, where the walls exhibit different lengths and appear stacked.

In addition to the conventional SWNTs and MWNTs, CNTs can also exhibit unique structures such as nano horns and nanofibres. Nano horns are carbon cones that resemble the end caps of nanotubes. They are formed through the high-temperature treatment of fullerene soot. On the other hand, nanofibres are carbon fibres that can be either hollow or solid, with lengths in the micron scale and widths ranging from tens to approximately 200 nanometres. Unlike SWNTs and MWNTs, nanofibres lack the characteristic cylindrical structure. Instead, they comprise a mixture of carbon forms, including graphite layers stacked at various angles and amorphous carbon, which lacks a large-scale regular structure.

8.3 PROPERTIES

CNTs possess an exceptional surface area and exhibit a significant aspect ratio. These distinctive characteristics give rise to extraordinary properties, including outstanding mechanical strength, high electrical conductivity and efficient thermal conductivity. These properties make CNTs highly relevant for a wide range of

TABLE 8.1

Comparison of the Physical Properties between MWNT and SWNT

Physical Properties	MWNT	SWNT
1. Diameter	5–100 nm	1–2 nm
2. Length	15000 nm	100–1000 nm
3. Thermal conductivity (at 300 K)	2000–3000 W/mK	3000–6000 W/mK
4. Band gap	2.9–3.7 eV	0–0.5 eV
5. Density	1600 ρ (kg/m3)	2600 ρ (kg/m3)
6. Melting point	3527°C	3527°C
7. Tensile strength	11–63 GPa	22.2 ± 2.2 Gpa

applications. To fully harness the potential of CNTs, it is crucial to have a comprehensive understanding of their physical and chemical properties.

8.3.1 PHYSICAL PROPERTIES

CNTs are a unique and highly versatile form of carbon known for their extraordinary physical properties. Although significant progress has been made in the controlled fabrication of CNTs, characterizing them accurately still remains a complex task. MWNTs and SWNTs exhibit remarkable electrical and thermal conductivity and exceptional mechanical strength. The anisotropic nature of CNTs, which demonstrate different property values along different directions, has fueled extensive research into their potential applications. CNTs have been extensively investigated for various uses, including microelectronic interconnects, heat sinks, and structural composites.

SWNTs and MWNTs possess different physical properties, owing to their structural dissimilarities. MWNTs consist of multiple concentric cylinders of SWNTs, resulting in thicker dimensions than SWNTs. Moreover, the outer tubes in MWNTs provide shielding to the inner ones. Table 8.1 summarizes the comparison of physical properties between SWNTs and MWNTs. Double-walled carbon nanotubes are unique structures composed of two concentric nanotubes. They share similar properties and morphology with single-walled carbon nanotubes but possess enhanced chemical resistance. The diameter difference between the two nanotubes plays a crucial role in their interaction, allowing modifications to be made to the outer nanotube without compromising the properties of the inner nanotube.

DWNTs exhibit exceptional thermal and electrical stability as well as flexibility. They combine the synthetic advantages of SWNTs and MWNTs (41). Structural defects in SWNTs can significantly impact their electrical and mechanical properties, and functionalized SWNTs are more susceptible to breakage (refer to Figure 8.5). In the following section, we will outline the fundamental physical properties of CNTs.

8.3.1.1 Electrical Property

The electronic structure of CNTs can be derived from two-dimensional graphite (43). However, due to their one-dimensional nature, CNTs exhibit electronic properties directly influenced by their diameter and helicity. These structures possess meagre

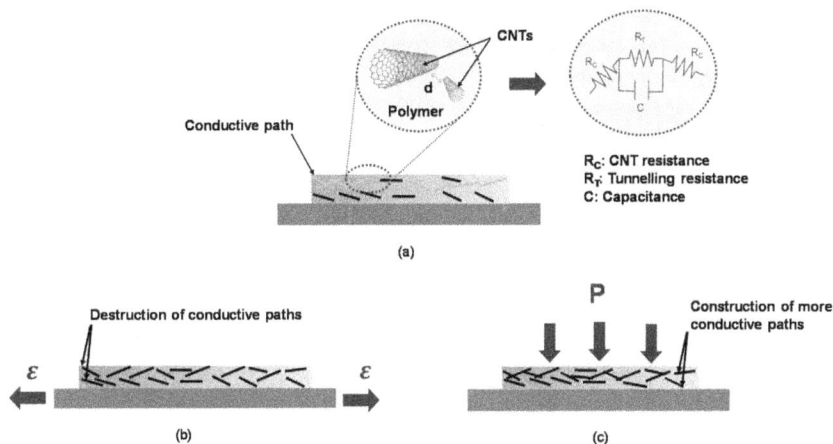

FIGURE 8.5 The conduction mechanisms of a polymer/CNTs sensor can be described as follows: (a) in the absence of a load, (b) when subjected to strain, and (c) when subjected to pressure (42).

electrical resistance, as resistance only occurs when electrons collide with defects in the crystal structure along their path. Defects can arise from impurities in atoms or crystal lattice, causing electron deflection. The small diameter and high aspect ratio of CNTs make electron scattering less likely (44). CNTs have a significantly high current-carrying capacity, surpassing that of superconductors.

CNTs possess various electrical properties, which can be categorized as either metallic or semiconducting based on the combinations of the indices "n" and "m." Specifically, armchair nanotubes are inherently metallic, while other nanotubes can exhibit either metallic or semiconducting behaviour. Semiconducting behaviour arises when the sum of "n" and "m" is a multiple of 3. This unique property facilitates the creation of semiconductor-semiconductor and semiconductor-metal junctions, which are valuable in device fabrication. However, it is important to note that, in small-diameter tubes with high curvature, the electrical properties can deviate significantly, leading to exceptions to these general rules. Moreover, it should be noted that the conductivity characteristics of MWNTs are more complex than their single-walled counterparts. The electrical properties of CNTs are significantly influenced by their treatment methods and aggregation state (45). In theory, CNTs with ideal characteristics demonstrate ballistic conduction, allowing efficient electron transport over micron-scale distances. This phenomenon arises from the one-dimensional confinement of electrons and the principles of energy conservation and momentum conservation, leading to rapid electron conduction without any resistance (11, 28, 29, 46, 47). One potential application of CNTs lies in electrical wiring. Unlike conventional copper and aluminium wires, CNTs present several advantages that can overcome existing challenges. These challenges include high weight, which is particularly problematic in aerospace applications, and the skin effect that limits their use in telecommunications. Additionally, conventional wires exhibit low mechanical performance – critical in overhead power lines and low electrical performance – leading to damage in microscopic wires used in electronic applications. CNTs offer a

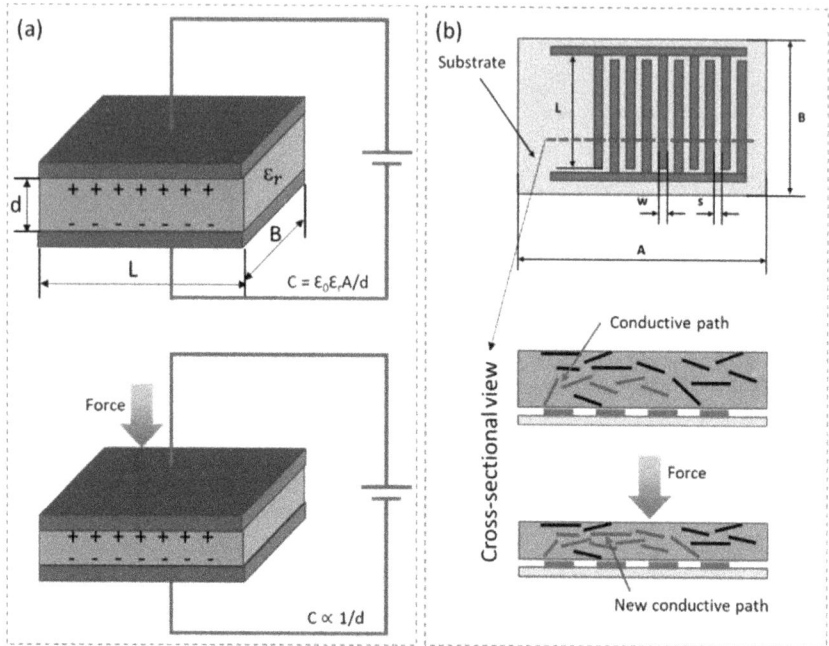

FIGURE 8.6 The conduction mechanisms of polymer/CNT pressure sensors, operating on the principle of piezocapacitance, can be categorized into two types: (a) sensors employing parallel plate electrode structures and (b) sensors utilizing interdigital electrode structures (42).

promising solution to address these issues. Figure 8.6 depicts the conduction mechanism in a CNTs/polymer-based sensor under the influence of external pressure.

Furthermore, CNTs can be utilized to develop impressive electron guns, which play a crucial role in creating thin, high-brightness, low-energy and lightweight displays known as field emission displays (FEDs). A single nanotube can function as a diode by connecting nanotubes of different diameters end-to-end. This suggests the exciting possibility of constructing electronic computer circuits entirely out of nanotubes.

8.3.1.2 Thermal Property

Traditionally, diamonds were renowned for their exceptional thermal conductivity until the discovery of CNTs. These nanotubes have demonstrated thermal conductivity that surpasses that of diamonds by at least twofold. The remarkable thermal properties of CNTs can be directly attributed to their distinctive structure and minute size. Their small size allows quantum effects to play a significant role, making CNTs ideal for studying low-dimensional phonon band structures and thermal management on both macro and micro scales (48). The thermal properties of CNTs, such as specific heat and thermal conductivity, are primarily governed by phonons. CNTs exhibit remarkable resilience to high temperatures, withstanding up to 750 °C under normal atmospheric pressures and up to 2800 °C in vacuum conditions, thanks to the strength of the atomic bonds within the tubes. The thermal conductivity of CNTs is influenced by several factors, such as the atomic arrangement, diameter, and length of the tubes, structural defects, impurities, and overall morphology (49).

Effective cooling poses a significant hurdle in the microelectronics, transportation, manufacturing, and metrology industries. When it comes to CNTs, the inter-tube coupling within single-walled nanotube ropes and the inter-shell coupling within multi-walled nanotubes lead to specific heat at low temperatures compared to that of three-dimensional graphite. According to theoretical predictions, it is suggested that nanotubes possess an impressive thermal conductivity of 6600 W/m K at room temperature. However, experimental measurements have demonstrated that, when it comes to bulk samples of SWNTs, the thermal conductivity exceeds 200 W/m K at room temperature. Moreover, individual MWNTs demonstrate an extraordinary thermal conductivity exceeding 3000 W/m K. In contrast, copper, a widely employed thermal conductor, possesses a thermal conductivity of only 385 W/m K. The remarkable characteristics of CNTs offer significant potential in tackling the thermal management and dissipation challenges associated with downsizing future microprocessors, which is crucial for a wide range of electrical applications (32, 44, 48, 50–52).

8.3.1.3 Mechanical Property

Measuring the mechanical properties of nanotubes is a complex task, primarily because it is challenging to acquire pure, homogeneous, and uniform samples. Nevertheless, theoretical predictions and experimental observations have consistently demonstrated the extraordinary mechanical characteristics of CNTs. CNTs exhibit incredible stiffness, surpassing even that of diamonds, with a maximum Young's modulus of 1.4 TPa and an impressive tensile strength of approximately 100 GPa (44). MWNTs offer superior protection against chemical interactions and exhibit enhanced tensile strength compared to their single-walled counterparts. The presence of sp^2 bonds in CNTs contributes to their exceptional tensile strength, surpassing that of materials such as Kevlar and steel (34). These sp^2 bonds are considerably stronger than the sp^3 bonds found in diamonds. Theoretically, SWNTs have the potential to possess a tensile strength hundreds of times greater than that of steel, and experimental studies have demonstrated Young's modulus of 1,002 GPa (32, 53).

CNTs possess another remarkable property: elasticity. Without sustaining damage, these nanotubes can withstand axial compressive forces, enduring bending, twisting, kinking, and buckling. They exhibit a high elasticity, enabling them to revert to their original structure. Nevertheless, their elasticity does have limitations, as they may undergo temporary deformation when subjected to significant forces. CNTs offer substantial mechanical advantages over other materials when comparing Young's modulus to tensile strength. Consequently, CNTs are promising for various mechanical applications, including CNT-polymer composites. However, it is worth noting that structural defects can undermine the mechanical properties of CNTs (29, 34, 46, 53, 54).

8.3.1.4 Vibrational Property

"Phonons" refer to lattice vibrations that occur when atoms or molecules within a lattice oscillate uniformly at a distinct frequency. Like photons, they are discrete and quantized entities carrying vibrational energy. Phonons significantly impact many phenomena in condensed matter systems, such as thermal, transport, and mechanical properties (43). Quasiparticles, known as phonons, play a significant role in the quantum mechanical quantization of vibrational modes in elastic structures formed by interacting particles. These phonons represent excited states within the system. There are typically 3N phonon branches in a crystal, where "N" refers to the number of atoms present in the unit cell. Among these branches, three are acoustic modes and possess zero frequency. They correspond to the uniform displacement of the entire crystal. The remaining branches involve the out-of-phase motion of atoms and are referred to as optical phonons. In particular, these optical phonons can directly interact with light in polar crystals.

In contrast to electrons, phonons belong to the category of bosons and do not adhere to the Pauli Exclusion Principle. They are pivotal in electron transport phenomena, including thermal conduction, heat capacity, and the equilibrium between electrons and the lattice structure (43). Phonons also significantly contribute to electron transport and optical properties, as they serve as the primary mechanism for the relaxation of excited electrons or other quasiparticles to lower energy states. Moreover, phonons constitute the exclusive source of electrical resistance in an ideal crystal at finite temperatures.

Graphene displays a phonon dispersion branch structure that is relatively simple and possesses higher symmetry compared to SWNTs. However, accurately measuring the phonon dispersion in CNTs poses a challenge due to the requirement of single crystals with identical chirality. SWNTs exhibit crucial vibration modes, including acoustic and optical phonons. Acoustic phonons involve uniform mass distribution among all components within the crystal lattice. The acoustic phonon modes in SWNTs can be categorized into three types: longitudinal acoustic (LA) mode, where atomic displacement occurs parallel to the nanotube axis, and two transverse acoustic (TA) modes, which are doubly degenerate and involve atomic displacement perpendicular to the nanotube axis (54, 55).

8.4   CNTs-BASED SENSORS IN INDUSTRIAL APPLICATIONS

CNTs have shown great potential for various industrial applications, particularly in the field of sensors. CNT-based sensors offer unique properties that suit various industrial sensing applications. Here are some insights on CNT-based sensors for industrial applications:

8.4.1   SENSING MECHANISM

CNTs-based sensors leverage the exceptional properties of CNTs to detect and quantify a wide range of physical and chemical parameters. When exposed to specific target analytes, these sensors detect alterations in the electrical, mechanical, or optical characteristics of the nanotubes. CNTs possess tremendous potential in sensing applications across diverse domains, including mechanical, chemical, biological, optical, and electrical interactions. CNTs find utility in sensing devices from individual single-walled nanotubes (SWNTs) and multi-walled nanotubes (MWNTs) to CNT networks, composites, and bulk CNT materials, such as forests or bundles.

Using CNTs in sensing devices confers numerous advantages, including integrating nanoscale devices for large-scale integration and a high surface-to-volume ratio that enhances sensitivity (16, 56–60).

CNTs have a high aspect-ratio, which increases the exposure of the sensor surface to the analyte. The structure of CNTs, characterized by π-electron conjugation, offers numerous active sites where analyte molecules can induce changes in electrical properties. This property enables the detection of multiple gases. Furthermore, the curvature of CNTs plays a crucial role in enhancing gas sensitivity by reducing the energy barrier required for adsorption of molecules onto the CNT surface. Importantly, CNT-based devices and sensors demonstrate low power consumption, addressing a significant concern in the field. CNTs exhibit high photoabsorption capability, efficient thermal conductivity in metallic CNTs, and a photovoltaic mechanism in semiconducting CNTs, making them promising for practical photo devices. Moreover, the decoration of aligned quantum dots further enhances external quantum efficiency. To adjust the bandgap of CNTs and enable the generation of more electron-hole pairs (excitons) upon photon incidence, chemical doping is employed.

CNTs, known for their remarkable mechanical strength and small size, have emerged as promising alternatives to conventional materials in various applications, such as atomic force microscopy (AFM) and scanning, tunnelling microscope tips. CNTs function as force sensors in these applications due to their exceptional properties. Furthermore, CNTs exhibit polarization under electrical and magnetic fields, making them suitable for actuators and resonators. To summarize, utilizing CNTs in sensing applications provides distinct advantages, including their seamless integration at the nanoscale level, heightened sensitivity owing to their high surface-to-volume ratio, and versatility across multiple sensing modalities. Ongoing research is focused on further exploration and optimization of CNT-based sensing devices to fully unleash their potential in diverse technological and scientific investigations (61). Here is an overview of the sensing mechanism of CNT-based sensors:

8.4.1.1 Electrical Sensing Mechanism

CNTs exhibit excellent electrical conductivity, and this property forms the basis of many CNT-based sensors. When CNTs are exposed to target analytes or environmental changes, such as gases, chemicals, or biomolecules, it alters their electrical properties. This can be attributed to charge transfer, doping, adsorption, or changes in the electronic structure of the CNTs. The resulting changes in electrical conductance or resistance can be measured and correlated with the concentration or presence of the target analyte. CNT-based electrical sensors are widely used for gas detection, chemical analysis, biosensing, and electronic skin applications (16, 24, 62, 63).

8.4.1.2 Mechanical Sensing Mechanism

CNTs possess exceptional mechanical properties, including high tensile-strength and flexibility. These properties enable their use in mechanical sensing applications. CNT-based sensors can detect mechanical deformations, strains, vibrations, or stresses by measuring the changes in the mechanical properties of the nanotubes. This can be achieved through various techniques, such as integrating CNTs into composite materials or using them as strain gauges. The mechanical changes induce alterations in the electrical conductivity, resistance, or optical properties of the CNTs, which can be measured to quantify the sensed mechanical parameter. CNT-based mechanical sensors find applications in structural health monitoring, robotics, aerospace, and automotive industries.

8.4.1.3 Optical Sensing Mechanism

CNTs also exhibit unique optical properties that can be harnessed for sensing applications. The interaction of CNTs with target analytes can result in changes in their optical properties, such as absorbance, fluorescence, or Raman scattering. These changes can be detected and quantified using optical techniques, including spectroscopy or imaging. CNT-based optical sensors offer advantages such as high sensitivity, label-free detection, and the potential for multiplexed analysis. They are used in environmental monitoring, biomedical sensing, and chemical analysis applications.

8.4.1.4 Hybrid Sensing Mechanisms

In some cases, CNT-based sensors combine multiple sensing mechanisms to enhance sensitivity, selectivity, or versatility. For example, a sensor may utilize both the electrical and mechanical properties of CNTs to detect a target analyte. It can provide complementary information and improve performance by integrating different sensing mechanisms. It's important to note that the sensing mechanism can vary, depending on the specific design, functionalization, and configuration of the CNT-based sensor. Functionalization techniques, such as chemical modifications or surface coatings, can enhance selectivity and sensitivity towards particular analytes. The development of CNT-based sensors continues to explore new materials, structures, and functionalization strategies to enhance their performance and expand their applications in various fields.

8.4.2 CNT-Based Gas Sensors

Gas sensing is a field that greatly benefits from the exceptional properties of CNTs. These nanotubes have garnered considerable attention due to their remarkable characteristics, making them an ideal choice for chemical sensors capable of detecting various organic and inorganic species in various environments. Gas detection using CNTs primarily relies on two fundamental mechanisms: physisorption and chemisorption. Physisorption occurs via weak van der Waals forces. At the same time, chemisorption involves the transfer of charges between gas molecules and the surface of CNTs. These mechanisms collectively contribute to the reliable and efficient detection of gases using CNTs.

Physisorption enables a rapid response when exposed to the gas, with subsequent quick desorption in the absence of gas (refer to Figure 8.7). This process is driven by the interaction of gas molecules with low-energy binding sites on the surface of CNTs, allowing for easy attachment and detachment of analyte molecules. In contrast, chemisorption necessitates overcoming an activation energy barrier to facilitate charge transfer. High-energy binding sites, such as structural defects and oxygen functional groups, actively facilitate chemisorption. Several factors, including the collision rate, surface area, availability of adsorption sites, and the adhesion coefficient,

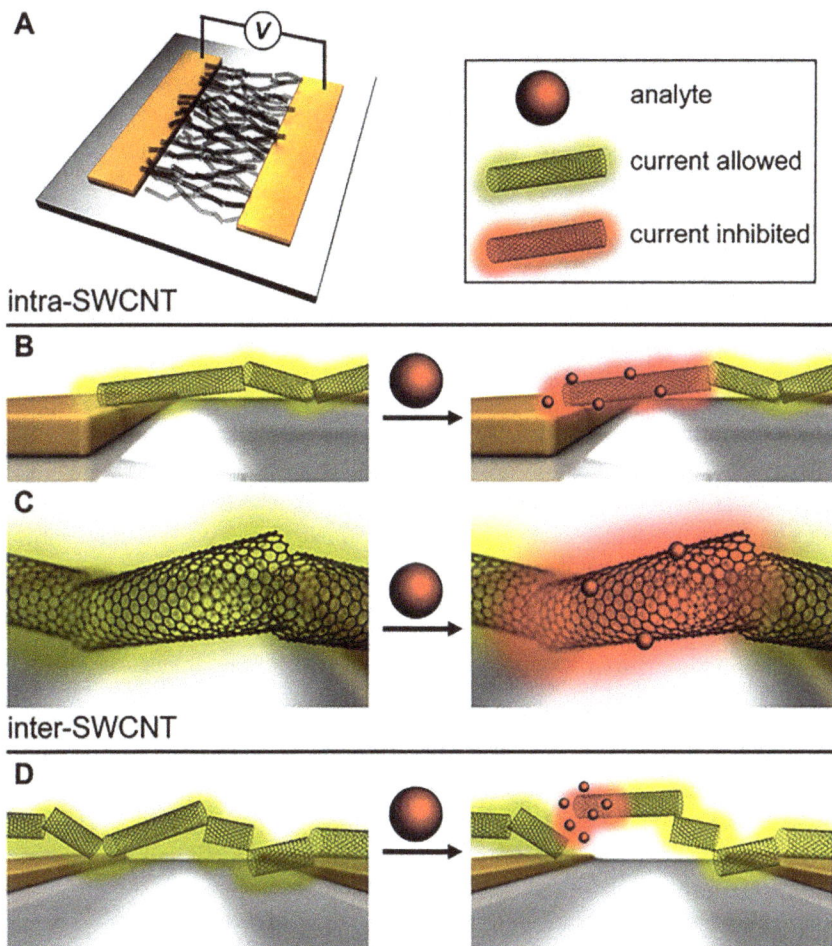

FIGURE 8.7 (a) Illustrates the sensing mechanisms involved in resistive sensors based on SWNTs networks. The schematic highlights different regions: (b) the interface between the metallic electrode and the SWNTs, known as the Schottky barrier; (c) the sidewall or along the length of the SWCNT, referred to as intra-SWNTs; and (d) the interface between SWNTs, referred to as inter-SWNTs (64).

determine the rate of adsorption. On the other hand, the desorption rate is influenced by the quantity of adsorbed particles and the average duration of their contact with the surface. Additionally, the adsorption mechanism is affected by the presence of metallic or semiconducting nanotubes. For example, metallic nanotubes contribute to chemisorption, whereas semiconducting nanotubes lead to physisorption.

Different gases induce varying changes in the conductance of CNTs. For example, nitrogen dioxide (NO_2) causes increased conductivity due to electron depletion, making CNTs more p-type. Conversely, ammonia (NH_3) decreases conductance by promoting hole recombination. Functionalizing CNTs specific to a particular gas is necessary to distinguish between different gases. Various methods have been explored, including the decoration of metal nanoparticles or oxides and the formation of hybrid structures using CNTs and polymers. However, these modifications have drawbacks related to operational temperature, sensitivity to humidity, and temperature-induced changes in the sensing properties of the materials.

Carbon nanotube field-effect transistors (CNTFETs) are preferred over chemo-resistive sensors to achieve more precise sensing responses. CNTFETs exhibit changes in conductivity upon exposure to different gas concentrations. The conductance response of gas sensors is influenced by the ability of the gases to donate or accept electrons. In the context of CNTFETs, the interaction between the CNTs and metal contacts is vital in sensing behaviour. Different configurations of these contacts result in diverse responses.

Further research has focused on elucidating the exact mechanisms of charge transfer and current changes in CNTFETs upon gas adsorption. The nature of contact between CNTs and metals and diffusion-assisted changes in conductance have been investigated to understand the sensing behaviour better. In summary, CNTs offer significant potential for gas-sensing applications, employing both physisorption and chemisorption mechanisms. Understanding the interactions between gases and CNT surfaces is crucial for developing selective and sensitive gas sensors. Further advancements in CNT-based gas sensing technologies, particularly in designing CNTFETs, hold promise for future gas detection and analysis applications.

8.4.3   CNT-BASED ORGANIC VAPOR SENSORS

The interaction between organic molecules and CNTs during adsorption is governed by several factors. These factors include the hydrophobic nature of the CNT surface, the high surface area of the CNT, and the electronic polarizability of the molecules. Modified CNTs introduce additional variations in shape, size, morphology, and impurities, such as attached chemical groups or metals. Nitroaromatic compounds like nitrobenzene exhibit stronger adsorption than nonpolar aromatic compounds like benzene, toluene, and chlorobenzene. Nitroaromatic compounds are known for their high polarity and strong electron-accepting properties. These compounds interact favourably with adsorbents that exhibit high electron polarizability. On the other hand, most organic compounds are usually nonpolar and do not readily accept electrons.

The adsorption of organic chemicals on CNTs typically exhibits heterogeneous adsorption and hysteresis phenomena. Heterogeneous adsorption occurs due to high-energy adsorption sites, such as defects in CNTs, functional groups, and the interstitial spaces between CNT bundles. Additionally, the surface influences heterogeneous adsorption and capillary condensation of liquid analytes. The thermodynamics and kinetics, which depend on concentration, indicate that organic molecules initially occupy high-energy adsorption sites before settling in lower-energy sites. On the other hand, hysteresis is observed as a deviation from the adsorption to desorption curve. It may be attributed to several factors, including the solid π-π coupling between the CNT surface and benzene-ring chemicals. Furthermore, hysteresis could also result from structural alterations or reorganization of the adsorbent following the adsorption of organic molecules.

The penetration of small molecules into the inner regions of CNTs restricted by a slow diffusion process, leading to significantly low diffusivity. Conversely, larger molecules demonstrate faster absorption rates as they experience reduced adsorption within the inner pores of CNTs. These larger molecules can adjust their conformation to match the curvature of CNTs, resulting in the formation of stable complexes. The adsorption of organic molecules onto CNTs is influenced by several factors, including their polarity, electron-accepting properties, high-energy adsorption sites, and the size and shape of the molecules. An understanding of these interactions is crucial for the development of efficient adsorption processes and the creation of customized, CNT-based adsorbents for organic compounds.

8.4.4 CNT-Based Optical Sensors

CNTs have emerged as highly promising optical sensors, specifically in infrared sensing, owing to their remarkable absorption capacity, reaching an astounding 99.98% (65). Numerous comprehensive investigations have been carried out on vertically aligned MWNTs that are synthesized through a plasma-enhanced chemical vapour deposition (PECVD) technique. The growth of MWNTs on silicon substrates involve the utilization of different templates; namely, Co/Ti and Co/Ti/NbTiN. These templates yielded distinct outcomes: the former resulted in reflective surfaces, while the latter produced dense, dark arrays of MWNTs. Notably, the Co/Ti/NbTiN template demonstrated a significantly reduced contact resistance due to the presence of the metal nitride-MWCNT interface. Examining the optical reflection properties revealed a direct correlation between the thickness of the catalyst and the observed phenomena. In particular, increased catalyst thickness corresponded to enhanced absorption and minimal backscattering. This behaviour can be attributed to weak electron coupling within the grown MWNTs.

However, the photocurrent generation mechanism in SWNTs differs significantly from that in bulk materials. To gain insight into this mechanism, a controlled experiment was conducted using scanning photocurrent microscopy (SPCM) to investigate the electronic structure of suspended metallic and semiconducting CNTs (66). SPCM is a highly effective technique to explore photocurrent generation in nanostructures and provides valuable insights into their behaviour. The study revealed distinct behaviours in metallic CNTs forming p-n junctions. These CNTs exhibited multiple polarity reversals and generated photothermal current along the nanotube axis due to the difference in Seebeck coefficients between the two ends of the tube. Conversely, semiconducting CNTs with a uniform doping profile displayed constant Seebeck coefficients along the nanotube, eliminating the possibility of photo thermal current. In the case of semiconducting CNTs, an electric field at the junction facilitated exciting formation, leading to photocurrent generation as the laser traversed from one end to the other. This behaviour resembled the photovoltaic effect. These findings unveiled distinct mechanisms for CNTs, depending on carrier generation and recombination processes.

The exceptional optical absorption abilities of CNTs have positioned them as promising candidates for terahertz detection applications (refer to the sensor shown in Figure 8.8). Terahertz optics, an emerging field, holds great potential for biomedical

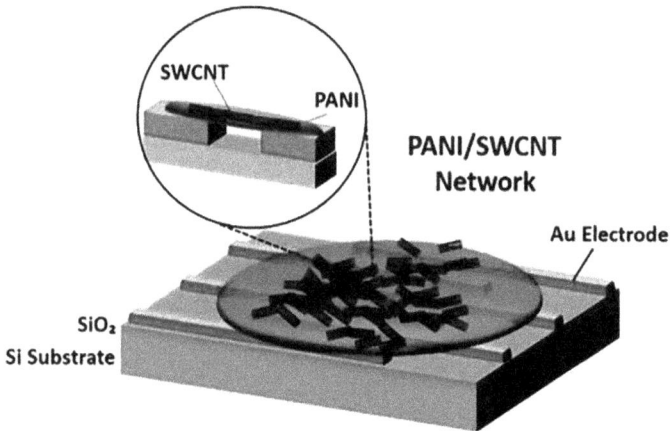

FIGURE 8.8 The diagram illustrates the arrangement of interdigitated electrodes with a coating of PANI (Polyaniline) layered with SWCNT–COOH (Single-Walled Carbon Nanotubes with Carboxylic Acid Functionalization) (69).

imaging, security, and environmental monitoring advancements. By chemically doping a film of n-type CNTs to create a p-n junction, researchers have achieved remarkable sensitivity to the polarization of terahertz beams. Previous studies have confirmed the significant role the Seebeck effect plays in the response, supported by optothermal and thermoelectric measurements (67). In addition, investigations have been conducted to elucidate the photocurrent mechanism in different CNT morphologies, including random assemblies (68). These studies have revealed that diffusion primarily contributes to photocurrent generation, with nanotube-metal contacts displaying the highest photocurrent due to Schottky barrier modulation, which aligns with earlier reports.

Furthermore, introducing metal oxides onto MWNTs has unveiled distinct photodetection mechanisms. For instance, Zhu et al. (70) demonstrated the synergistic combination of ZnO's three-photon absorption properties with MWNTs' saturable absorption properties, leading to ultra-fast optical detection suitable for optical switches. To summarize, CNTs have demonstrated exceptional optical sensing capabilities, particularly in the infrared and terahertz ranges, with the specific mechanisms varying depending on the type of CNT and the experimental setup. These findings have paved the way for further advancements and applications of CNT-based optical sensors.

8.4.5 CNT-Based Resonators or Resonant Sensors

Resonant sensors are utilized to identify alterations in mechanical stress or strain by measuring variations in resonant frequency caused by external forces. CNTs exhibit exceptional qualities that make them highly suitable for ultrasensitive resonators. These qualities include a high Young's modulus, a small cross-section, and robust tensile strength. For such measurements, a preferred approach is to employ a cantilever configuration or suspend the nanotube between two supports. The first experiment on multi-walled CNTs used a cantilever configuration to study field-induced vibrations, demonstrating their elastic behaviour even under high electric-fields.

Due to their nanoscale dimensions, classic mechanics concepts are inadequate when modelling CNT resonators. Atomistic modelling techniques, on the other hand, are preferred. In a particular study, the simulation of covalent bonding was conducted between carbon atoms in SWNTs by representing them as structural beams. Three stiffness parameters were evaluated: tensile resistance, flexural rigidity, and torsional stiffness. The fundamental frequencies of both cantilever and bridged configurations were calculated, assuming free vibration of the SWCNT as a space-frame-like structure. The findings revealed that the fundamental frequency decreases as the nanotube length increases while maintaining a constant aspect ratio. Additionally, an increase in diameter was observed to result in a higher fundamental frequency. When comparing the resonant properties of MWNTs and SWNTs, it was determined that those of MWNTs exhibit greater reproducibility. This can be attributed to their enhanced rigidity and resistance to deformation, thereby making them more suitable for a wide range of applications, including signal processing.

8.4.6 CNT-Based Piezoresistive Sensors

Piezoresistive sensors can detect changes in electrical resistance when subjected to strain. CNTs exhibit favourable mechanical and electrical properties, making them intriguing piezoresistive materials (refer to Figure 8.9). The ability of CNTs to detect even negligible strain has been demonstrated using atomic force microscopy (AFM). Tombler et al. (71) conducted an experiment employing a bridge configuration on a SiO_2/Si substrate with metal electrodes. To position the suspended SWNTs, an Atomic Force Microscope (AFM) tip was employed at its centre, pushing it down towards a trench. As a result of this procedure, the AFM tip experienced deflection, leading to a measurable change in the resistance of the SWNT. By correlating the

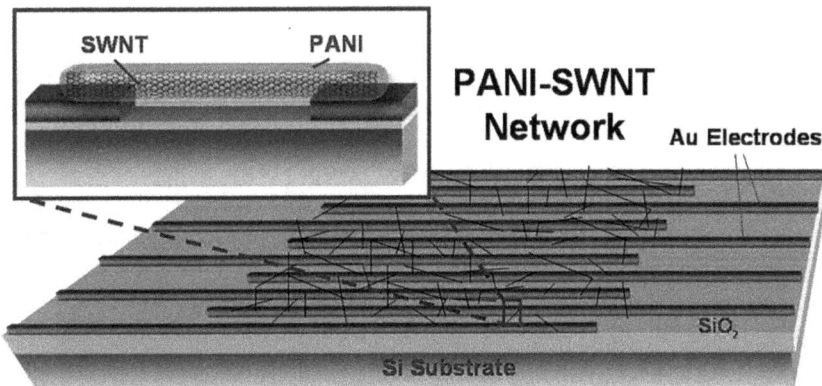

FIGURE 8.9 Illustrates the proposed interaction mechanism between NH_3 and the PTh/MWNTs nanocomposite (72).

deflection of the AFM tip with the movement of SWNT, the actual response of the SWNT to applied strain was obtained at various angles. Notably, the sample exhibited robustness by retracting to its initial position, even after repeated pushing with the AFM tip. It was observed that the conductance of the SWNT decreased as it bent under the influence of the AFM. The rate of decrease in conductance accelerated with the increase in bending angle, and at maximum bending, the decrease in conductance reached two orders of magnitude. Simulations were conducted to understand the underlying mechanism behind this conductance decrease.

8.5 ADVANTAGES OF CNT

CNTs offer numerous sensing advantages, including high sensitivity, fast response time, wide detection range, small size, mechanical robustness, low power-consumption, versatility, and compatibility. These properties make CNTs ideal for developing susceptible and efficient sensors for various applications.

- High Sensitivity: CNTs exhibit exceptional sensitivity to various physical and chemical stimuli, enabling them to detect subtle environmental changes. This increased sensitivity makes them suitable for applications requiring precise and accurate measurements.
- Wide Sensing Range: CNTs-based sensors can detect various parameters, including temperature, pressure, strain, humidity, gas composition, and biological molecules. This versatility allows for diverse industrial applications across different sectors.
- Miniaturization: CNTs have nanoscale dimensions, allowing for the development of miniaturized sensors. Their small size enables integration into compact devices and systems, making them ideal for space-limited applications.
- Fast Response Time: CNTs-based sensors exhibit rapid response and recovery times. They can quickly detect environmental changes and provide real-time data, which is crucial for dynamic industrial processes and monitoring systems.
- Mechanical Strength: CNTs possess exceptional mechanical strength and resilience. They can withstand harsh conditions, such as high temperatures, extreme pressures, and chemical exposure, making them suitable for robust and durable sensors in industrial environments.
- Low Power Consumption: CNT-based sensors often require less power, leading to energy-efficient sensing systems. This characteristic is particularly advantageous for battery-powered or portable devices, enhancing their overall performance and longevity.
- Chemical Selectivity: Functionalization of CNTs allows for selective sensing of specific target molecules or gases. By modifying the surface of CNTs, sensors can be tailored to detect and differentiate between different analytes, enhancing their selectivity and reducing false readings.
- Long-term Stability: CNTs exhibit excellent stability over extended periods, maintaining their sensing properties without significant degradation. This

stability ensures reliable and consistent performance of the CNTs-based sensors in industrial applications.

- Compatibility: CNTs can be easily integrated into various substrates and materials commonly used in industrial settings, such as polymers, metals, and ceramics. This compatibility facilitates the incorporation of CNT-based sensors into existing industrial systems and manufacturing processes.
- Cost-effectiveness: With advancements in manufacturing techniques, the production of CNTs has become more cost-effective. As a result, CNT-based sensors offer an attractive balance between performance and affordability, making them economically viable for industrial applications.

Overall, the unique properties of CNTs make them highly advantageous for industrial sensing applications, enabling enhanced sensing capabilities, improved efficiency, and increased reliability in diverse industrial processes and monitoring systems.

8.6 DISADVANTAGES OF CNTS

While CNTs possess many desirable properties, they have certain limitations that hinder their commercial utilization, one of which is their hydrophobicity and chemical inertness that makes it challenging to suspend them in solution due to strong van der Waals attractions between the nanotubes. Moreover, the aggregation of nanotubes in solution is a common issue that diminishes the unique mechanical and electrical properties of the individual tubes. To overcome these challenges, the surface of CNTs can be modified through covalent or non-covalent methods. Various techniques can be employed to enable effective dissolution in water, such as using surfactants or polymer wrapping. Surfactants like sodium dodecyl sulfate, sodium dodecylbenzene sulfonate, cetyltrimethylammonium bromide, and polyethene glycol are commonly employed. These techniques play a crucial role to ensure the suitability of materials for biomedical applications and biophysical processing schemes.

Defects in CNTs represent a notable drawback in their practical applications. During the growth process of CNTs, various structural defects are generated, which have a discernible impact on the physical and chemical properties of the nanotubes. These defects can be categorized into three groups:

- Topological defects occur when ring sizes other than hexagons, such as pentagons and heptagons, are introduced into the graphene sheet. As a result, local deformations in the nanotube's width can occur, leading to potential changes in its helicity.
- Rehybridization defects: Carbon atoms in graphene can undergo rehybridization between sp, sp^2, and sp^3 states. This rehybridization allows for out-of-plane flexibility in the graphene structure. When the graphene sheet is bent out of plane, it acquires some sp^3 character, forming defect lines with strong sp^3 character in the folds.
- Incomplete bonding and other defects: This category includes vacancies, dislocations, impurity attachments, and other defects commonly associated with graphite. Point defects like vacancies or dislocations can also occur

in nanotubes, although dislocations are less prevalent in high-temperature-formed nanotubes compared to catalytically grown ones.

The presence of defects significantly influences the observed properties of nanotubes. Although the chemical reactivity of nanotubes is relatively inert, subjecting them to annealing and purification processes before conducting physical measurements is advisable. This precaution ensures the consistency and unambiguous interpretation of experimental results. In conclusion, addressing the issue of defects in CNTs is crucial for their successful utilization. Understanding the nature and impact of these defects is essential in order to optimize the properties of nanotubes and advance their applications in various fields. In summary, the challenges associated with CNTs include their hydrophobicity, tendency to aggregate, and the presence of defects. However, surface modifications and careful consideration of defects can mitigate these issues, enabling more consistent and reliable utilization of CNTs in industrial applications.

8.7 CONCLUSIONS

CNTs have garnered significant attention as highly promising sensing application materials, primarily due to their extraordinary mechanical, electrical, and chemical properties. With their exceptional characteristics, CNTs-based sensors present several advantages, such as high sensitivity, rapid response times, and the ability to detect diverse analytes. This chapter provides a comprehensive overview of the key aspects associated with CNTs-based sensors. Firstly, the unique mechanical properties of CNTs make them excellent candidates for sensing applications. Their high Young's modulus and tensile strength enable the detection of minute mechanical deformations induced by external forces. Resonant sensors utilizing CNTs as ultrasensitive resonators have been demonstrated, where changes in the resonant frequency correlate with external stimuli. Secondly, CNTs exhibit remarkable electrical properties, making them ideal for electrical sensing. Piezoresistive sensors utilize the change in electrical resistance of CNTs, upon applying strain. The strain-induced modulation of electrical resistance can be exploited to detect mechanical deformation, pressure, or strain in various applications. Additionally, CNTs can function as excellent field-effect transistors, enabling electrical sensing of gases, biomolecules, and other analytes.

Surface modifications can be applied to overcome challenges associated with CNTs, such as hydrophobicity and aggregation. Covalent or non-covalent modifications enhance solubility and dispersibility in solvents, enabling their integration into various sensing platforms. Moreover, defects in CNTs, which can influence their properties, should be carefully considered and minimized through annealing and purification processes. CNTs offer numerous advantages for sensing applications, including high sensitivity, fast response times, and a wide detection range. Their mechanical and electrical properties and surface modifications provide a versatile platform for developing sensors across various domains, including environmental monitoring, biomedical diagnostics, and industrial applications. Continued research and development in this field hold significant potential for advancing sensor technology and addressing real-world challenges.

REFERENCES

1. Morris JE, Iniewski K. *Graphene, carbon nanotubes, and nanostructures: techniques and applications*. Boca Raton, USA: Routledge, Taylor and Francis Group, 2013.
2. Soldano C, Mahmood A, Dujardin E. Production, properties and potential of graphene. Carbon. 2010;48(8):2127–50.
3. Sheehan JE, Buesking K, Sullivan B. Carbon-carbon composites. Annual Review of Materials Science. 1994;24(1):19–44.
4. Windhorst T, Blount G. Carbon-carbon composites: a summary of recent developments and applications. Materials & Design. 1997;18(1):11–15.
5. Shenderova O, Zhirnov V, Brenner D. Carbon nanostructures. Critical Reviews in Solid State and Material Sciences. 2002;27(3–4):227–356.
6. Suzuki A. Carbon-carbon bonding made easy. Chemical Communications. 2005(38):4759–63.
7. Henning T, Salama F. Carbon in the universe. Science. 1998;282(5397):2204–10.
8. Hirlekar R, Yamagar M, Garse H, Vij M, Kadam V. Carbon nanotubes and its applications: a review. Asian Journal of Pharmaceutical and Clinical Research. 2009;2(4):17–27.
9. Belin T, Epron F. Characterization methods of carbon nanotubes: a review. Materials Science and Engineering: B. 2005;119(2):105–18.
10. Endo M, Iijima S, Dresselhaus MS. *Carbon nanotubes*. Elsevier. 1997.
11. Thostenson ET, Ren Z, Chou T-W. Advances in the science and technology of carbon nanotubes and their composites: a review. Composites Science and Technology. 2001;61(13):1899–912.
12. Prasek J, Drbohlavova J, Chomoucka J, Hubalek J, Jasek O, Adam V, et al. Methods for carbon nanotubes synthesis. Journal of Materials Chemistry. 2011;21(40):15872–84.
13. Dervishi E, Li Z, Xu Y, Saini V, Biris AR, Lupu D, et al. Carbon nanotubes: synthesis, properties, and applications. Particulate Science and Technology. 2009;27(2):107–25.
14. Varshney K. Carbon nanotubes: a review on synthesis, properties and applications. International Journal of Engineering Research and General Science. 2014;2(4):660–77.
15. Ibrahim KS. Carbon nanotubes? Properties and applications: A review. Carbon Letters. 2013;14(3):131–44.
16. Zaporotskova IV, Boroznina NP, Parkhomenko YN, Kozhitov LV. Carbon nanotubes: sensor properties. A review. Modern Electronic Materials. 2016;2(4):95–105.
17. Tasis D, Tagmatarchis N, Bianco A, Prato M. Chemistry of carbon nanotubes. Chemical Reviews. 2006;106(3):1105–36.
18. Heidarinejad Z, Dehghani MH, Heidari M, Javedan G, Ali I, Sillanpää M. Methods for preparation and activation of activated carbon: a review. Environmental Chemistry Letters. 2020;18:393–415.
19. Gibson RF, Ayorinde EO, Wen Y-F. Vibrations of carbon nanotubes and their composites: a review. Composites science and technology. 2007;67(1):1–28.
20. Gong K, Yan Y, Zhang M, Su L, Xiong S, Mao L. Electrochemistry and electroanalytical applications of carbon nanotubes: a review. Analytical Sciences. 2005;21(12):1383–93.
21. Eletskii AV. Carbon nanotubes. Physics-Uspekhi. 1997;40(9):899.
22. Purohit R, Purohit K, Rana S, Rana R, Patel V. Carbon nanotubes and their growth methods. Procedia Materials Science. 2014;6:716–28.
23. Kruss S, Hilmer AJ, Zhang J, Reuel NF, Mu B, Strano MS. Carbon nanotubes as optical biomedical sensors. Advanced Drug Delivery Reviews. 2013;65(15):1933–50.
24. Nag A, Alahi MEE, Feng S, Mukhopadhyay SC. IoT-based sensing system for phosphate detection using graphite/PDMS sensors. Sensors and Actuators A: Physical. 2019;286:43–50.
25. Qian D, Wagner A, Gregory J, Liu WK, Yu M-F, Ruoff RS. Mechanics of carbon nanotubes. Applied Mechanics Reviews. 2002;55(6):495–533.

26. Baddour CE, Briens C. Carbon nanotube synthesis: a review. International Journal of Chemical Reactor Engineering. 2005;3(1).
27. Dai H. Carbon nanotubes: synthesis, integration, and properties. Accounts of Chemical Research. 2002;35(12):1035–44.
28. Meyyappan M, Delzeit L, Cassell A, Hash D. Carbon nanotube growth by PECVD: a review. Plasma Sources Science and Technology. 2003;12(2):205.
29. Nessim GD. Properties, synthesis, and growth mechanisms of carbon nanotubes with special focus on thermal chemical vapor deposition. Nanoscale. 2010;2(8):1306–23.
30. Jariwala D, Sangwan VK, Lauhon LJ, Marks TJ, Hersam MC. Carbon nanomaterials for electronics, optoelectronics, photovoltaics, and sensing. Chemical Society Reviews. 2013;42(7):2824–60.
31. Sharma R, Sharma AK, Sharma V. Synthesis of carbon nanotubes by arc-discharge and chemical vapor deposition method with analysis of its morphology, dispersion and functionalization characteristics. Cogent Engineering. 2015;2(1):1094017.
32. Lan Y, Wang Y, Ren Z. Physics and applications of aligned carbon nanotubes. Advances in Physics. 2011;60(4):553–678.
33. Gupta N, Gupta SM, Sharma S. Carbon nanotubes: Synthesis, properties and engineering applications. Carbon Letters. 2019;29:419–47.
34. Eatemadi A, Daraee H, Karimkhanloo H, Kouhi M, Zarghami N, Akbarzadeh A, et al. Carbon nanotubes: properties, synthesis, purification, and medical applications. Nanoscale Research Letters. 2014;9:1–13.
35. Aqel A, Abou El-Nour KM, Ammar RA, Al-Warthan A. Carbon nanotubes, science and technology part (I) structure, synthesis and characterisation. Arabian Journal of Chemistry. 2012;5(1):1–23.
36. Kuzmany H, Kukovecz A, Simon F, Holzweber M, Kramberger C, Pichler T. Functionalization of carbon nanotubes. Synthetic Metals. 2004;141(1–2):113–22.
37. Shah KA, Tali BA. Synthesis of carbon nanotubes by catalytic chemical vapour deposition: a review on carbon sources, catalysts and substrates. Materials Science in Semiconductor Processing. 2016;41:67–82.
38. Fan Z, Advani SG. Rheology of multiwall carbon nanotube suspensions. Journal of Rheology. 2007;51(4):585–604.
39. Salah LS, Ouslimani N, Bousba D, Huynen I, Danlée Y, Aksas H. Carbon nanotubes (CNTs) from synthesis to functionalized (CNTs) using conventional and new chemical approaches. Journal of Nanomaterials. 2021;2021:4972770.
40. Norizan MN, Moklis MH, Demon SZN, Halim NA, Samsuri A, Mohamad IS, et al. Carbon nanotubes: functionalisation and their application in chemical sensors. RSC Advances. 2020;10(71):43704–32.
41. Bhatt A, Jain A, Gurnany E, Jain R, Modi A, Jain A. Carbon nanotubes: a promising carrier for drug delivery and targeting. Nanoarchitectonics for Smart Delivery and Drug Targeting. 2016:465–501.
42. Kanoun O, Bouhamed A, Ramalingame R, Bautista-Quijano JR, Rajendran D, Al-Hamry A. Review on conductive polymer/CNTs nanocomposites based flexible and stretchable strain and pressure sensors. Sensors. 2021;21(2):341.
43. Dresselhaus MS, Dresselhaus G, Saito R, Jorio A. Raman spectroscopy of carbon nanotubes. Physics Reports. 2005;409(2):47–99.
44. Khare R. Carbon nanotube based composites: a review. Journal of Minerals and Materials Characterization and Engineering. 2005;4(1):31.
45. Lekawa-Raus A, Patmore J, Kurzepa L, Bulmer J, Koziol K. Electrical properties of carbon nanotube based fibers and their future use in electrical wiring. Advanced Functional Materials. 2014;24(24):3661–82.
46. Bernholc J, Brenner D, Buongiorno Nardelli M, Meunier V, Roland C. Mechanical and electrical properties of nanotubes. Annual Review of Materials Research. 2002;32(1):347–75.

47. Tang Q, Shafiq I, Chan Y, Wong N, Cheung R. Study of the dispersion and electrical properties of carbon nanotubes treated by surfactants in dimethylacetamide. Journal of Nanoscience and Nanotechnology. 2010;10(8):4967–74.
48. Hone J. Carbon nanotubes: thermal properties. Dekker Encyclopedia of Nanoscience and Nanotechnology. 2004;7:603–10.
49. Kumanek B, Janas D. Thermal conductivity of carbon nanotube networks: a review. Journal of Materials Science. 2019;54(10):7397–427.
50. Guo J. Uric acid monitoring with a smartphone as the electrochemical analyzer. Analytical chemistry. 2016;88(24):11986–9.
51. Ruoff RS, Lorents DC. Mechanical and thermal properties of carbon nanotubes. Carbon. 1995;33(7):925–30.
52. Eastman JA, Phillpot S, Choi S, Keblinski P. Thermal transport in nanofluids. Annual Review of Materials Research. 2004;34:219–46.
53. Yu M-F, Files BS, Arepalli S, Ruoff RS. Tensile loading of ropes of single wall carbon nanotubes and their mechanical properties. Physical Review Letters. 2000;84(24):5552.
54. Jishi R, Venkataraman L, Dresselhaus M, Dresselhaus G. Phonon modes in carbon nanotubules. Chemical Physics Letters. 1993;209(1–2):77–82.
55. Singh R, Gupta S. *Introduction to nanotechnology*. Oxford University Press. 2016.
56. Zhao Q, Gan Z, Zhuang Q. Electrochemical sensors based on carbon nanotubes. Electroanalysis: An International Journal Devoted to Fundamental and Practical Aspects of Electroanalysis. 2002;14(23):1609–13.
57. Wang Y, Yeow JT. A review of carbon nanotubes-based gas sensors. Journal of Sensors. 2009;2009.
58. Balasubramanian K, Burghard M. Biosensors based on carbon nanotubes. Analytical and Bioanalytical Chemistry. 2006;385:452–68.
59. Wang Q, Arash B. A review on applications of carbon nanotubes and graphenes as nano-resonator sensors. Computational Materials Science. 2014;82:350–60.
60. Sinha N, Ma J, Yeow JT. Carbon nanotube-based sensors. Journal of Nanoscience and Nanotechnology. 2006;6(3):573–90.
61. Mahar B, Laslau C, Yip R, Sun Y. Development of carbon nanotube-based sensors: a review. IEEE Sensors Journal. 2007;7(2):266–84.
62. Alahi MEE, Xie L, Mukhopadhyay S, Burkitt L. A temperature compensated smart nitrate-sensor for agricultural industry. IEEE Transactions on Industrial Electronics. 2017;64(9):7333–41.
63. Alahi MEE, Mukhopadhyay SC, Burkitt L. Imprinted polymer coated impedimetric nitrate sensor for real-time water quality monitoring. Sensors and Actuators B: Chemical. 2018;259:753–61.
64. Fennell Jr JF, Liu SF, Azzarelli JM, Weis JG, Rochat S, Mirica KA, et al. Nanowire chemical/biological sensors: status and a roadmap for the future. Angewandte Chemie International Edition. 2016;55(4):1266–81.
65. Kaul PB, Bifano MF, Prakash V. Multifunctional carbon nanotube-epoxy composites for thermal energy management. Journal of Composite Materials. 2013;47(1):77–95.
66. Adamson A, Gast A. Adsorption of gases and vapors on solids. In *Physical chemistry of surfaces*. Wiley. 1997. pp. 599–676.
67. He X, Fujimura N, Lloyd JM, Erickson KJ, Talin AA, Zhang Q, et al. Carbon nanotube terahertz detector. Nano Letters. 2014;14(7):3953–8.
68. Sarker BK, Arif M, Stokes P, Khondaker SI. Diffusion mediated photoconduction in multiwalled carbon nanotube films. Journal of Applied Physics. 2009;106(7).
69. Zhang T, Nix MB, Yoo BY, Deshusses MA, Myung NV. Electrochemically functionalized single-walled carbon nanotube gas sensor. Electroanalysis: An International Journal Devoted to Fundamental and Practical Aspects of Electroanalysis. 2006;18(12):1153–8.

70. Zhu Y, Elim HI, Foo YL, Yu T, Liu Y, Ji W, et al. Multiwalled carbon nanotubes beaded with ZnO nanoparticles for ultrafast nonlinear optical switching. Advanced Materials. 2006;18(5):587–92.
71. Tombler TW, Zhou C, Alexseyev L, Kong J, Dai H, Liu L, et al. Reversible electro-mechanical characteristics of carbon nanotubes underlocal-probe manipulation. Nature. 2000;405(6788):769–72.
72. Husain A, Ahmad S, Mohammad F. Electrical conductivity and ammonia sensing studies on polythiophene/MWCNTs nanocomposites. Materialia. 2020;14:100868.

9 Carbon Nanotubes-Based Sensors for Environmental Applications

Alivia Mukherjee

9.1 INTRODUCTION

Modern society has benefited from advances in science and technology through the expansion of industrial manufacturing, which has resulted in higher living standards, sustained economic growth, and enhanced business competitiveness. Despite the economic miracles and urbanization, the environment pays a higher price due to the pervasive pollution issues that plague communities around the world [1, 2]. The challenges that lie ahead of us, such as escalating depletion of natural resources (fossil fuels), growing energy demand, global environmental pollution, and CO_2 emissions, pose a significant and imminent threat to the eco-system [3, 4]. The emission of various detrimental pollutants, such as toxic gases, volatile organic compounds, dyes, and heavy metals, into the surrounding ecosystem is of global concern and can have significant impacts on human beings. However, the extent of the impact primarily depends upon the duration of the exposure to the pollutants [1]. For instance, the issue of deteriorating air quality resulting from industrial emissions and automobile exhaust has emerged as a significant concern in several urban environments worldwide [5]. Furthermore, the quality of freshwater resources is rapidly degraded by the discharge of pollutants into the aquatic environment from major industrial processes [2]. Indeed, the exploration and utilization of green, renewable resources to promote sustainable development and tackle environmental remediation is crucial amidst rapid global economic and social advancement.

Carbon can be used to revolutionize device applications for removing environmental contaminants and for green engineering [3]. In particular, sp2 hybridization is alluring for the scientific community because it produces notable structures such as fullerene, carbon nanotubes (CNTs), graphene, and its derivatives (graphene oxide and reduced graphene oxide) that have displayed exceptional and distinct physiochemical characteristics. Among other carbon-based nanomaterials (CNMs), nanotubes have garnered significant attention in the field of nanotechnology, attracting interest from both fundamental research and an advanced perspective [6, 7]. Notably, Ijima's 1991 discovery of CNTs sparked a surge in global scientific research and technological focus, fueling the aspirations of revolutionizing numerous frontiers in

DOI: 10.1201/9781003376071-9

the realm of nanotechnology [8, 9]. The exceptional mechanical strength, distinctive physical and chemical characteristics, and high thermal conductivity of CNTs have stimulated an immense impetus within the scientific community [7, 10]. CNTs grasp a noticeable position as a carbon allotrope due to the rich science underpinning their highly requisite real-time, cutting-edge technical applications such as, e.g., sensors, energy, catalysis, environment, electromagnetic shielding, and other fields, combined with exceptional characteristics. There are several studies that highlight the various environmental uses of CNTs, which encompass proactive measures such as avoiding environmental deterioration and optimizing energy efficiency as well as retrospective approaches, including remediation, wastewater treatment, and pollutant transformation [11]. As a consequence of this, the capacity for the manufacturing of CNTs is increasing to thousands of tons annually [12]. This increased interest has stimulated efforts to identify their potential applications and advance the field of CNTs.

Mirroring this endeavor, a diverse array of applications, including CNTs in the domain of sensors, is now being developed. This review is self-reliant and intended to contribute to the recent progress in the emerging field of CNTs in environmental remediation applications. Herein, we have discussed the applicability and recent advancements of numerous CNTs-based chemical sensors in environmental monitoring and effluent treatment, with relevant examples. Following a brief introduction, we reviewed and provided an overview of the state-of-the-art applicability of CNTs in environmental monitoring and their suitable implantations to remove pollutants from air and water: (1) in the realm of environmental remediation, detection of toxic gases; hydrogen sulfide (H_2S), nitrogen dioxide (NO_2), and ammonia (NH_3); (2) in environmental monitoring and detection of benzene, toluene, and xylene (BTX) vapor from the environment; and (3) in wastewater treatment for monitoring and detection of cationic and anionic dyes and heavy transition metal ions in aquatic environment.

9.2 CARBON NANOTUBES

Carbon stands out as the most intriguing element worldwide from the last few decades, due to their abundant availability, remarkable inherent characteristics to form diverse structures through intermolecular reactions, and environmental friendliness [13]. CNTs possess cylindrical structures composed of sp2 carbon units arranged in hexagonal networks on honeycomb lattices, measuring several nanometers in diameter and micron axial length, and are analogous to seamless, rolled graphene sheets [3, 10]. CNTs can be classified into two primary groups based on their shell structure: single-walled carbon nanotubes (SWCNTs) and multi-walled carbon nanotubes (MWCNTs) [7]. SWCNTs can be viewed as rolled graphene sheet into a cylinder with capped ends, and MWCNTs, on the other hand, consist of multiple graphene sheets layered in concentric cylinders [9, 14]. For both SWCNTs and MWCNTs, the diameters are in the range of nanometers, but their lengths can extend up to several centimeters [8]. CNTs can be synthesized by three primary methodologies, including chemical vapor deposition (CVD), arc discharge, and laser deposition techniques [15]. In comparison to the existing methodologies, the low-temperature CVD process demonstrates economic viability, practical feasibility, and the potential for large-scale manufacturing [15, 16]. MWCNTs are more easily produced from a wide range of hydrocarbons undergoing low-temperature CVD technique. On the other hand, SWCNTs can only be formed by high-temperature CVD processes (ranging from 900 to 1200°C), utilizing specific hydrocarbons methane (CH_4), and carbon monoxide (CO) [16].

CNTs demonstrate phenomenal physical and chemical properties, including a high surface area, remarkable thermal, mechanical, and electrical properties, attributed to their unique inherent architecture [9]. For instance, CNTs demonstrate remarkable physical properties, such as tensile strength that surpasses steel by a factor of one hundred, thermal conductivity that is equivalent to the purest form of diamond, and electrical conductivity comparable to copper but with the ability to transport far higher currents [8]. Due to their distinctive physical and chemical features, CNTs have garnered significant attention within the scientific community.

The primary factors that determine the overall chemical reactivity of CNTs are structural defects; namely, vacancies and Stone-Wales defects. These defects lead to the creation of localized double bonds between carbon atoms, thereby augmenting the local chemical reactivity. Furthermore, CNTs' functionality and different properties can be precisely controlled and altered by modifying their surfaces and side chains/walls. This can be achieved by altering the surface chemistry, size distribution, surface area, aggregation, and structure of the CNTs by chemical modification and re-engineering. For instance, modifying CNTs through cap removal or chemical treatment techniques involving covalent and/or non-covalent functionalization can have a substantial influence on their detection/adsorption performance. Also, the accessible, high, specific surface-area-to-volume ratio and porous structure of CNTs have an impact on their adsorption properties [2]. In the field of environmental remediation, CNTs have gained significant attention in recent years. This is mostly owing to their notable characteristics, such as greater condensation pressures, lower temperatures of adsorption, and improved adsorption qualities. These attributes make CNTs particularly suitable for performing remediation processes under ambient conditions.

9.3 ENVIRONMENTAL MONITORING BY CARBON NANOTUBES

The utilization of low-power, economically viable sensors has played a pivotal role in propelling research efforts toward chemical sensors that utilize CNTs. As a result, these sensors have found extensive usage in several domains of environmental remediation, such as the passive monitoring of air and water quality. In the present scenario, sensors developed using CNTs have emerged as a promising area of research, owing to the exceptional performance of CNTs in environmental remediation systems. The CNTs-based sensors have demonstrated exceptional long-term stability and the capacity to function well at normal temperatures under ambient conditions and in the presence of moisture in the background. Additionally, the virtually one-dimensional characteristic of these nanomaterials is responsible for the significant surface-to-volume ratio and exceptional charge-transport capabilities. Their effectiveness in pollutant removal processes, particularly adsorption, has contributed

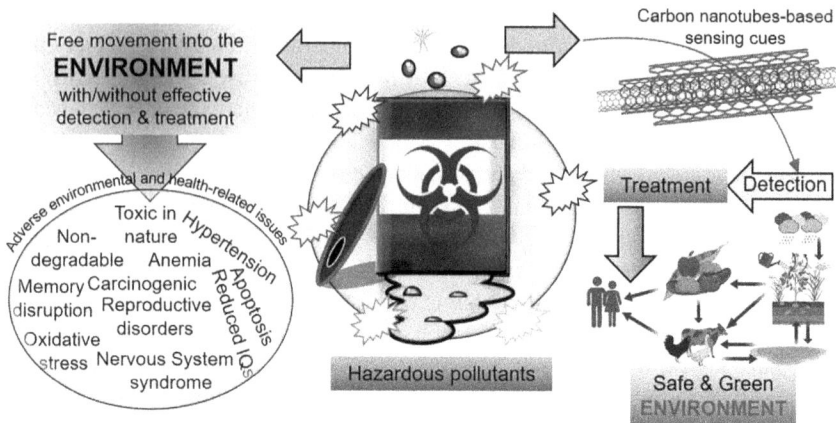

FIGURE 9.1 A schematic illustration of detrimental environmental concerns associated with hazardous pollutants and their carbon-nanotubes-assisted detection to develop a safe and environmentally friendly environment (reprinted from the reference [18], with permission from Elsevier. Copyright (2019)).

to their wide range of usage in this field [17]. In this context, pristine and functionalized CNTs layers are extensively employed in the development of sensors to detect the pollutants and to develop an environmentally friendly environment. Figure 9.1 presents a graphical illustration of the detrimental health and environmental concerns associated with hazardous contaminants, along with the use of CNTs for their detection. The objective is to promote the establishment of a sustainable and safe environment [18].

It is noteworthy to state that CNTs are promising candidates for the development of advanced sensors and tools for environmental monitoring and aqueous environmental assessment, owing to their remarkable inherent properties and characteristics, as stated: (1) CNTs exhibit exemplary structural, functional, electronic, and optical attributes and multi-walled nature make them highly suitable to use for pollutant removal; (2) CNTs have a faster response, higher sensitivity and selectivity, ambient operating temperature, and a wider spectrum of detectable pollutants (in gaseous or aqueous medium); and (3) CNTs provide novel strategies for complete integration and present exceptional opportunities for further miniaturization. These attributes are essential for the development of sensors with high sensitivity and target selectivity to the pollutants (gas, volatile and/or liquid) [19]. Nevertheless, there has been a persistent effort to enhance the range of selectivity and sensitivity by altering the functional moieties attached to the surface and side chains/side-walls of CNTs. The successful monitoring of air and water quality in an urban environment using CNTs requires the ability to accurately quantify trace amounts of harmful substances (air and water) within the low concentration range, often in parts per billion (ppb) to parts per million (ppm). Hence, the primary aim of this study is to examine the potential of CNTs as sensing agents/sensors for the detection and identification of environmentally relevant toxic chemicals in both gas and liquid phases.

9.4 DETECTION OF GASEOUS TOXICANTS USING CARBON NANOTUBES

The identification and detection of hazardous gases are of the utmost significance with regard to the health and safety of human beings. In the realm of environmental monitoring and remediation, specifically pertaining to commercial electrochemical sensors that predominantly employ metal oxides (MOs), CNTs demonstrate exceptional physical and chemical stability. The phenomenon of CNTs displaying interactions with diverse hazardous gas molecules has been observed [20]. The nature of these interactions is dependent on the reducing or oxidizing properties of the gas molecules, which can facilitate the introduction or removal of electrons from the CNTs. As a result, the transfer of electrons leads to the emergence of a detectable electrical signal. The aforementioned effects possess the potential to be harnessed in the advancement of compact and portable gas sensors engineered for the identification of toxic or environmentally damaging compounds. Furthermore, CNTs exhibit enhanced electrical conductivity toward the electrodes and present a wide range of possible architectures and operational properties, making them a unique material for gas sensing applications. Numerous CNTs-based gas sensor devices have been investigated and tested so far, revealing that CNTs possess considerable potential for the advancement of gas sensor technology [21–23]. This review will primarily address the feasibility of utilizing CNTs-based gas sensors for the purpose of monitoring concentrations of hydrogen sulfide (H_2S), nitrogen dioxide (NO_2), and ammonia (NH_3) in the urban environment and reveal the mechanisms of molecular sensing. The choice of a combination of small reducing (H_2S, and NH_3) and oxidizing gas molecules (NO_2) as prototype polluting gases is rooted in the objective of mitigating risks to ecosystems, human health, aquatic life, and safety.

9.4.1 DETECTION OF HYDROGEN SULFIDE (H_2S)

Hydrogen sulfide (H_2S) is a common, flammable, highly corrosive, and water-soluble toxic gas. It possesses hazardous properties, even when present in low quantities measured in ppm. It has a typical odorous egg smell and colorless hazardous vapor and is the source of the deterioration of the environment [24]. Naturally occurring sources of H_2S include fossil fuels, volcanic emissions, and decomposed organic material. The detection of H_2S gas is of considerable significance in various anthropogenic activities, including coal mining, oil and gas extraction, natural gas purification, wastewater treatment, animal husbandry, and several other industrial sectors [25]. H_2S gas is recognized for its significant detrimental effects on human health due to its toxic nature and pronounced flammability that remains potent even at minimal quantities. Long-term exposure of H_2S content over 1,000 ppm can result seriously endanger human health and even lead to death [26]. On the other hand, in daily life, exposure to this chemical is potentially fatal and acts as a neurotoxin and creates life-threatening consequences impacting the respiratory tract and eyes (refer to Table 9.1). Considering its corrosiveness and toxicity, the detection and removal of H_2S have been regarded as crucial [27].

TABLE 9.1

Summary of Toxic Gases, Their Uses, Primary Sources of Emissions, Permissible Exposure Level Limit (PEL), and Their Impacts on Human Beings [14, 37, 40, 41]

Toxic Gases	Uses	Primary Harmful Sources	Permissible Exposure Limit (PEL)	Effects	Reference
Hydrogen sulfide (H_2S)	H_2S has been utilized for the manufacturing of sulfuric acid as well as other inorganic sulfides	Release from natural gases, coal and petroleum factories/refineries	TWA[1] of 10 ppm for 8-h exposure (OSHA[2])	Low concentration causes severe eye, nose, throat, and respiratory tract irritation. High concentration (beyond 1000 ppm) causes death. Act as neurotoxin.	[14, 40]
Nitrogen dioxide (NO_2)	Used as an inhibitor for polymerization, for producing explosives and used as fuel for rocket	Combustion activities from chemical industries, vehicle fumes, and other by-products	TWA of 5 ppm (OSHA)	Leads to asthma and respiratory inflammations and causes coughs. Long exposure may cause lung-related chronic diseases and lead to smell impairment.	[14, 37, 41]
Ammonia (NH_3)	Utilized in the industrial production of explosives, plastics, textiles, and used in refrigerators	Agricultural activities, applications of fertilizers and animal husbandry	TWA of 25–50 ppm (OSHA)	High concentration (500–1000 ppm) causes immense nose and throat irritation. Lower concentration leads to eye, respiratory and skin irritation.	[14, 37, 41]

[1]TWA: Total weight average; [2]OSHA: Occupational Safety and Health Administration

The existing gas sensors for H_2S detection utilize materials such as metal oxides, metal-doped oxides, composite material of metal oxides (SnO_2-CuO multilayered), and Ag/Ag_2S electrodes, which entail significant expenses and consume substantial amounts of electricity [28]. Hence, it is imperative to conduct research on a novel material that exhibits enhanced characteristics like accelerated reaction time, heightened sensitivity, target selectivity, and decreased operational temperature, in comparison to current gas sensor materials [29]. In this context, CNTs have the potential to serve as a cost-effective and promising candidate for the detection of H_2S gas [12]. In addition to the existence of pristine CNTs, literature contains a wide range of studies documenting the application of functionalized CNTs for the aim of detecting and monitoring H_2S gas [28].

To improve the selectivity for H_2S over SO_2, Zhang et al. [30] developed dielectric barrier discharge (DBD) air plasma-modified MWNTs containing nitrogen- and carboxyl-functional groups [30]. They observed that the MWNTs modified by DBD air plasma technique exhibited favorable selectivity toward H_2S and demonstrated significant promise in the field of H_2S detection using modified MWCNTs over SO_2.

Several literatures have demonstrated that the presence of copper (Cu) and copper oxide (CuO) on the surface of pristine CNTs may improve the sensitivity of H_2S sensors [26, 27, 31]. Musayeva and co-workers [27] fabricated nanocomposites-based MWCNTs and functionalized the surface with a thin layer of copper (Cu) films by physical means (electron beam evaporation) and chemical means (electrochemical deposition). The objective of their investigation was to compare and investigate the sensing capabilities of the nanocomposites prepared by both the physical and chemical methods toward H_2S gas [27]. All the prepared sensors exhibited notable sensitivity and selectivity to H_2S in ambient temperature conditions. The researchers observed that both physically and chemically modified nanocomposites – namely, f-MWCNTs/Cu – exhibited comparable performance in terms of selectivity. Additionally, they found that the functionalization of CNTs with oxygenated groups improved the sensing performance. Therefore, it may be utilized as a cost-efficient gas sensor under ambient conditions. Liu and co-workers [26] presented a study that focused on the development and fabrication of a surface plasmon resonance (SPR) gas sensor, as schematically represented in Figure 9.2, designed specifically to detect H_2S in the range of 10–100 ppm. This was achieved through the use of fiber-optic SPR technology. The sensing sheet utilized in the experiment was obtained by coating Ag/APTES/Cu-MWCNT composite film on single-mode fiber (SMF). The sensor exhibits exceptional selectivity and demonstrates favorable stability in terms of response time, humidity, and temperature. Consequently, it holds promise for possible use in the realm of trace H_2S gas monitoring [26].

Similarly, in another study, Liu et al [32] prepared a nanofluid (NF) system, which includes the use of deep eutectic solvent (DES) and copper (Cu). The optimized, NF-based DES solution and Cu (NF@Cu-1%) demonstrates a notable improvement in desulfurization efficiency when compared to its original form of solution containing only DES. The utilization of NF systems using both DES and Cu as absorbents shows promise for the removal of H_2S.

Mubeen et al. synthesized SWCNTs-based conductometric H_2S-detecting sensors decorated with Au nanoparticles using the electrodeposition technique. At ambient

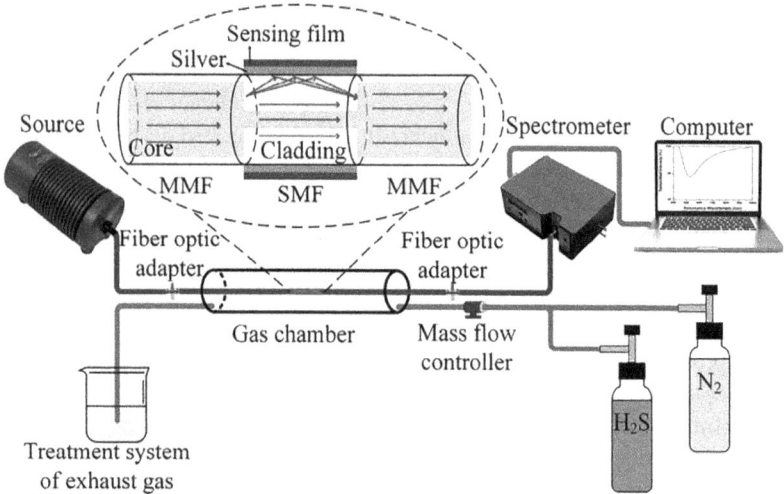

FIGURE 9.2 Schematic diagram of H_2S-sensing experimental system consisting of sensing element (reprinted from the reference [26], with permission from Elsevier. Copyright (2023)).

temperature, the prepared sensor with discrete gold nano-deposits has a detection limit of 3 ppb for H_2S and displayed enhanced recovery time [33]. However, the main obstacle in the production of functionalized CNTs sensors lies in the irreversible reactions caused by the high attraction between metallic nanoparticles (Au or Ag) and sulfur-containing compounds [34]. In this context, several efforts have been dedicated to the advancement of H_2S sensors. These efforts involve the replacement of gold nanoparticles (Au) with alternative metal or metallic nanoparticles (MNPs). The objective of this substitute is to tackle the main challenges posed by organo-metallic nanoparticles. For instance, SnO_2 has been identified as a novel substitute for achieving the reversible functionality of the CNTs surface. Mendoza et al. [35] and Dai et al. [36] presented a CNT-based nanocomposite sensor with SnO_2 coating, prepared using the CVD technique and wet-chemical method, respectively. Under normal conditions, the exhibited H_2S sensors can detect H_2S in the ppm range [35, 36]. Additionally, the functionalized sensors exhibit a recovery time of 1 minute with relative ease and rapid self-recovery [35].

9.4.2 DETECTION OF NITROGEN DIOXIDE (NO_2)

Nitrogen dioxide (NO_2) is a major air pollutant that arises as a by-product of coal combustion and petroleum refining processes. Indeed, the emission of NO_2 from motor vehicles is well recognized as a significant contributor to the occurrence of photochemical smog and acid rain in the atmosphere [37]. Severe exposure of NO_2 beyond permissible exposure limit can cause asthma and lung-related chronic respiratory issue of the human body as highlighted in Table 9.1. The permissible time-weight-average (TWA) contact limit of NO_2 for 8 h is 5 ppm (refer to Table 9.1) [14]. Therefore, NO_2 has an injurious and acute impact on human health and the eco-system, and its accurate detection is crucial.

Sedelnikova et al. [38] synthesized sulfur-coated SWCNTs that were treated with sulfur for the purpose of detecting gaseous NO_2. The researchers noted that the sensing capabilities of sulfur-modified SWCNTs are significantly affected by surface and volumetric processes as well as interface effects. The sensitivity of the sulfur-modified SWCNTs to detect NO_2 within a concentration range of 1 ppb to 10 ppm is remarkable. This can be attributed to the participation of sulfur species in facilitating charge transfer between the nanotubes and NO_2 molecules [38]. Penza and co-workers [39] used noble metal catalysts (Au and Pt) to functionalize the surface of CNTs. The functionalized MWCNTs film was prepared by plasma-enhanced CVD technique for the detection of NO_2 and NH_3 at a temperature range of 25–100°C. They made an observation that the choice of catalysts (Au and Pt) influences the gas sensitivity. This influence was shown to be substantial, mostly due to the spillover effects of the catalysts toward NO_2 and NH_3 [39]. The metal-functionalized gas sensors displayed high gas-sensitivity, fast response and recovery time, selectivity in the sub-ppm range, and repeatability, although various research groups are working on the utilization of pristine CNTs network-based sensors to detect facilely NO_2 at lower concentrations (ppb) at room temperature under ambient conditions.

A study conducted by Ammu et al. [42] documented the fabrication of lightweight, all-organic chemiresistors, using SWCNTs as the active material. The SWCNTs were deposited as thin films onto both paper and cloth substrates. The researchers employed the inkjet-printing technique to apply a layer of CNTs onto cellulosic paper, specifically chosen for its 100% acid-free composition. The purpose of this research was to investigate the suitability of these all-organic chemiresistors for the detection and identification of aggressive oxidizing vapor (NO_2) under ambient conditions. The chemiresistors demonstrate full spontaneous reversibility, achieving a limit of detection (LOD) of 125 ppb for NO_2 without requiring a vapor concentrator. These results were obtained at room temperature, under ambient circumstances. The research team conducted additional analysis on the concept of total recovery and observed a complete retrieval due to the limited interaction between CNTs and NO_2 [42].

9.4.3 Detection of Ammonia (NH_3)

The emission of ammonia (NH_3) into the atmosphere, which has negative impacts on human health, is mostly attributed to agronomic improvements, including the use of fertilizers and animal products. Simultaneously, the assessment of NH_3 concentrations in soil (e.g., from fertilizer) has paramount importance due to the potential to disrupt the nutrient cycle and ecological balance. This disruption arises from the ability of an excessive amount of NH_3 to alter the acidity of the soil [43]. Moreover, a notable impetus for the enhancement of NH_3 sensing lies in the recognition that the existence of NH_3 in drinking water or food might potentially give rise to hazards to human welfare. The time-weight-average (TWA) exposure limit of NH_3, as prescribed by OSHA for the time frame of 8h, is between the range of 25 and 50 ppm (refer to Table 9.1) [12]. A compelling alternative approach for NH_3 detection entails the development of chemical sensors capable of providing straightforward and effective identification of the toxicant. Rigoni and co-workers [44] developed NH_3 gas sensor devices based on CNTs through sonication and dielectrophoresis techniques for the purpose of detecting NH_3 concentrations over the ppb thresholds. The group examined the operational efficiency of SWCNTs-based chemiresistor gas sensors to detect NH_3 operated at ambient temperature conditions. They achieved a detection limit in the range of 3–20 ppb, which falls within the typical range of NH_3 concentrations found in urban environment. This detection limit is also significantly lower than the sensitivities previously documented for pristine, non-functionalized SWCNTs operating at room temperature and ambient conditions [44]. The enhanced performance of the SWCNTs-based chemiresistor gas sensor may be ascribed to two primary factors. Firstly, the manufacturing technique results in a higher quality SWCNTs bundle layer. Secondly, the identification and accurate monitoring of a rapid dynamics channel during the desorption mechanism of the NH_3 molecules [44].

The presence of strong sp2 carbon-carbon bonds makes pristine CNTs chemically stable. However, this characteristic poses a challenge when attempting to establish strong bonds with the majority of gas molecules, including NH_3 molecules. Nevertheless, the prevailing technique to improve the selectivity and sensitivity of the probe for NH_3 involves functionalizing the surface and side chains/side walls of CNTs with metal-based nanoparticles (NPs). This entails the creation of functionalized layers of CNTs that incorporate metal NPs [45]. Accordingly, Valentini's group [43] developed composite paste electrodes composed of CNTs and copper (Cu) by an electrochemical method. These electrodes (CNTs/Cu) were designed to effectively detect and quantify NH_3. The optimized MWCNT/Cu nanocomposite paste electrodes showed comparatively better NH_3 detection performance in terms of sensitivity, response time, and reproducibility than pristine CNTs or Cu paste electrodes, owing to the catalytic action of the CNTs/Cu composite material [43]. Nguyen et al. [46] fabricated MWCNTs/Pt onto alumina substrates using the CVD technique for NH_3 gas-sensing applications. However, to enhance the gas sensitivity and selectivity MWCNTs-sensor at room temperature, the MWCNTs/Pt were modified by sputtering cobalt (Co) nanoparticles on the surface. They observed improved recovery rates, selectivity, gas sensitivity, and repeatability of the Co-functionalized MWCNTs/Pt sensors [46]. Similarly, Cui and co-workers [45] developed functionalized MWCNTs with Ag nanocrystals (Ag NC–MWCNTs) for selective detection of NH_3 at room temperature. The use of Ag nanocrystals significantly increased the sensors' sensitivity at ambient temperature. The elevated sensitivity and selectivity observed in this particular context can be attributed to the presence of the hollow sites on oxidized silver's (AgO) surface.

Numerous studies have reported the utilization of a range of semiconducting oxide materials, including TiO_2, SnO_2, and ZnO, to develop gas sensors for the detection of hazardous gas molecules [47–49]. Hoa and coworkers fabricated a simple, semiconducting metal-oxide-based NH_3 gas sensor. The research team employed a nanocomposite material consisting of tin oxide (SnO_2) and SWCNTs. The walls of CNTs were adorned with tin (Sn) by the application of the arc discharge technique. The fabricated SNO_2/SWCNTs-based sensor displayed remarkable selectivity and desirable sensitivity, as it effectively detects NH_3 with a detection limit of 10 ppm [49]. Similarly, Ghaddab et al. [50] devised a multi-gas sensing system (specifically targeting NH_3 and O_3) at ambient temperature conditions, utilizing a hybrid material composition. The gas-sensitive composite thin films were fabricated by the utilization of sol-gel and dip-coating methodologies, including SnO_2 in conjunction with

SWCNTs. The sensor, which is based on SnO_2/SWCNTs, has the capability to detect NH_3 at a concentration as low as 1 ppm [50].

In general, sensors based on metal nanoparticles (NPs) or CNTs with metal oxide coatings demonstrate improved sensitivity, target selectivity, rapid response times (typically less than 10 seconds), and complete removal of pollutants from the surface under room temperature and ambient conditions. These enhancements are attributed to the presence of functional moieties and defect densities, which are lacking in sensors utilizing pristine CNTs.

9.5 BENZENE, TOLUENE AND XYLENE

VOCs represent the most detrimental consequences arising from the synergistic impact of both biogenic emissions, such as those originating from algae, marine microorganisms, and plants as well as anthropogenic sources, including transportation fuels, landfills, fires, household items, personal care products, and construction materials [51, 52]. Figure 9.3 illustrates the many categories of VOCs, along with the corresponding examples [53].

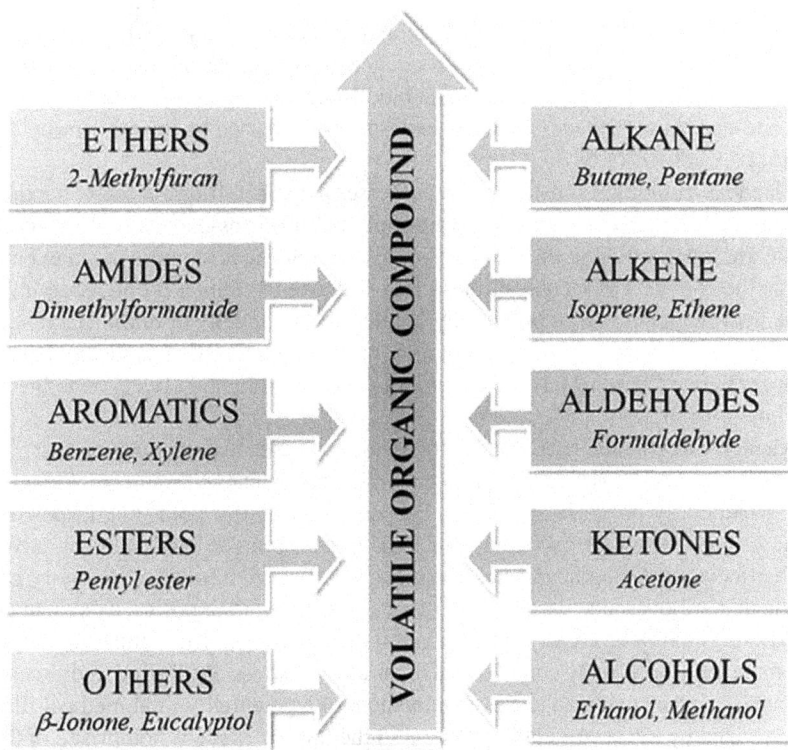

FIGURE 9.3 Categories of volatile organic compounds with pertinent examples (reprinted with permission from the reference [53], Elsevier Copyright (2023)).

Among the large spectrum of VOCs shown in Figure 9.3, aromatic hydrocarbons such as benzene (C_6H_6), toluene (C_7H_8), xylene (C_8H_{10}), often referred to as BTX [54], and ethylbenzene ($C_{10}H_{14}$), collectively known as BTEX [52], are ubiquitous in the environment [55]. BTX, which stands for benzene, toluene, and xylene, is a subset of VOCs that find widespread application in the manufacturing sector. They are utilized as essential components in the production of petrochemical goods as well as solvents for diverse chemical reactions. Additionally, BTX is employed in the maintenance of building interiors and the cleaning of industrial machinery [14]. Despite being highly sought after in industrial sectors, BTX has attracted significant interest among the scientific community due to its potential long-term, negative effects on human health when individuals are exposed to it for longer periods of time [56]. The identification and quantification of VOCs are crucial in urban environments due to their significance in environmental surveillance, occupational safety, and public health considerations. In order to safeguard human health, the acceptable exposure limits established by OSHA for BTX/BTEX are in the range of 1–200 ppm in an 8-hour work day [57, 58]. Among the BTX group, benzene and toluene are illustrative examples of the most hazardous organic solvents that warrant particular scrutiny, owing to their profound carcinogenic properties and severe acute respiratory impacts on human well-being [59, 60].

Benzene is an odorless chemical that is colorless in appearance. It possesses a relatively low density of 0.878 gr cm^{-3} at standard ambient temperature and exhibits a boiling point of 80.1°C [61]. It is widely recognized as a substance with carcinogenic properties in humans. Prolonged and frequent exposure to elevated levels of benzene has the potential to impact the blood and blood-forming organs. The development of leukemia as a consequence of benzene exposure is supported by the most compelling data [62].

Toluene is a colorless substance and possesses a distinct, fragrant scent. It exhibits a density of 0.866 gr cm^{-3} at a temperature of 20°C and possesses a boiling point of 110.7°C [63]. According to the OSHA, the PEL for toluene is recommended to be 100 ppm as the 8-hour TWA concentration [61]. Toluene exhibits a pronounced impact on the neurological system, potentially leading to disruptions in brain function and impairments in various sensory and motor abilities such as balance, vision, hearing, and speech. The chemical has the potential to elicit detrimental effects on kidneys or liver functions [62].

Xylene is an aromatic VOC that is colorless, possesses toxicity, and emits a pleasant fragrance. The OSHA has established a PEL for xylene within occupational settings, particularly prescribing a TWA concentration of 100 ppm [61]. Exposure to xylene at concentrations exceeding the PEL of 50 ppm can lead to many adverse health effects, including headache, dizziness, exhaustion, tremors, respiratory complications, cardiovascular issues, and kidney dysfunction. Exposure to xylene concentrations equal to or above 200 ppm can induce pulmonary irritation, resulting in symptoms such as chest discomfort and respiratory distress. Prolonged and excessive exposure of xylene can lead to the development of a potentially fatal medical illness characterized by the accumulation of fluid in the lungs. Xylene is comprised of three isomers; namely, ortho-xylene, meta-xylene, and para-xylene, all of which share the identical chemical formula C_8H_{10}. The primary differentiating characteristic among the three isomers is in the spatial configuration of the methyl groups ($-CH_3$) on the aromatic ring. P-xylene, which is one of the multiple isomers, is widely utilized in various industrial applications, particularly in the production of polyester [64].

9.5.1 CARBON NANOTUBES FOR THE DETECTION OF BTX

The concentration of BTX is normally minimal, posing a challenge in their detection due to their low levels, commonly measured in ppb (v/v) [65]. Within the framework of BTX identification, numerous detection strategies have been extensively documented in the existing literature. These identification techniques include gas chromatography, coupled with mass spectrometry (GC-MS) and flame ionization detection (FID) techniques [51, 66]. However, to achieve BTX detection at meaningful concentrations, the current BTX detection techniques require high power input, time, and labor. Furthermore, the task of conducting real-time monitoring and targeted evaluation of these hazardous chemicals at their lowest detectable levels in both indoor and outdoor settings continues to provide significant challenges. In recent times, there has been a growing need for the development of novel chemical gas sensors to address the challenges involved with the selective monitoring of BTX compounds. In the context of this framework, the effectiveness of gas sensors that utilize innovative approaches is contingent upon their ability to fulfill the criteria of environmental monitoring, as depicted in Figure 9.4 [53]. Specifically, these requirements can be obtained from the introduction of tailored nanomaterials, which can aid in achieving target selectivity and sensitivity at levels as low as in the ppb (v/v) range, robust interaction with the target absorbents, and facile detection of the polluting absorbents in the presence of a moist background. The scientific community has identified the use of nanoparticles as active components in gas sensors as a crucial approach to enhance their detection capabilities. In this context, the sensing performance of these nanomaterials is a consequence of their nanoscale features, high surface-area-to-volume ratio, high physical and chemical stability at various environmental conditions, exceptional mechanical strength, unique thermal properties, fast recovery and response, ease of composite formation, and low fabrication cost [67]. The nanomaterials-based gas sensors have the potential to serve as either an independent detection system or as an integral part of a more comprehensive sensing system. All these features of nanomaterials make them the most promising materials for gas-sensing applications, including the possibility of using low-cost CNTs (SWCNTs, MWCNTs) to detect target analytes. Hence, this chapter examines several sensors based on CNTs that are utilized for the purpose of selectively and easily detecting these ubiquitous contaminants. Along with defective and pristine CNTs, the literature contains numerous evidence of CNTs utilized for the purpose of environmental remediation of BTX compounds.

9.5.2 PORPHYRINS FUNCTIONALIZED CARBON NANOTUBES FOR THE DETECTION OF BTX/BTEX

Existing BTX detection systems often consist of intricate, multi-component devices or need substantial power input to reliably detect BTX at significant concentrations under ambient conditions. However, this established approach may be modified by including tailor-made nanomaterials. Porphyrins, a relatively recent class of

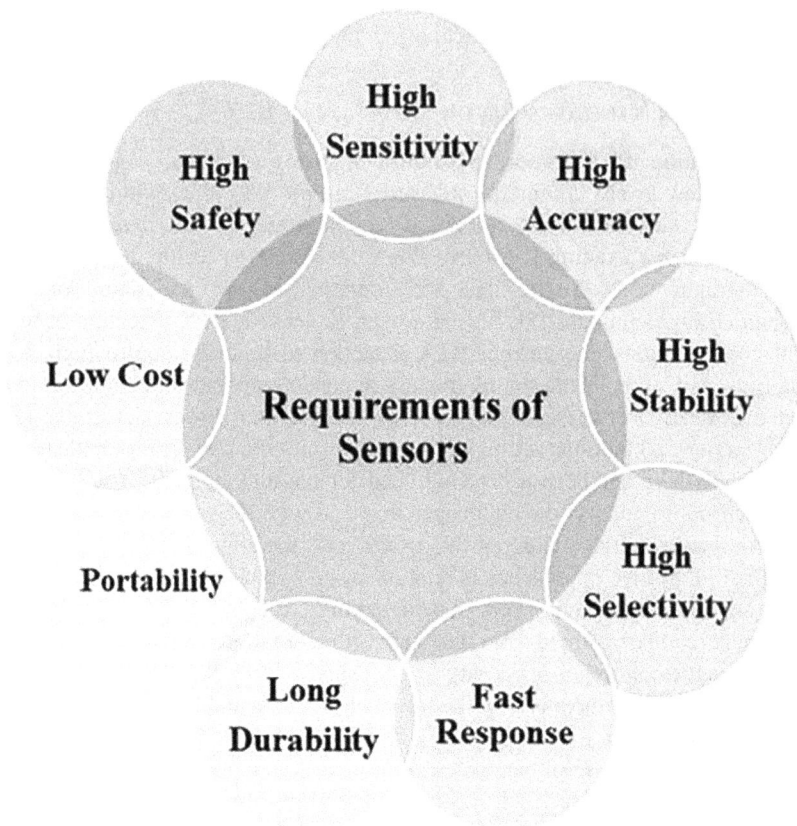

FIGURE 9.4 Essential requirements of a gas sensor for the detection of volatile organic compounds in the environment (reprinted with permission from the reference [53], Elsevier Copyright (2023)).

compounds characterized by an extensive-electronic network, have attracted considerable interest as highly desirable, functional groups for BTX vapor detection [63]. The surface modification of CNTs using porphyrins has been proven to be effective in the identification of BTX vapor, as this approach allows for the manipulation of charge transfer by combining porphyrins with CNTs. The use of a diverse range of porphyrins for the noncovalent functionalization of CNTs has been investigated as a viable material for the construction of chemiresistors in BTX sensing applications. Nanocomposites have been employed in the fabrication of highly responsive sensors for the detection of BTX vapors by several researchers [55, 65, 68]. The use of tetraphenylporphyrin to electrochemically modify SWNTs has been recognized as a notably efficient approach by Sarkar and Srinives [55] for the identification of BTEX vapor. The electrochemical functionalization of SWNTs resulted in a notable and quantifiable reaction to BTEX vapor, even when present at the lowest concentration of 1.25 ppm. Nevertheless, the experimental results indicated that there was no notable reaction, save for the presence of background noise, even when the concentration of BTEX vapor reached its maximum value of 15 ppm. They reported that the successful exhibition of highly sensitive hybrid fabrication via electrochemical methods, coupled with the accessibility of diverse freebase and metalloporphyrin variants that offer varying sensitivity towards different analytes, presents an opportunity for the advancement of a dependable sensor system based on an array configuration [55]. In another study conducted by Rushi et al. [65], they employed a non-covalent functionalization approach to modify SWNTs with iron and cobalt tetraphenyl porphyrins. This fabrication was achieved using a drop-casting technique and was reported to be an excellent approach in detecting organic pollutants. The researchers have documented a clear preference for detecting toluene in the functionalized sensors, which may be attributed to the enhanced π-delocalization occurring at the interfacial locations between metalloporphyrins and SWNTs. Significantly, the functionalized SWNTs produced using iron tetraphenyl porphyrin (FeTPP) exhibited remarkable sensitivity toward toluene in the concentration range of 500 ppb to 10 ppm, when compared to sensors infused with cobalt tetraphenyl porphyrin (CoTPP) at ambient temperature. The key factor that confers target selectivity of the sensors is the variance in electron donating-accepting mechanism displayed by FeTPP over CoTPP [65]. Saxena and co-workers [68] developed functionalized MWCNTs using metal-tetraphenylporphyrins (M-TPP) via a non-covalent functionalization technique. In this study, they have modified the sensors using cobalt- and copper-tetraphenylporphyrins on MWCNTs using the blending and ultrasonification techniques in toluene solution. Different morphological techniques (SEM and TEM) reveal the formation of nanosized clusters of aggregated M-TPP molecular species around the MWNTs surface. The resistance, responsiveness, and recovery periods of the fabricated sensors exhibited variation upon exposure to BTX vapor at normal temperature. They reported that the sensors exhibited resistivity within the kiloohms range. These sensors demonstrated response and recovery durations of a few seconds at normal temperature under ambient conditions. In addition to their ease of fabrication, rapid response, and fast recovery capabilities, these sensors, functionalized with M-TPP, provide distinct advantages.

9.5.3 Functionalized Carbon Nanotubes for the Detection of BTX

Carbon nanotubes (CNTs) are commonly used in gas sensors, encompassing both SWCNTs and MWCNTs. The sensors are equipped with an active layer that exhibits high sensitivity and experiences alterations in electrical resistance upon exposure to certain target analytes. The observed modifications are a result of molecular-level interactions between the CNTs and the analytes. The nature of these interactions, whether they are characterized by either chemisorption, defined by strong covalent bonding, or physisorption, are characterized by weak intermolecular forces. The performance parameter includes its sensitivity, response, and recovery time as well as its detection range. Unlike metal-oxide-based gas sensors, sensors utilizing CNTs demonstrate effective performance under ambient conditions, owing to their low activation energy. Consequently, these characteristics hold promise for the advancement of economically viable gas sensor technologies in the commercial sector. However, they suffer from certain limitations, including their low target selectivity, partial recovery, and extended response-recovery duration. To address these challenges, several strategies have been proposed, including but not limited to chemical functionalization using oxygenated-functional groups, inexpensive acid and amide treatment, and decorating with hybrid metal nanoparticles (gold, rhodium, nickel, or palladium) [54, 69, 70].

The work conducted by Bohli and co-workers [54] focused on the investigation of a gas sensor utilizing MWCNTs for the purpose of detecting toluene and benzene at ambient temperature. The fabrication of the gas sensor involved the application of gold nanoparticles (Au-MWCNTs) and functionalized with a long-chain thiol, 1-hexadecanethiol (HDT) (HDT/Au-MWCNT). The use of Au-MWCNT and HDT/Au-MWCNT sensors has demonstrated the capability to detect aromatic vapors at concentrations as low as in the ppm range. It was reported that the presence of layers and the functionalization contributed toward the improved sensitivity of the sensors by up to 17 times toward selectivity. The use of waste resources to manufacture environmentally friendly products is considered a key approach in the pursuit of sustainable development. The research conducted by Le et al. [69] centered around the production of functionalized MWCNTs generated from rice husks. The study investigated the benzene and toluene detection ability of the oxygen functionalized MWCNTs, which were treated with different oxygenated functional groups, including HNO_3/H_2SO_4, NaOCl, and H_2O_2. Empirical findings have demonstrated that porosity plays a pivotal role in guiding the adsorption of benzene and toluene onto AC. However, the degree of functionalization and the surface chemical properties of the MWCNTs synthesized in advance are the decisive elements influencing their adsorption capacity. While the MWCNTs oxidized using the mixed acid exhibit the greatest quantity of oxygen-functional groups, the MWCNTs oxidized using sodium hypochlorite (NaOCl) exhibit the highest degree of functionalization. Moreover, the mechanism of adsorption was clarified, revealing that the process of oxidative functionalization of MWCNTs results in an augmentation of their negative charge and an enrichment of their π-π electron orbit. The examination of the point-of-zero charge indicates that the absorbates exhibit a positive charge, whereas the negatively charged surface of the functionalized MWCNTs readily engages in electrostatic interactions. The π-π electron-donor-acceptor mechanism is most likely to dominate during the adsorption process, consequently facilitating the interaction between the adsorbates (benzene and toluene) and rice husk-derived oxygen-functionalized MWCNTs, as depicted in Figure 9.5. It is worth noting that the adsorption of toluene exhibits a higher degree of favorability onto carbon-based materials like MWCNTs compared to that of benzene [69].

Specifically, Hodul and co-workers [70], in their research, demonstrated a feasible chemical treatment approach to develop a novel, resonant mass-sensor platform that is capable of selectively detecting aromatic BTX analytes. The methodology employed in their study entails the utilization of SWCNTs that have undergone treatment with inexpensive hydrochloric acid (HCl) and hydroxylamine hydrochloride (HHCl). The researchers made observations regarding the distinctive features of the surface chemistry and tailored material of the nanostructure of SWCNTs that provide a robust interaction with the target gas analytes and contribute to their capacity to induce a selective and unique response when exposed to each BTX analyte.

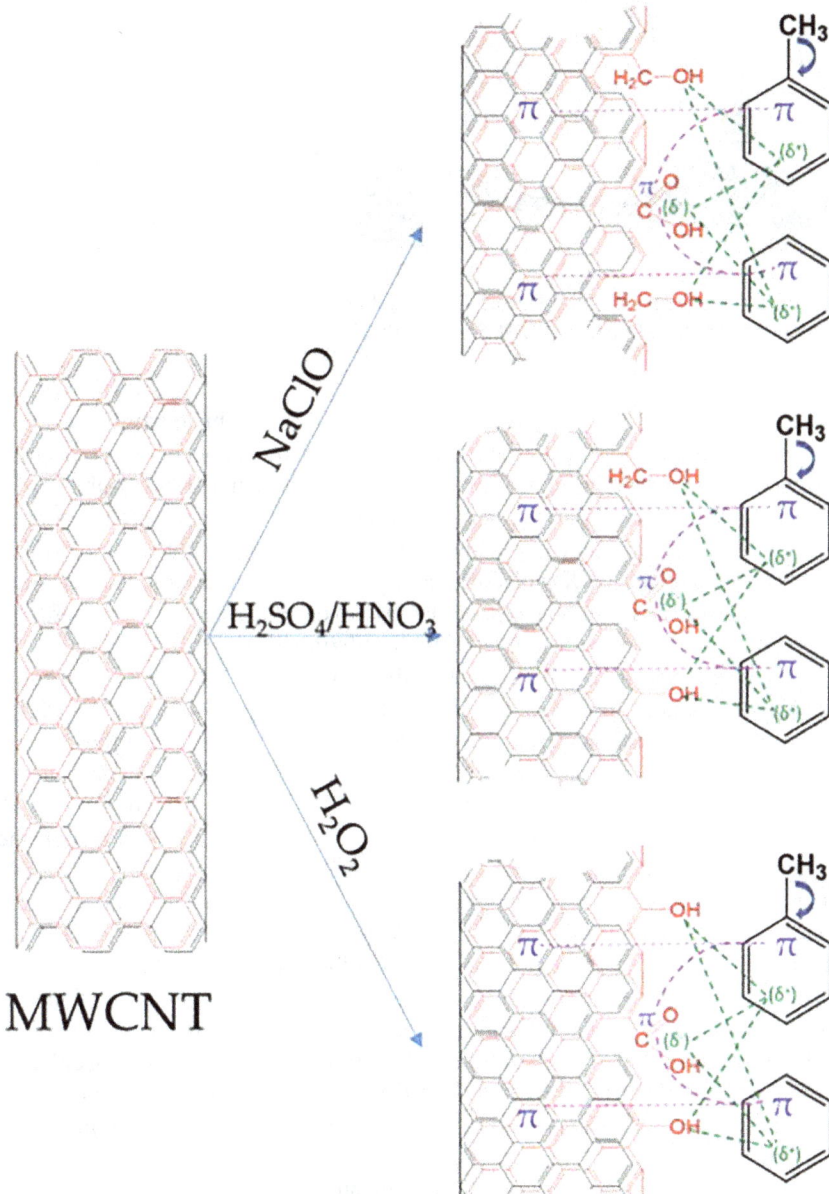

FIGURE 9.5 The adsorption mechanism of the aromatic chemical compounds on oxygen-functionalized (NaClO, H_2SO_4/HNO_3, H_2O_2) MWCNTs (reprinted with permission from the reference [69], Elsevier Copyright (2023)).

Upon treatment with hydrochloric acid, the SWCNTs undergo a process in which the surface iron oxide is efficiently eliminated. Subsequently, the surface iron (III) chloride undergoes reduction to iron (II) by the use of cost-effective chemicals. The use of this chemical treatment results in the production of nitrous oxide gas, which

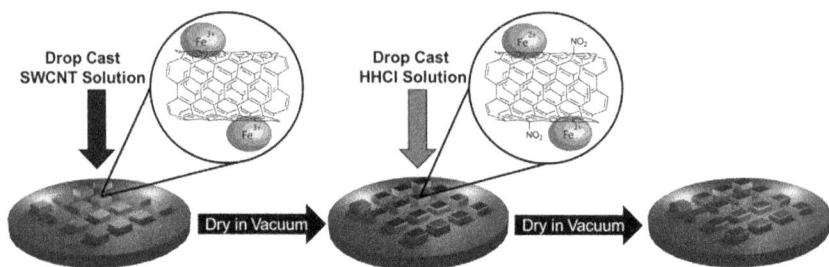

FIGURE 9.6 Schematic of the Device Chemical Functionalization Protocol (reprinted with permission from reference [70], ACS Copyright (2020)).

aids in the in-situ modification of the surface of SWCNTs. This modification allows for the preferential adsorption of aromatic analytes with an abundance of electrons, as schematically shown in Figure 9.6. Consequently, these materials exhibit specific interactions and distinct reactions toward the BTX analytes. This approach offers an integration of surface chemistry with resonant mass sensor technology capable of effectively monitoring BTX molecules, along with many other aromatic chemicals [70]. The study offers a durable sensor platform to develop a cost-effective, efficient, and reusable sensor for the detection of BTX utilizing SWCNTs.

Furthermore, researchers have functionalized pristine CNTs with acid and amide chemical compounds to detect the trace of benzene [58]. Janudin and co-workers [58] developed cost-effective acid and amide-functionalized CNTs for the detection of benzene gas at room temperature and under a controlled-humidity environment. These functionalized CNTs were denoted as CNT-carboxylic and CNT-amide, respectively. The traces of carboxylic acid and dodecylamine were further confirmed through various sophisticated analytical techniques (surface morphology, elemental composition, and FTIR spectroscopy). The successful attachment of the functional groups was confirmed through the morphology study, as depicted in Figure 9.7. As can be seen in Figure 9.7, the pristine CNTs exhibited a rather uniform and smooth morphology in contrast to the rough surface of the acid and amide functionalized CNTs (CNT-carboxylic and CNT-amide). Based on the findings, they reported that CNT-carboxylic and CNT-amide provide an extra surface area, facilitating enhanced interaction between the gas analyte and CNTs. Consequently, this augmentation leads to an amplified response and heightened sensitivity of the sensing material at room temperature.

In another study by Pireaux et al [71], they created a micro-sensor array by employing MWCNTs that were decorated with hybrid metal nanoparticles, such as gold (Au), rhodium (Rh), nickel (Ni), or palladium (Pd). The purpose of developing this functionalized, MWCNTs-based array was to detect and identify minute amounts of benzene vapors with exceptional sensitivity and specificity under ambient conditions. The MWCNTs were functionalized by the utilization of plasma treatment technology. Additionally, the integrated sensor arrays were developed using microsystems technology. According to their findings, the use of micro-sensor arrays adorned with hybrid metal nanoparticles (Rh, Ni, or Pd) enables the detection of benzene present,

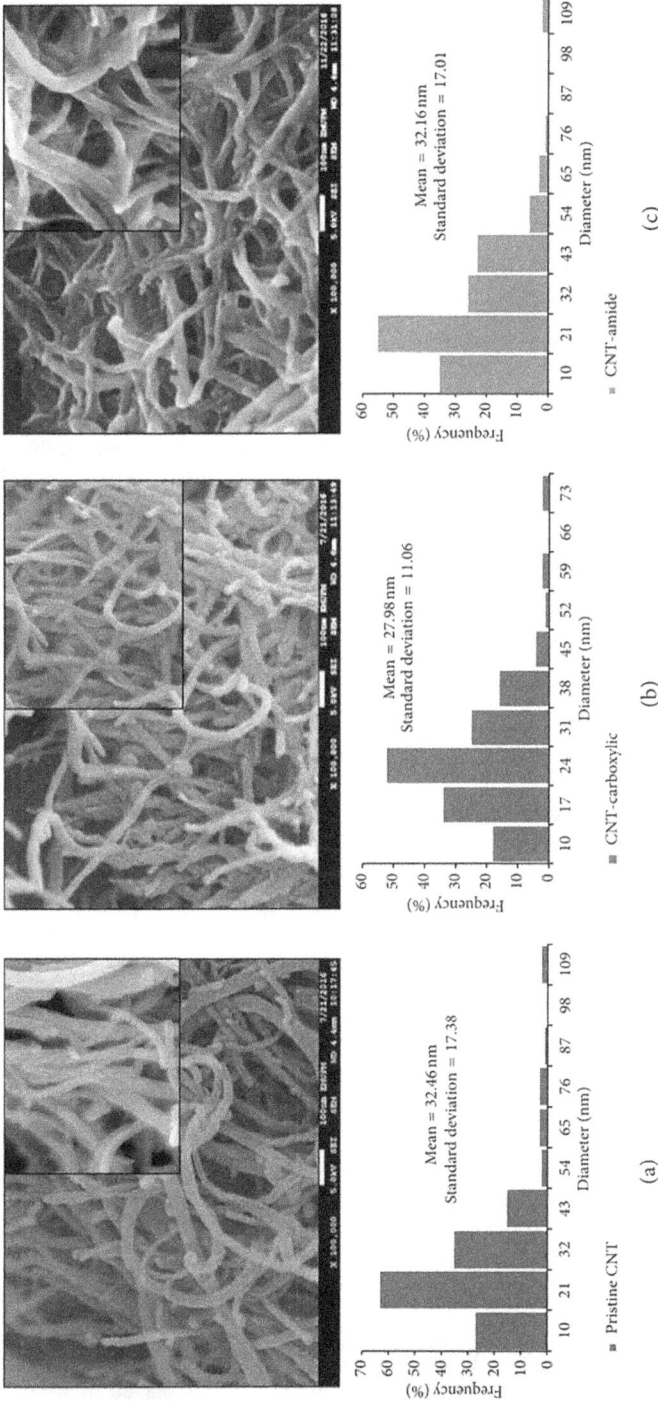

FIGURE 9.7 Micrographs of (a) pristine CNT, (b) CNT-carboxylic, and (c) CNT-amide (reprinted from the reference [58], Hindawi Copyright (2018)).

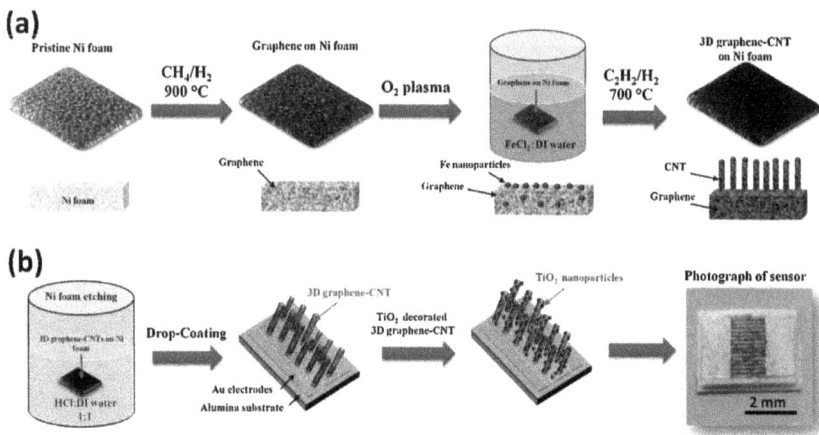

FIGURE 9.8 Schematic diagram of (a) 3D G-CNT nanostructure synthesis and (b) 3D TiO$_2$/G-CNT nanostructure gas sensor fabrication process (reprinted from the reference [72], with permission from Elsevier Copyright (2023)).

even at trace levels within the ppb range (specifically, 50 ppb). The enhanced sensing capability of the hybrid nanomaterials can be ascribed to the varying efficiency of electron transmission between distinct metal nanoclusters and MWCNTs.

The manufacture of poly(ethylene glycol)-grafted MWCNTs (PEG-g-MWCNTs) was explored by Sarafraz-Yazdi et al. [52] using the covalent functionalization approach. The objective of this synthesis was to facilitate the identification of exceedingly low levels (ultra trace) of BTEX in aqueous phase (water samples). The researchers noticed and documented improvements in the porous surface morphology, performance, selectivity, sensitivity, longevity, and thermal stability of the novel fiber. The combination of distinct properties of CNTs and the inherent benefits of the sol-gel coating technique introduced the desirable traits in PEG-g-MWCNTs required for routine ultra-trace analysis of BTEX in aqueous medium [52].

In addition, researchers have functionalized CNTs with oxygenated functional groups to develop low-cost gas sensors for the detection of toluene at trace level at room temperature [72, 73]. Seekaew et al. [72] produced a gas sensor with great sensitivity for detecting toluene at ambient temperature conditions. The sensor was constructed using a three-dimensional structure of titanium dioxide/graphene-carbon nanotubes (3D TiO$_2$/G-CNT), which was synthesized by CVD and sparking procedures. Figure 9.8 illustrates the methodologies employed by the researchers to fabricate three-dimensional graphene-carbon nanotube (G-CNT) nanostructures. The authors conducted a comparative analysis of the toluene detection capabilities of 3D TiO$_2$/G-CNT structures in relation to 3D G-CNT, TiO$_2$-CNT, pristine graphene, and pristine CNTs toward 500 ppm toluene vapor at room temperature. The researchers made the observation that the best three-dimensional (3D) TiO$_2$/G-CNT material had a significantly enhanced toluene response under the same conditions. Within a low concentration range of 50–500 ppm, the 3D G-CNT displayed improved sensitivity and selectivity toward the target compound compared to TiO$_2$-CNT, pristine

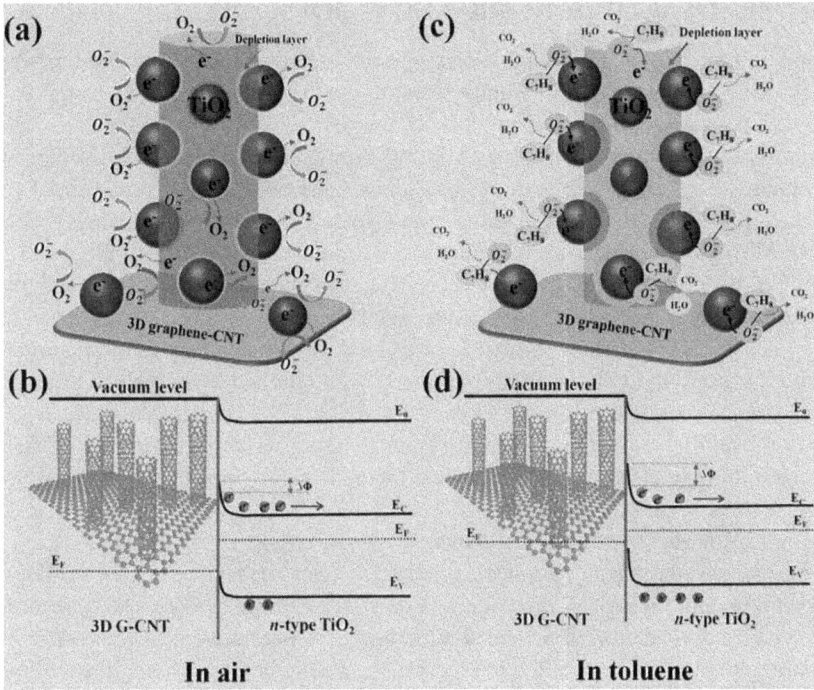

FIGURE 9.9 Proposed toluene-sensing mechanism of 3D TiO$_2$/G-CNT gas sensors in air (a, b) and in toluene (c, d) (reprinted with permission from the reference [72], Elsevier Copyright (2019)).

graphene, and pristine CNTs. The outstanding performance exhibited by the 3D TiO$_2$/G-CNT composite can be attributed to the creation of Schottky M-S junctions between the metallic 3D G-CNT nanostructures and the n-type TiO$_2$ nanoparticles by either low-temperature reduction reactions or direct charge transfer processes, as illustrated in Figure 9.9 [72].

Furthermore, Acosta and co-workers performed the initial functionalization of CNTs with the introduction of oxygenated functional groups by the utilization of oxygen plasma functionalization technique, which is recognized as a rapid, environmentally friendly, and very effective method. The technique described previousy showed a significant achievement of 22.1% atomic concentration of oxygen within a brief 5-minute time of functionalization. Subsequently, it was empirically shown that the oxygen functionalities that were grafted onto the basal plane of carbon nanotubes (CNTs) served as nucleation sites for the formation of palladium (Pd) and nickel-palladium (Ni-Pd) nanoparticles. In this study, the performance of Ox-CNTs decorated with Pd and Ni-Pd as active layers for toluene sensing was assessed. The experimental sample, consisting of bimetallic nanoparticles composed of 5-Ni-Pd, exhibited enhanced performance, indicating a potential synergistic effect arising from the inherent features of the nickel metal [73].

9.6   REMOVAL OF CATIONIC AND ANIONIC DYES USING CNTs

Rapid industrialization, urbanization, and the steady population surge generate wastewater discharge, which pollutes water and is considered a consequential environmental issue. Dyes are a major class of aromatic chemical compounds that contaminate surface and ground water, causing significant harm to human beings and aquatic life because of their toxicity. Nowadays, more than 100,000 commercial synthetic dyes (CSDs) are known, and the yearly production of commercial dyes exceeds $7*10^5$ MT [74, 75]. However, to enhance the aesthetic appearance of the products, dyes are widely exploited by numerous industries like textiles; for instance, dyestuffs, pharmaceutical drugs, cosmetics, petrochemicals, plastics, textiles, leather, and paints [76]. Furthermore, textile industries are one of the gigantic consumers of dyes, exceeding more than 10,000 tons per year, and indiscriminate discharge of these dye-laden effluents are considered as aesthetically unpleasant concern [77]. The removal of dyes from water bodies is challenging, as they have a complex structure, but it is essential to develop efficient methods for the elimination of these dyes from wastewater, as their presence in even a trace amount in water can cause skin irritation, allergic responses, and cancer [78]. Hazardous dyes have been removed from wastewater using physical, chemical, electrochemical, photo-catalysis, adsorption, and biological approaches. The adsorption process is widely regarded as a viable approach for water remediation because of several advantages, including its straightforward design and implementation, cost-effectiveness, and the ability to utilize versatile and reusable adsorbents and generate no harmful by-products. Dye detection advanced spectroscopic techniques, such as GC–MS and LC-MS, yield fast and accurate outcomes but are expensive, require complex instrumentation, and involve organic chemicals. In the present day, it is imperative for technology to possess attributes such as high speed, precision, and affordability. The current emphasis of researchers is mostly directed toward nanomaterials, owing to their distinctive structural characteristics and the ability to adjust their dimensions. Nanoparticles, including graphene nano-sheets, carbon nanotubes, and polymeric nanoparticles, have been utilized in the removal of diverse harmful substances, such as dyes and medicines, from wastewater. In this context, MWCNTs garner attention because of their hollow morphology, which enhances the surface area, improves the adsorption capacity, and can be modified easily. However, pristine MWCNTs have a limited adsorption capacity compared to functionalized MWCNTS [79].

9.6.1   DYES' ADSORPTION MECHANISM ONTO CNTs

The active sites and defects present on the surface of CNTs have a significant influence on the process of dye adsorption onto CNTs. The active sites of CNTs bundles include the exterior surface, interstitial and groove regions formed between the CNTs, and the interior pores of the tubes, as seen in Figure 9.10 [80]. The adsorption of dyes onto CNTs is frequently affected by three primary factors; specifically, the conditions of the solution, the attributes of the adsorbent, and the features of the adsorbates. In addition, other factors play a role in the adsorption process, such as

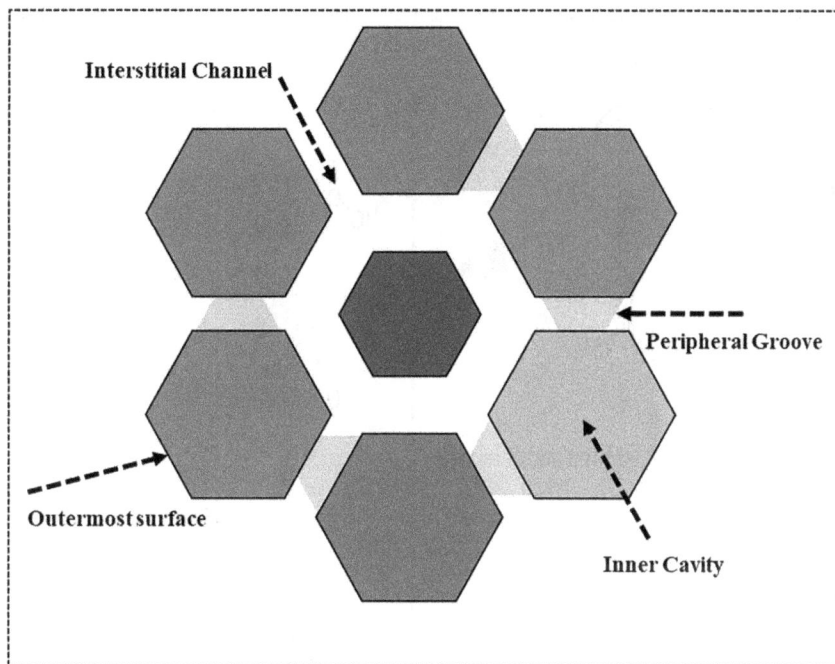

FIGURE 9.10 Schematic structure of CNTs bundle (modified from the reference [80]).

the size of the adsorbent and adsorbates, the charge properties of the adsorbent surface and adsorbates, as well as the pH and temperature of the solution, among other variables [81].

The primary processes by which CNTs absorb various dye compounds are dependent upon the nature of the dyes; namely, whether they are cationic or anionic. The adsorption prediction of dyes on CNTs is often complex, with several hypothesized interactions between the two entities. The primary mechanisms responsible for the adsorption of organic molecules on carbon nanoparticles are commonly attributed to hydrophobic interactions, π-π bonding, hydrogen bonding, as well as covalent and electrostatic interactions [82, 83]. Figure 9.11 depicts a schematic representation of the possible interactions between dyes (for example methylene blue) and CNTs. The hydrophobic nature of the side walls of CNTs can be attributed to the elevated electron-density of sp2 carbon atoms. Dyes have the potential to engage in hydrophobic interactions with the side walls of CNTs [84].

The phenomenon of π-π bonding frequently occurs when there is an interaction between the extended π system of CNTs and dye molecules that possess either C-C or benzene ring structures. The primary functional groups responsible for hydrogen-bond formation between dye molecules and CNTs are -COOH, -OH, and -NH$_2$. Furthermore, the adsorption of ionic substances is predominantly governed by electrostatic interactions due to the presence of charged surfaces of CNTs [85, 86].

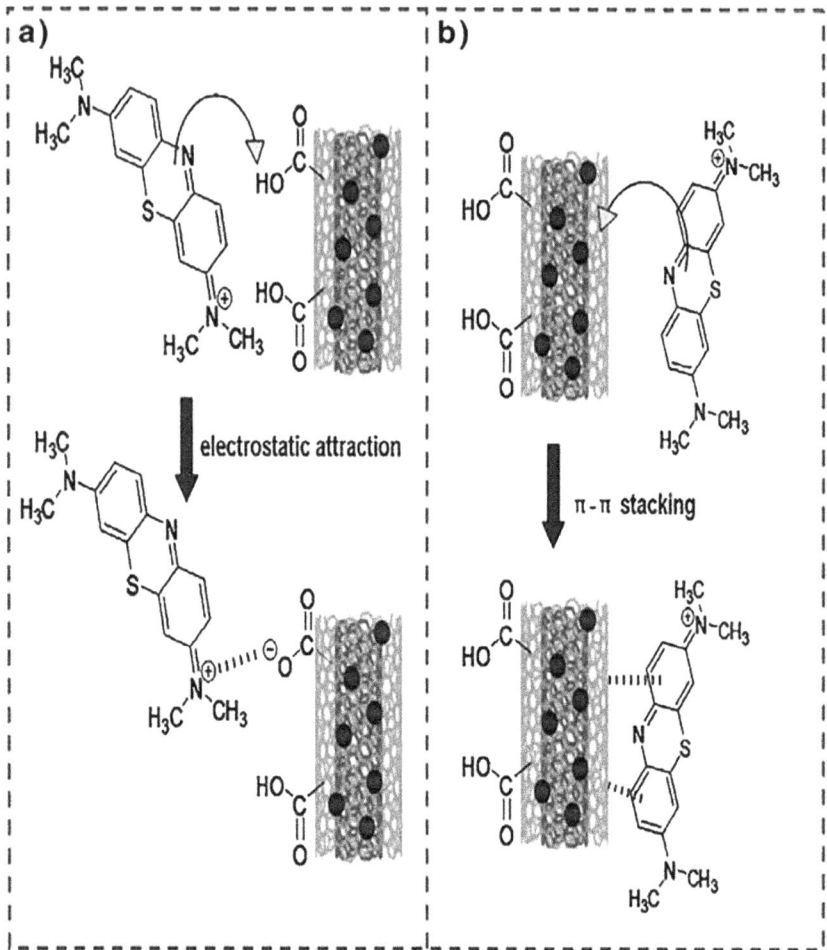

FIGURE 9.11 Possible interaction between MWCNTs and methylene blue dye: (a) electrostatic attraction and (b) π–π stacking [85] (reprinted with permission from the reference [85], Elsevier Copyright (2013)).

9.6.2 REMOVAL OF CATIONIC DYE (METHYLENE BLUE)

Methylene Blue (MB) is a synthetic dye and has wide applications in the dyeing industries, food, chemical, plastic, leather, textile, cosmetic industries, biology, as well as in medical science. It is also named as Basic Blue 9, cationic in nature, and has the chemical formula $C_{16}H_{18}N_3SCl$, with a molecular weight of 319.85 g/mol [87]. Hypertension, skin allergies, vomiting, nausea, and generation of mental confusion ensue with acute MB exposure [88]. Its high water-solubility makes it difficult to remove using typical water treatment techniques. Since MB dye is a non-biodegradable and long-lasting cationic dye, its aromatic structure is a further challenge, the researchers struggle to remove it, and it has been selected as a model cationic dye for removal from the aqueous solution in this study. There are numerous techniques developed for dye removal such as adsorption, ion exchange, oxidation, electrochemical and photocatalytic degradation methods [89, 90]. However, there are still significant issues with some of these technologies. The oxidation process has the potential to generate hazardous by-products, even from biodegradable dyes in wastewater. While the electrochemical method has been subject to limited investigation, mostly because of its elevated operational cost, the adsorption technique exhibited high efficiency and showed complete removal of the pollutant with simplicity, high efficiency, and low energy consumption, while avoiding the generation of any by-products.

9.6.2.1   Magnetic Nanomaterials for the Removal of Methylene Blue

Nanomaterials, such as CNTs mainly MWCNTs, have undergone thorough investigation and demonstrated considerable potential as adsorbents due to their distinctive attributes, which encompass a high surface-area, chemical stability, an elongated π-electron structure, and a distinctive network, including sp2 carbon atoms. However, the potential enhancement in selectivity and adsorptive ability of CNTs toward polar chemicals, such as MB, can be achieved by functionalization of the CNTs by introducing additional functionalities onto the surface of their walls. Furthermore, the enhancement of water solubility in CNTs necessitates the implementation of surface changes. In comparison to SWCNTs, MWCNTs are frequently employed in a wider range of environmental remediation applications. Magnetic nanoparticles that include MWCNTs are typically synthesized by the chemical deposition of γ-Fe_2O_3 and magnetite (Fe_3O_4) onto MWCNTs that have been covalently modified. Numerous investigations are dedicated to investigating the use of functionalized FWCNTs in the development of novel nanocomposite adsorbents. The integration of CNT composites with magnetic nanoparticles leads to the emergence of a distinctive category of adsorbents. Magnetite (Fe_3O_4) and maghemite (γ-Fe_2O_3) are widely recognized as prominent examples of magnetic nanoparticles. Furthermore, these materials exhibit significant magnetic responsiveness and mechanical robustness, hence facilitating their efficient separation and augmenting their durability for chemical regeneration following treatment. The use of a magnetically recoverable and reusable adsorbent for the solid-phase extraction of dyes is of great significance in terms of sustainability and environmental considerations. Several studies have been conducted to present an efficient method for synthesizing magnetic iron oxide functionalized nanoparticles with high scalability [91].

Boukhalfa et al. [90] effectively synthesized magnetic nanocomposite beads using the process of encapsulating citrate-coated maghemite nanoparticles and acid-functionalized MWCNTs into calcium alginate beads. A schematic representation of A-F-Fe_2O_3 beads preparation is illustrated in Figure 9.12. The synthesized beads were designated as magnetic alginate beads (A-Fe_2O_3) and magnetic alginate functionalized MWCNTs beads (A-F-Fe_2O_3), respectively. In comparison to A-Fe_2O_3 beads, A-F-Fe_2O_3 beads exhibited superior adsorption capabilities for the removal of MB, demonstrating over a large domain of pH (4–10), with a maximum monolayer adsorption capacity of 905.5 mg/g.

The adsorption process exhibited a dependence on both the ionic strength and the starting dye concentration. These magnetic nanocomposite beads display a strong

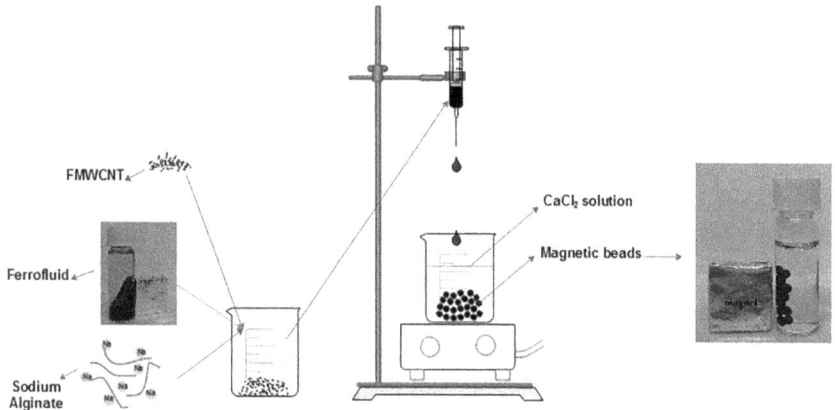

FIGURE 9.12 Schematic representation of magnetic beads preparation technique [90] (reprinted with permission from the reference [90], Elsevier Copyright (2019)).

affinity and reusability for MB dye while also allowing for convenient separation following treatment, owing to their super magnetic properties.

In another study, Bhakta and colleagues [88] have developed a unique and effective method for synthesizing maghemite nanocrystals that are attached to MWCNTs. They reported that the method adapted was simple and worked well because it solved the problems that have come up over the last 10 years when trying to decorate CNTs using maghemite nanoparticles. The synthesis process involved the utilization of infrared (IR) irradiation and diazonium chemistry. The resulting nanocomposites exhibited enhanced magnetic and electrical properties. The synergistic impact arising from the combination of the catalytic and magnetic capabilities of MWCNTs with maghemite nanocrystals results in the development of novel nanocomposites exhibiting enhanced features. The efficacy of p-MWCNTs/MC as catalysts for the removal of MB has been demonstrated to be superior to that of pure MWCNTs. The aforementioned material exhibits compatibility with environmental and many nano technological applications [88]. Magnetic MWCNTs nanocomposites (MMWCNTs) were synthesized using commercial MWCNTs and iron oxide nanoparticles for the detection of multiple cationic dyes, including MB [92]. Compared to the present adsorption techniques, the magnetic nanocomposites (MMWCNTs) displayed a significant advantage in terms of separation convenience.

9.6.2.2 Functionalized Carbon Nanotubes for the Removal of Methylene Blue

A distinctive approach was utilized by Yadav et al [93] to fabricate a novel adsorbent; namely, CNTs with eucalyptus-derived activated carbon (CNT-EBAC). They investigated the effectiveness of a particular adsorbent material in the removal of MB from wastewater. The effectiveness of the adsorbent in removing MB from wastewater has been evaluated by considering the influence of numerous parameters, including the pH, adsorbent dosage, contact duration, and initial concentration of the adsorbates. The adsorbent demonstrates a peak adsorption efficiency of 49.61 mg/g for the

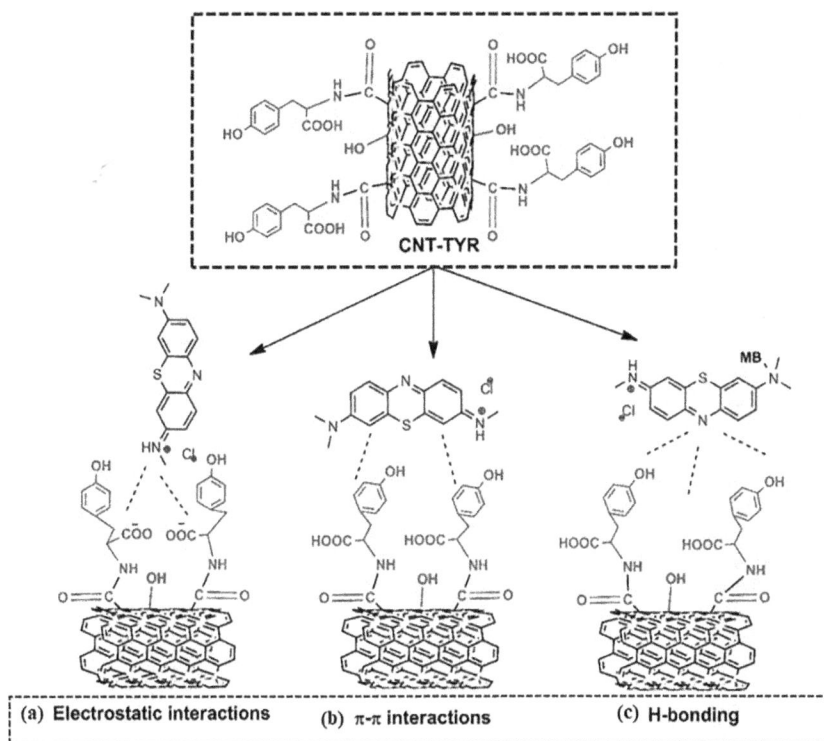

FIGURE 9.13 Schematic of possible interaction mechanisms of CNT-TYR with MB dye in wastewater: (a) electrostatic interactions, (b) π–π interactions, and (c) H-bonding [79] (reprinted with permission from the reference [79], Elsevier Copyright (2020)).

elimination of MB. The utilization of CNTs-based electrically conductive adsorbents (CNT-EBAC) exhibits significant promise in the realm of removal of colorants from wastewater and water purification endeavors [93].

Saxena et al. [79] presented the utilization of tyrosine-functionalized MWCNTs as proficient nanomaterials for the efficient removal of MB dye from aqueous solutions. The researchers performed an optimization study on the key MB adsorption parameters, which encompassed the influence of stirring speed (ranging from 200 to 800 rpm), contact duration (ranging from 0 to 150 minutes), the quantity of nanoadsorbents (ranging from 5 to 25 mg), pH (ranging from 2 to 10), temperature (25 to 65°C), and initial dye concentration (10 to 300 mg/L). The maximum adsorption capacity attained under ideal conditions was 440 mg/g, exceeding the capabilities of other nano-adsorbents documented in the literature [79]. The significant adsorption capacity of CNT-TYR may be attributed to several types of interactions, encompassing electrostatic forces, π–π interactions, and hydrogen bonding. These interactions, illustrated in Figure 9.13, provide insight into the adsorption process involving MB. The results indicate that CNT-TYR has significant effectiveness in the removal of various colors and dangerous contaminants present in water.

The successful functionalization of CNTs with amine groups, also known as compound 1, was accomplished with high efficiency by Saleh et al. [94]. Following this, two prophyrine derivatives with poor symmetry were chemically linked to the functionalized nanotubes, leading to the creation of compound 2 and compound 3. The objective of this conjugation method was to augment the active sites and surface area of the nanotubes. The researchers conducted a series of batch adsorption experiments to assess the suitability of the synthesized materials for removing MB from aqueous solutions. In a wide range of pH values, it was revealed that Compound 3 exhibited superior adsorption effectiveness in comparison to Compound 1 and Compound 2, respectively. Compound 3 demonstrated a significant improvement in adsorption efficiency for MB when the dose of adsorbent was raised from 0.001 to 0.06 g. The research findings indicate that the process of MB adsorption is a complex mechanism that encompasses both non-electrostatic and electrostatic interactions. The results provided in the study indicate that the produced material has shown efficacy as an adsorbent for the adsorption of MB from wastewater discharged from industries.

9.6.3 REMOVAL OF ANIONIC DYE (CONGO RED)

Congo red (CR, sodium salt of benzidinediazobis-1-naphthyl-amine-4-sulfonic acid) is classified as a benzidine anionic azo dye, characterized by its chemical formula $C_{32}H_{22}N_6Na_2O_6S_2$. The aromatic group of CR includes a central biphenyl group with two azo (R–N=N–R') groups and is water-soluble in nature [75, 95]. The utilization of CR in modern dyeing techniques is prevalent due to its cost-effectiveness, remarkable solubility, and distinctive coloring-characterized properties. However, CR has a comparatively elevated degree of toxicity in relation to the majority of other dyes and pigments. The higher toxicity of this aromatic compound compared to other dyes can be attributed to the significant solubility of the dye that negatively impacts the clarity of the water and the limited biodegradability inside wastewater systems. The CR dye possesses the ability to degrade carcinogenic, aromatic amines through the reduction mechanism. Upon being activated, these carcinogenic, aromatic amines possess the ability to alter the structure of human DNA and trigger the progression of fatal cancer [96]. The choice of CR as the representative anionic dye in this study was made due to its complex chemical structure, considerable solubility in water-based solutions, and its capacity to persist in the environment after being released.

In order to mitigate the issue of CR contamination in aquatic medium, a range of methodologies has been utilized, including activated carbon adsorption, CNT-based adsorption, chemical catalysis, biological degradation, and photo-catalytic oxidation, among others [80, 97]. The lack of selectivity, slow rate of deterioration, and high energy consumption are only a few of the shortcomings associated with these techniques. Therefore, the selective removal of CR from wastewater remains a substantial obstacle. Among the several options considered for the removal of organic and inorganic dyes from aqueous solutions, the utilization of adsorption with an environmentally acceptable and cost-effective adsorbent is widely acknowledged as a viable approach. Therefore, the selection of readily obtainable, affordable, environmentally sustainable, and efficient materials that are widely accessible, cost-effective, ecologically friendly, and efficient. Several researchers have studied the influence of several factors influencing the CR adsorption process in the aqueous medium using CNTs [80, 98].

In a study conducted by Kamil et al. [80], the researchers examined the effectiveness of MWCNTs as a treatment method for the removal of CR dye from aqueous solutions, specifically, in the context of wastewater treatment. A comprehensive investigation was conducted to analyze the impact of several variables (mass of the adsorbent, initial concentration of the solution, ionic strength, pH, and temperature) on the CR adsorption process. The researchers observed that the adsorption capacity of the CR dye on MWCNTs exhibited an upward trend when the starting concentration of CR rose. The elimination percentage of CR exhibited an increase in correlation with the rise in salinity, which can be attributed to the alteration of electrostatic forces. Additionally, the experimental findings indicate that the adsorption process exhibited spontaneity and endothermic characteristics.

Similarly, in another study conducted by Sheibani et al. [98], they evaluated the efficacy of oxidized and nitric acid-functionalized MWCNTs in the removal of CR from an aqueous solution. Furthermore, they studied that the efficacy of the CR adsorption process on the surface of functionalized MWCNTs. They reported that HNO_3-treated MWCNTs are a promoting and environmentally friendly approach, as evidenced by their significant adsorption capacity of 357.14 mg/g, achieved under optimal circumstances, including all pertinent variables.

9.7 DETECTION OF HEAVY METAL IONS THROUGH CNTs-ASSISTED SENSORS

The term "heavy metal" is utilized to categorize metallic elements and metalloids that possess atomic densities above 5 g/cm^3. Heavy metals have a significant role in polluting the biosphere as a consequence of their intrinsic toxicity and persistent characteristics [99]. Trace amounts of heavy metals like lead (Pb), arsenic (As), zinc (Zn), mercury (Hg), cadmium (Cd), and chromium (Cr) are heavy metals that possess high toxicity levels, owing to their non-degradable and non-destructible nature [100]. Owing to the rise in anthropogenic activities carried out by humans, including mining, industrial operations, agriculture, and metallurgical processes, there is a buildup of heavy transition metals (HTMs) in the surrounding environment. HTMs can also be released into water bodies via rain, soil, weathering, and rock erosion, accelerating hazardous metal discharge [101]. The hazardous and toxic substances emitted from these anthropogenic and natural sources tend to build in the atmosphere, contaminate drinking water, and deposit on soil. Indeed, heavy metal ions can bio-accumulate in both plants and animals subjected to environmental pollution, particularly in the aquatic environment [102]. HTMs ions possess a prolonged half-life, potential accumulation in different parts of the body, and resistance to biodegradation. Consequently, these entities have gained much attention because of serious environmental issues and numerous perilous health hazards. The impact of HTMs' toxicity is more pronounced in children than in adults [103]. For instance,

TABLE 9.2

The Table Highlights the List of Heavy Transition Metals, Their Permissible Exposure Limit, Primary Sources of Heavy Transition Metal Emission, and Impacts on Human Beings upon Exposure [2, 100, 103]

Heavy Metals	Permissible Limit Exposure[a] (mg/L)	Primary Sources	Effects
Zinc (Zn)	3.0	Electroplating, smelting, acid mine drainage, effluents from chemical processes, discharge of untreated domestic sewage, cosmetics, and pigments	Depression, stomach cramps, neuronal disorder, and prostate cancer risks
Copper (Cu)	2.0	Fertilizers, tanning and sewage effluent	Liver damage, insomnia, allergies, anemia, and arthritis
Silver (Ag)	0.1	Refining metals like copper, zinc, gold, nickel, and jewelry	Gastroenteritis, mental fatigue, neuronal disorders, and rheumatism
Chromium (Cr)	0.05	Leather industry, textile industry, and electroplating industries	Diarrhea, skin and systemic effects involving kidney, liver, gastrointestinal tract, and circulatory system
Arsenic (As)	0.01	Coloring agent in textile, toy-making industries, pesticides, and fertilizer	Causes skin damage, cancer, neuro behavior sickness, effects on central (CNS) and peripheral nervous system (PNS), cardiovascular and pulmonary diseases, and pigmentation
Lead (Pb)	0.01	Gasoline, plastic, paint, PVC pipes in sanitation, gasoline, lead acid batteries, etc.	Effect circulatory and nervous system, impaired growth in children, and induce learning disabilities
Cadmium (Cd)	0.003	Electroplating industry, battery, plastics, synthetic rubber	Renal toxicity, hypertension, kidney damage, and cancer
Mercury (Hg)	0.006	Thermometer, combustion of coal, and municipal solid-waste incineration	Rheumatoid arthritis, effects circulatory, and nervous system

[a] *Permissible limit exposure (PEL) as prescribed by the World Health Organization (WHO), 2017*

an increased dosage of lead in children has the potential of inducing organ damage and facilitating neurotoxicity. However, it should be noted that certain heavy metals, such as chromium (hexavalent Cr) and cadmium (Cd), possess carcinogenic properties, as referred to in Table 9.2. The prescribed maximum concentrations of the metal ions, primary sources of emission, and associated health effects of common heavy metals are provided in Table 9.2. As observed from the Table 9.2, the permissible levels of copper (Cu), zinc (Zn), chromium (Cr), and mercury (Hg) are in the range of 0.001–5 mg/L, as announced by WHO. Furthermore, Table 9.2 illustrates that the primary source of release is mainly from industrial operations or from establishments with inadequate wastewater monitoring and treatment infrastructure. Consequently, the removal of toxic metals from contaminated water and the issue of HTM pollution is imperative, as it poses a potential threat to human health, plants, and aquatic ecosystems [104].

The present techniques employed for the identification of heavy metals comprise spectroscopic techniques, which exhibit a high degree of sensitivity and selectivity [105]. Nevertheless, the use of these techniques necessitates sophisticated equipment and proficient employees and incurs substantial costs for operation and maintenance. Hence, it is imperative to devise a straightforward, highly responsive, specific, and easily transportable detection system for the detection of heavy metals. For the detection, monitoring, and quantification of these non-degradable, heavy toxicants in the aquatic environment, various modern sensing techniques such as chemical sensors and electrochemical sensors are utilized. In view of their excellent physicochemical properties (enhanced conductivity, stability, large surface area, small dimensions, and simple functionalization), carbon nanomaterials are employed in chemical and electrochemical sensors to increase the sensing performance of HTM in contaminated water [19].

Morton et al. [99] utilized an innovative and a sensitive voltammetric technique to detect trace quantities of heavy metal ions (specifically Pb^{2+} and Cu^{2+}) by employing chemically modified CNTs as electrode surfaces. The surface of CNTs were covalently functionalized using cysteine, a class of amino acids. The sensor based on modified MWCNTs has the capability to detect trace levels of copper ions (Cu^{2+}) and lead ions (Pb^{2+}) in the aquatic environment, with detection limits of 15 ppb and 1 ppb, respectively. Barati et al. [106] developed a novel MWCNT-based-paste electrochemical sensor containing the synthesized N-(1-(4-bromophenyl)-3-oxo-3-phenylpropyl) acetamide. The authors observed that the modified carbon-paste electrode has great chemical stability, selectivity, and an enhanced detection limit for chromium (III) ions. The enhanced selectivity and detection ability were attributed to the ability of the ionophore to form a stable chelate with Cr(III) ions. The CNT-encapsulated composite electrodes have proved to be excellent materials for the simultaneous detection of multiple heavy metal ions in complex aquatic environment. Bao et al. [107] proposed a novel composite electrode based on a polyaniline (PANi) framework doped with bismuth nanoparticle@graphene oxide MWCNT (Bi NPs@GO-MWCNTs) for the detection of mercury ion (Hg(II)) and copper ion (Cu(II)). The electrochemical mechanisms involved in demonstrating the sensing capabilities of the proposed composite electrodes were cyclic voltammetry and differential pulse voltammetry. The identification of Hg(II) and Cu(II) ions was demonstrated using modified PANi-Bi NPs@GO-MWCNT. Wu et al. [108] devised a novel composite material – namely, chitosan-CNTs (Chit-CNT) – for the purpose of adsorbing copper, cadmium, and lead ions. The Chit-CNT composite exhibits a wide range of functional groups, which might potentially lead to various adsorption processes when interacting with heavy metal ions [108].

9.8 CONCLUSION

The present study has examined the emerging technologies pertaining to nano-materials based on carbon nanotubes (CNTs), with a particular emphasis on their application in environmental remediation. In summary, this study has presented a comprehensive examination of the significant research undertaken on chemical sensors based on CNTs. The chapter has, furthermore, provided a comprehensive overview of the environmental remediation associated with the utilization of these CNTs-based sensors for the purpose of monitoring a diverse array of toxic contam-inants, both in gaseous and aqueous media. The contaminants consist of hazard-ous gas molecules as well as vapors and liquid pollutants. Additionally, they include heavy transition metal ions (HTMs) that are released into the environment, primarily as a result of human activities, particularly within industrial sectors. Based on the review of relevant literature in the preceding sections and sub-sections, it is evident that there is a pressing requirement to establish a robust platform that can accurately and efficiently identify and quantify the risks associated with the discussed contam-inants. The findings indicate that, among nanomaterials, mainly CNTs exhibit their effectiveness in reducing several environmental pollutants in laboratory settings, owing to their unique physical and chemical properties. However, further research is required in order to modify and enhance this technology for more extensive environ-mental applications under ambient conditions. Significant emphasis is made on the advancement of innovative nanomaterials that possess remarkable capacity, selec-tivity, sensitivity, recovery time, stability, reproducibility, and efficacy in industrial columns.

REFERENCES

[1] K. Panchamoorthy, G. Dai, V. N. Vo, and D. Gnana, "Environmental applications of car-bon-based materials: a review," *Environ. Chem. Lett.*, vol. 19, no. 1, pp. 557–582, 2021, doi: 10.1007/s10311-020-01084-9.

[2] A. Tuan, and S. Ni, "Chemosphere Heavy metal removal by biomass-derived car-bon nanotubes as a greener environmental remediation: a comprehensive review," *Chemosphere*, vol. 287, no. June 2021, 2022.

[3] C. Wei, T. Kok, H. Tan, and Y. Thai, "Energy and environmental applications of carbon nanotubes," *Environ. Chem. Lett.*, pp. 265–273, 2012, doi: 10.1007/s10311-012-0356-4.

[4] D. Miyashiro, R. Hamano, and K. Umemura, "A review of applications using mixed materials of cellulose, nanocellulose and carbon nanotubes," *Nanomaterials*, vol. 10, 2020.

[5] M. Meyyappan, "Carbon nanotube-based chemical sensors," *Small*, vol. 12, no. 16, pp. 2118–2129, 2016, doi: 10.1002/smll.201502555.

[6] S. Rathinavel, K. Priyadharshini, and D. Panda, "A review on carbon nanotube: an overview of synthesis, properties, functionalization, characterization, and the applica-tion," *Mater. Sci. Eng. B Solid-State Mater. Adv. Technol.*, vol. 268, no. December 2020, p. 115095, 2021, doi: 10.1016/j.mseb.2021.115095.

[7] R. Shoukat and M. I. Khan, "Carbon nanotubes: a review on properties, synthesis meth-ods and applications in micro and nanotechnology," *Microsyst. Technol.*, vol. 27, no. 12, pp. 4183–4192, 2021, doi: 10.1007/s00542-021-05211-6.

[8] M. N. Norizan, H. Moklis, and Z. Ngah, "Carbon nanotubes: functionalisation and their," *RSC Advances*, pp. 43704–43732, 2020, doi: 10.1039/d0ra09438b.

[9] J. Peng, Y. He, C. Zhou, S. Su, and B. Lai, "The carbon nanotubes-based materials and their applications for organic pollutant removal: a critical review," *Chinese Chem. Lett.*, vol. 32, no. 5, pp. 1626–1636, 2021, doi: 10.1016/j.cclet.2020.10.026.

[10] S. V. Sawant, A. W. Patwardhan, J. B. Joshi, and K. Dasgupta, "Boron doped carbon nanotubes: synthesis, characterization and emerging applications – a review," *J. Chem. Eng.*, vol. 427, no. July 2021, 2022.

[11] M. S. Mauter, and M. Elimelech, "Environmental applications of carbon-based nanomaterials: critical review," *Am. Chem. Soc.*, vol. 42, no. 16, pp. 5843–5859, 2008.

[12] T. Rasheed, F. Nabeel, M. Adeel, K. Rizwan, M. Bilal, and H. M. N. Iqbal, "Carbon nanotubes-based cues: A pathway to future sensing and detection of hazardous pollutants," *J. Mol. Liq.*, vol. 292, 2019, doi: 10.1016/j.molliq.2019.111425.

[13] O. Zaytseva, and G. Neumann, "Carbon nanomaterials: production, impact on plant development, agricultural and environmental applications," *Chem. Biol. Technol. Agric.*, pp. 1–26, 2016, doi: 10.1186/s40538-016-0070-8.

[14] R. B. Onyancha *et al.*, "A systematic review on the detection and monitoring of toxic gases using carbon nanotube-based biosensors," *Sens. Bio-Sensing Res.*, vol. 34, p. 100463, 2021, doi: 10.1016/j.sbsr.2021.100463.

[15] N. Anzar, R. Hasan, M. Tyagi, N. Yadav, and J. Narang, "Carbon nanotube – a review on synthesis, properties and plethora of applications in the field of biomedical science," *Sensors Int.*, vol. 1, no. December 2019, p. 100003, 2020, doi: 10.1016/j.sintl.2020.100003.

[16] J. Prasek *et al.*, "Methods for carbon nanotubes synthesis – review," *J. Mater. Chem.*, vol. 21, no. 40, pp. 15872–15884, 2011, doi: 10.1039/c1jm12254a.

[17] D. Liu *et al.*, "Recent advances in MOF-derived carbon-based nanomaterials for environmental applications in adsorption and catalytic degradation," *Chem. Eng. J.*, vol. 427, no. May 2021, p. 131503, 2022, doi: 10.1016/j.cej.2021.131503.

[18] T. Rasheed, M. Adeel, F. Nabeel, M. Bilal, and H. M. N. Iqbal, "TiO2/SiO2 decorated carbon nanostructured materials as a multifunctional platform for emerging pollutants removal," *Sci. Total Environ.*, vol. 688, pp. 299–311, 2019, doi: 10.1016/j.scitotenv.2019.06.200.

[19] G. Drera *et al.*, "Exploring the performance of a functionalized CNT-based sensor array for breathomics through clustering and classification algorithms: from gas sensing of selective biomarkers to discrimination of chronic obstructive pulmonary disease," *RSC Adv.*, vol. 11, no. 48, pp. 30270–30282, 2021, doi: 10.1039/d1ra03337a.

[20] J. A. Robinson, E. S. Snow, Ş. C. Bădescu, T. L. Reinecke, and F. K. Perkins, "Role of defects in single-walled carbon nanotube chemical sensors," *Nano Lett.*, vol. 6, no. 8, pp. 1747–1751, 2006, doi: 10.1021/nl0612289.

[21] A. Goldoni, L. Petaccia, S. Lizzit, and R. Larciprete, "Sensing gases with carbon nanotubes: a review of the actual situation," *J. Phys. Condens. Matter*, vol. 22, no. 1, 2010, doi: 10.1088/0953-8984/22/1/013001.

[22] M. Penza, R. Rossi, M. Alvisi, M. A. Signore, and E. Serra, "Effects of reducing interferers in a binary gas mixture on NO2 gas adsorption using carbon nanotube networked films based chemiresistors," *J. Phys. D. Appl. Phys.*, vol. 42, no. 7, 2009, doi: 10.1088/0022-3727/42/7/072002.

[23] M. Penza, R. Rossi, M. Alvisi, and E. Serra, "Metal-modified and vertically aligned carbon nanotube sensors array for landfill gas monitoring applications," *Nanotechnology*, vol. 21, no. 10, 2010, doi: 10.1088/0957-4484/21/10/105501.

[24] Y. H. Chan *et al.*, "A state-of-the-art review on capture and separation of hazardous hydrogen sulfide (H2S): recent advances, challenges and outlook," *Environ. Pollut.*, vol. 314, no. August, p. 120219, 2022, doi: 10.1016/j.envpol.2022.120219.

[25] K. Jangam, Y. Y. Chen, L. Qin, and L. S. Fan, "Perspectives on reactive separation and removal of hydrogen sulfide," *Chem. Eng. Sci. X*, vol. 11, 2021, doi: 10.1016/j.cesx.2021.100105.

[26] Y. Liu, Y. Chen, C. Li, and X. Yang, "Copper-multiwalled carbon nanotubes decorated fiber-optic surface plasmon resonance sensor for detection of trace hydrogen sulfide gas," *Opt. Fiber Technol.*, vol. 76, no. November 2022, p. 103221, 2023, doi: 10.1016/j. yofte.2022.103221.

[27] N. Musayeva, H. Khalilova, B. Izzatov, G. Trevisi, S. Ahmadova, and M. Alizada, "Highly selective detection of hydrogen sulfide by simple Cu-CNTs nanocomposites," *C*, vol. 9, no. 1, p. 25, 2023, doi: 10.3390/c9010025.

[28] H. Wu *et al.*, "Stably dispersed carbon nanotubes covalently bonded to phthalocyanine cobalt(II) for ppb-level H2S sensing at room temperature," *J. Mater. Chem. A*, vol. 4, no. 3, pp. 1096–1104, 2016, doi: 10.1039/c5ta09213b.

[29] R. Jonuarti, M. Yusfi, and Suprijadi, "Energetics and stability of hydrogen sulphide adsorption on defective carbon nanotube," *Int. J. Comput. Mater. Sci. Surf. Eng.*, vol. 10, no. 1, pp. 62–72, 2021, doi: 10.1504/IJCMSSE.2021.116614.

[30] X. Zhang, B. Yang, X. Wang, and C. Luo, "Effect of plasma treatment on multi-walled carbon nanotubes for the detection of H2S and SO2," *Sensors (Switzerland)*, vol. 12, no. 7, pp. 9375–9385, 2012, doi: 10.3390/s120709375.

[31] V. Adavan Kiliyankil, B. Fugetsu, I. Sakata, Z. Wang, and M. Endo, "Aerogels from copper (II)-cellulose nanofibers and carbon nanotubes as absorbents for the elimination of toxic gases from air," *J. Colloid Interface Sci.*, vol. 582, pp. 950–960, 2021, doi: 10.1016/j.jcis.2020.08.100.

[32] X. Liu, B. Wang, X. Lv, Q. Meng, and M. Li, "Enhanced removal of hydrogen sulfide using novel nanofluid system composed of deep eutectic solvent and Cu nanoparticles," *J. Hazard. Mater.*, vol. 405, no. October 2020, p. 124271, 2021, doi: 10.1016/j. jhazmat.2020.124271.

[33] S. Mubeen *et al.*, "Sensitive detection of H2S using gold nanoparticle decorated single-walled carbon nanotubes," *Anal. Chem.*, vol. 82, no. 1, pp. 250–257, 2010, doi: 10.1021/ac901871d.

[34] D. W. H. Fam, A. I. Y. Tok, A. Palaniappan, P. Nopphawan, A. Lohani, and S. G. Mhaisalkar, "Selective sensing of hydrogen sulphide using silver nanoparticle decorated carbon nanotubes," *Sensors Actuators, B Chem.*, vol. 138, no. 1, pp. 189–192, 2009, doi: 10.1016/j.snb.2009.01.008.

[35] F. Mendoza, D. M. Hernández, V. Makarov, E. Febus, B. R. Weiner, and G. Morell, "Room temperature gas sensor based on tin dioxide-carbon nanotubes composite films," *Sensors Actuators, B Chem.*, vol. 190, pp. 227–233, 2014, doi: 10.1016/j.snb.2013.08.050.

[36] H. Dai, P. Xiao, and Q. Lou, "Application of SnO 2/MWCNTs nanocomposite for SF 6 decomposition gas sensor," *Phys. Status Solidi Appl. Mater. Sci.*, vol. 208, no. 7, pp. 1714–1717, 2011, doi: 10.1002/pssa.201026562.

[37] B. Timmer, W. Olthuis, and A. Van Den Berg, "Ammonia sensors and their applications – a review," *Sensors Actuators, B Chem.*, vol. 107, no. 2, pp. 666–677, 2005, doi: 10.1016/j.snb.2004.11.054.

[38] O. V. Sedelnikova *et al.*, "Role of interface interactions in the sensitivity of sulfur-modified single-walled carbon nanotubes for nitrogen dioxide gas sensing," *Carbon N. Y.*, vol. 186, no. 2, pp. 539–549, 2022, doi: 10.1016/j.carbon.2021.10.056.

[39] A. You, M. Be, and I. In, "Enhancement of sensitivity in gas chemiresistors based on carbon nanotube surface functionalized with noble metal (Au, Pt) nanoclusters □," *Appl. Phys. Lett.*, vol. 19, no. February 2012, 2016, doi: 10.1063/1.2722207.

[40] V. Schroeder, S. Savagatrup, M. He, S. Lin, and T. M. Swager, "Carbon nanotube chemical sensors," *Chem. Rev.*, vol. 119, no. 1, pp. 599–663, 2019, doi: 10.1021/acs. chemrev.8b00340.

[41] S. N. Behera, M. Sharma, V. P. Aneja, and R. Balasubramanian, "Ammonia in the atmosphere: a review on emission sources, atmospheric chemistry and deposition on terrestrial bodies," *Environ. Sci. Pollut. Res.*, vol. 20, no. 11, pp. 8092–8131, 2013, doi: 10.1007/ s11356-013-2051-9.

[42] S. Ammu *et al.*, "Flexible, All-Organic Chemiresistor for Detecting Chemically Aggressive Vapors," *J. Am. Chem. Soc.*, vol. 134, 2012.

[43] F. Valentini, V. Biagiotti, C. Lete, G. Palleschi, and J. Wang, "The electrochemical detection of ammonia in drinking water based on multi-walled carbon nanotube/copper nanoparticle composite paste electrodes," *Sensors Actuators, B Chem.*, vol. 128, no. 1, pp. 326–333, 2007, doi: 10.1016/j.snb.2007.06.010.

[44] F. Rigoni *et al.*, "Enhancing the sensitivity of chemiresistor gas sensors based on pristine carbon nanotubes to detect low-ppb ammonia concentrations in the environment," *Analyst*, vol. 138, no. 24, pp. 7392–7399, 2013, doi: 10.1039/c3an01209c.

[45] S. Cui *et al.*, "Fast and selective room-temperature ammonia sensors using silver nanocrystal-functionalized carbon nanotubes," *ACS Appl. Mater. Interfaces*, vol. 4, no. 9, pp. 4898–4904, 2012, doi: 10.1021/am301229w.

[46] L. Q. Nguyen, P. Q. Phan, H. N. Duong, C. D. Nguyen, and L. H. Nguyen, "Enhancement of NH3 gas sensitivity at room temperature by carbon nanotube-based sensor coated with Co nanoparticles," *Sensors (Switzerland)*, vol. 13, no. 2, pp. 1754–1762, 2013, doi: 10.3390/s130201754.

[47] S. K. Hazra, S. Roy, and S. Basu, "Growth of titanium dioxide thin films via a metallurgical route and characterizations for chemical gas sensors," *Mater. Sci. Eng. B*, vol. 110, no. 2, pp. 195–201, 2004, doi: 10.1016/j.mseb.2004.03.006.

[48] G. Sarala Devi, V. Bala Subrahmanyam, S. C. Gadkari, and S. K. Gupta, "NH3 gas sensing properties of nanocrystalline ZnO based thick films," *Anal. Chim. Acta*, vol. 568, no. 1–2, pp. 41–46, 2006, doi: 10.1016/j.aca.2006.02.040.

[49] N. D. Hoa, N. Van Quy, Y. S. Cho, and D. Kim, "Nanocomposite of SWNTs and SnO 2 fabricated by soldering process for ammonia gas sensor application," *Phys. Status Solidi Appl. Mater. Sci.*, vol. 204, no. 6, pp. 1820–1824, 2007, doi: 10.1002/pssa.200675318.

[50] B. Ghaddab *et al.*, "Detection of O 3 and NH 3 using hybrid tin dioxide/carbon nanotubes sensors: influence of materials and processing on sensor's sensitivity," *Sensors Actuators, B Chem.*, vol. 170, pp. 67–74, 2012, doi: 10.1016/j.snb.2011.01.044.

[51] Y. Jang, J. Bang, Y. Seon, D. You, and J. Oh, "Carbon nanotube sponges as an enrichment material for aromatic volatile organic compounds," *J. Chromatogr. A*, vol. 1617, p. 460840, 2020, doi: 10.1016/j.chroma.2019.460840.

[52] A. Sarafraz-Yazdi, A. Amiri, G. Rounaghi, and H. E. Hosseini, "A novel solid-phase microextraction using coated fiber based sol-gel technique using poly (ethylene glycol) grafted multi-walled carbon nanotubes for determination of benzene, toluene, ethylbenzene and oxylene in water samples with gas chromatogra," *J. Chromatogr. A*, vol. 1218, no. 34, pp. 5757–5764, 2011, doi: 10.1016/j.chroma.2011.06.099.

[53] R. Suresh, and R. Saravanan, Nanocomposites in detection of volatile organic compounds. In *Nanocomposites-advanced materials for energy and environmental aspects* (pp. 273–296). Cambridge, UK: Woodhead Publishing, 2023.

[54] N. Bohli, M. Belkilani, J. Casanova-Chafer, E. Llobet, and A. Abdelghani, "Multiwalled carbon nanotube based aromatic volatile organic compound sensor: sensitivity enhancement through 1-hexadecanethiol functionalisation," *Beilstein J. Nanotechnol.*, vol. 10, pp. 2364–2373, 2019, doi: 10.3762/bjnano.10.227.

[55] T. Sarkar, and S. Srinives, "Electrochemically functionalized single-walled carbon nanotubes for ultrasensitive detection of BTEX vapors," *Microelectron. Eng.*, vol. 247, no. March, p. 111584, 2021, doi: 10.1016/j.mee.2021.111584.

[56] S. P. Subin David, S. Veeralakshmi, J. Sandhya, S. Nehru, and S. Kalaiselvam, "Room temperature operatable high sensitive toluene gas sensor using chemiresistive Ag/Bi2O3 nanocomposite," *Sensors Actuators, B Chem.*, vol. 320, no. April, p. 128410, 2020, doi: 10.1016/j.snb.2020.128410.

[57] A. Rushi, K. Datta, P. Ghosh, A. Mulchandani, and M. D. Shirsat, "Iron tetraphenyl porphyrin functionalized single wall carbon nanotubes for the detection of benzene," *Mater. Lett.*, vol. 96, pp. 38–41, 2013, doi: 10.1016/j.matlet.2013.01.003.

[58] N. Janudin *et al.*, "Effect of functionalized carbon nanotubes in the detection of benzene at room temperature," *J. Nanotechnol.*, vol. 2018, 2018, doi: 10.1155/2018/2107898.

[59] H. Lahlou, R. Leghrib, E. Llobet, X. Vilanova, and X. Correig, "Development of a gas pre-concentrator based on carbon nanotubes for benzene detection," *Procedia Eng.*, vol. 25, pp. 239–242, 2011, doi: 10.1016/j.proeng.2011.12.059.

[60] B. Edokpolo, Q. J. Yu, and D. Connell, "Health risk assessment of ambient air concentrations of benzene, toluene and Xylene (BTX) in service station environments," *Int. J. Environ. Res. Public Health*, vol. 11, no. 6, pp. 6354–6374, 2014, doi: 10.3390/ijerph110606354.

[61] A. Mirzaei, J. H. Kim, H. W. Kim, and S. S. Kim, "Resistive-based gas sensors for detection of benzene, toluene and xylene (BTX) gases: a review," *J. Mater. Chem. C*, vol. 6, no. 16, pp. 4342–4370, 2018, doi: 10.1039/c8tc00245b.

[62] H. F. Frasch, and A. M. Barbero, "In vitro human epidermal permeation of nicotine from electronic cigarette refill liquids and implications for dermal exposure assessment," *J. Expo. Sci. Environ. Epidemiol.*, vol. 27, no. 6, pp. 618–624, 2017, doi: 10.1038/jes.2016.68.

[63] S. Saxena and A. L. Verma, "Metal-tetraphenylporphyrin functionalized carbon nanotube composites as sensor for benzene, toluene and xylene vapors," *Adv. Mater. Lett.*, vol. 5, no. 8, pp. 472–478, 2014.

[64] V. D. Martins, M. A. Granato, and A. E. Rodrigues, "Isobaric vapor-liquid equilibrium for binary systems of 2,2,4-trimethylpentane with o -xylene, m -Xylene, p -xylene, and ethylbenzene at 250 kPa," *J. Chem. Eng. Data*, vol. 59, no. 5, pp. 1499–1506, 2014, doi: 10.1021/je401057z.

[65] A. D. Rushi, K. P. Datta, P. S. Ghosh, and A. Mulchandani, "Selective discrimination among benzene, toluene, and xylene: probing metalloporphyrin-functionalized single-walled carbon nanotube-based field effect transistors," *J. Phys. Chem. C*, vol. 118, no. 41, pp. 24034–24041, 2014.

[66] K. A. Mirica, J. G. Weis, J. M. Schnorr, B. Esser, and T. M. Swager, "Mechanical drawing of gas sensors on paper," *Angew. Chemie – Int. Ed.*, vol. 51, no. 43, pp. 10740–10745, 2012, doi: 10.1002/anie.201206069.

[67] Y. Seekaew, A. Wisitsoraat, and D. Phokharatkul, "Sensors and actuators B: chemical room temperature toluene gas sensor based on TiO 2 nanoparticles decorated 3D graphene-carbon nanotube nanostructures," *Sensors Actuators B. Chem.*, vol. 279, no. August 2018, pp. 69–78, 2019, doi: 10.1016/j.snb.2018.09.095.

[68] P. Venkatesu, K. Ravichandran, and B. K. Reddy, "Morphological and impedance studies in diluted Mn-doped cds nanocomposites," *Adv. Mater. Lett.*, vol. 4, no. 10, pp. 786–791, 2013, doi: 10.5185/amlett.2013.2429.

[69] A. H.Q. Le, H. Y. Hoang, T. Le Van, T. Hoang Nguyen, and M. Uyen Dao, "Adsorptive removal of benzene and toluene from aqueous solutions by oxygen-functionalized multi-walled carbon nanotubes derived from rice husk waste: a comparative study," *Chemosphere*, vol. 336, no. March, p. 139265, 2023, doi: 10.1016/j.chemosphere.2023.139265.

[70] J. N. Hodul *et al.*, "Modifying the surface chemistry and nanostructure of carbon nanotubes facilitates the detection of aromatic hydrocarbon gases," *ACS Appl. Nano Mater.*, vol. 3, no. 10, pp. 10389–10398, 2020, doi: 10.1021/acsanm.0c02295.

[71] R. Leghrib, A. Felten, F. Demoisson, F. Reniers, J. J. Pireaux, and E. Llobet, "Room-temperature, selective detection of benzene at trace levels using plasma-treated metal-decorated multiwalled carbon nanotubes," *Carbon NY.*, vol. 48, no. 12, pp. 3477–3484, 2010, doi: 10.1016/j.carbon.2010.05.045.

[72] Y. Seekaew, A. Wisitsoraat, D. Phokharatkul, and C. Wongchoosuk, "Room temperature toluene gas sensor based on TiO2 nanoparticles decorated 3D graphene-carbon nanotube nanostructures," *Sensors Actuators, B Chem.*, vol. 279, no. June 2018, pp. 69–78, 2019, doi: 10.1016/j.snb.2018.09.095.

[73] S. Acosta, J. Casanova-Chafer, E. Llobet, A. Hemberg, M. Quintana, and C. Bitten-court, "Plasma-sputtered growth of Ni-Pd bimetallic nanoparticles on carbon nano-tubes for toluene sensing," *Chemosensors*, vol. 11, no. 6, p. 328, 2023, doi: 10.3390/chemosensors11060328.

[74] D. Pathania, S. Sharma, and P. Singh, "Removal of methylene blue by adsorption onto activated carbon developed from Ficus carica bast," *Arab. J. Chem.*, vol. 10, pp. S1445–S1451, 2017, doi: 10.1016/j.arabjc.2013.04.021.

[75] Y. Han *et al.*, "In situ synthesis of titanium doped hybrid metal-organic framework UiO-66 with enhanced adsorption capacity for organic dyes," *Inorg. Chem. Front.*, vol. 4, no. 11, pp. 1870–1880, 2017, doi: 10.1039/c7qi00437k.

[76] D. Balarak, M. Zafariyan, and C. Adaobi, "Adsorption of acid blue 92 dye from aqueous solutions by single-walled carbon nanotubes: isothermal, kinetic, and thermodynamic studies," *Environ. Process.*, pp. 869–888, 2021, doi: 10.1007/s40710-021-00505-3.

[77] F. Mashkoor, and A. Nasar, "Carbon nanotube - based adsorbents for the removal of dyes from waters: a review," *Environ. Chem. Lett.*, vol. 18, no. 3, pp. 605–629, 2020, doi: 10.1007/s10311-020-00970-6.

[78] S. Mallakpour, and S. Rashidimoghadam, "Poly(vinyl alcohol)/Vitamin C-multi walled carbon nanotubes composites and their applications for removal of methylene blue: advanced comparison between linear and nonlinear forms of adsorption isotherms and kinetics models," *Polymer (Guildf).*, vol. 160, no. November 2018, pp. 115–125, 2019, doi: 10.1016/j.polymer.2018.11.035.

[79] M. Saxena, N. Sharma, and R. Saxena, "Highly efficient and rapid removal of a toxic dye: adsorption kinetics, isotherm, and mechanism studies on functionalized multi-walled carbon nanotubes," *Surfaces and Interfaces*, vol. 21, no. June, p. 100639, 2020, doi: 10.1016/j.surfin.2020.100639.

[80] A. M. Kamil, H. T. Mohammed, A. F. Alkaim, and F. H. Hussein, "Adsorption of congo red on multiwall carbon nanotubes: equilibrium isotherm and kinetic studies," *Int. J. Chem. Sci.*, vol. 14, no. 3, pp. 1657–1669, 2016.

[81] R. Kumar, and R. Ahmad, "Biosorption of hazardous crystal violet dye from aqueous solution onto treated ginger waste (TGW)," *Desalination*, vol. 265, no. 1–3, pp. 112–118, 2011, doi: 10.1016/j.desal.2010.07.040.

[82] K. Yang, and B. Xing, "Adsorption of organic compounds by carbon nanomaterials in aqueous phase: Polanyi theory and its application," *Chem. Rev.*, vol. 110, no. 10, pp. 5989–6008, 2010, doi: 10.1021/cr100059s.

[83] M. Rajabi, K. Mahanpoor, and O. Moradi, "Removal of dye molecules from aqueous solution by carbon nanotubes and carbon nanotube functional groups: critical review," *RSC Adv.*, vol. 7, no. 74, pp. 47083–47090, 2017, doi: 10.1039/c7ra09377b.

[84] Y. Yan, M. Zhang, K. Gong, L. Su, Z. Guo, and L. Mao, "Adsorption of methylene blue dye onto carbon nanotubes: a route to an electrochemically functional nanostructure and its layer-by-layer assembled nanocomposite," *Chem. Mater.*, vol. 17, no. 13, pp. 3457–3463, 2005, doi: 10.1021/cm0504182.

[85] V. K. Gupta, R. Kumar, A. Nayak, T. A. Saleh, and M. A. Barakat, "Adsorptive removal of dyes from aqueous solution onto carbon nanotubes: a review," *Adv. Colloid Interface Sci.*, vol. 193–194, pp. 24–34, 2013, doi: 10.1016/j.cis.2013.03.003.

[86] F. Liu, S. Chung, G. Oh, and T. S. Seo, "Three-dimensional graphene oxide nanostruc-ture for fast and efficient water-soluble dye removal," *ACS Appl. Mater. Interfaces*, vol. 4, no. 2, pp. 922–927, 2012, doi: 10.1021/am201590z.

[87] W. Hassan, U. Farooq, M. Ahmad, M. Athar, and M. A. Khan, "Potential biosorbent, Haloxylon recurvum plant stems, for the removal of methylene blue dye," *Arab. J. Chem.*, vol. 10, pp. S1512–S1522, 2017, doi: 10.1016/j.arabjc.2013.05.002.

[88] A. K. Bhakta, S. Kumari, S. Hussain, and P. Martis, "Synthesis and characterization of maghemite nanocrystals decorated multi-wall carbon nanotubes for methylene blue dye removal," *J. Mater. Sci.*, vol. 54, no. 1, pp. 200–216, 2019, doi: 10.1007/s10853-018-2818-y.

[89] B. Maazinejad *et al.*, "Taguchi L9 (34) orthogonal array study based on methylene blue removal by single-walled carbon nanotubes-amine: adsorption optimization using the experimental design method, kinetics, equilibrium and thermodynamics," *J. Mol. Liq.*, vol. 298, p. 112001, 2020, doi: 10.1016/j.molliq.2019.112001.

[90] N. Boukhalfa, M. Boutahala, N. Djebri, and A. Idris, "Maghemite/alginate/functionalized multiwalled carbon nanotubes beads for methylene blue removal: adsorption and desorption studies," *J. Mol. Liq.*, vol. 275, pp. 431–440, 2019, doi: 10.1016/j.molliq.2018.11.064.

[91] D. Patra, B. Gopalan, and R. Ganesan, "Direct solid-state synthesis of maghemite as a magnetically recoverable adsorbent for the abatement of methylene blue," *J. Environ. Chem. Eng.*, vol. 7, no. 5, p. 103384, 2019, doi: 10.1016/j.jece.2019.103384.

[92] J. L. Gong *et al.*, "Removal of cationic dyes from aqueous solution using magnetic multi-wall carbon nanotube nanocomposite as adsorbent," *J. Hazard. Mater.*, vol. 164, no. 2–3, pp. 1517–1522, 2009, doi: 10.1016/j.jhazmat.2008.09.072.

[93] S. K. Yadav, S. R. Dhakate, and B. Pratap Singh, "Carbon nanotube incorporated eucalyptus derived activated carbon-based novel adsorbent for efficient removal of methylene blue and eosin yellow dyes," *Bioresour. Technol.*, vol. 344, no. PB, p. 126231, 2022, doi: 10.1016/j.biortech.2021.126231.

[94] T. A. Saleh, A. M. Elsharif, and O. A. Bin-Dahman, "Synthesis of amine functionalization carbon nanotube-low symmetry porphyrin derivatives conjugates toward dye and metal ions removal," *J. Mol. Liq.*, vol. 340, p. 117024, 2021, doi: 10.1016/j.molliq.2021.117024.

[95] R. Soltani, A. Marjani, M. R. S. Moguei, B. Rostami, and S. Shirazian, "Novel diamino-functionalized fibrous silica submicro-spheres with a bimodal-micro-mesoporous network: ultrasonic-assisted fabrication, characterization, and their application for superior uptake of Congo red," *J. Mol. Liq.*, vol. 294, p. 111617, 2019, doi: 10.1016/j.molliq.2019.111617.

[96] P. Zong *et al.*, "Eco-friendly approach for effective removal for Congo red dye from wastewater using reusable Zn-Al layered double hydroxide anchored on multiwalled carbon nanotubes supported sodium dodecyl sulfonate composites," *J. Mol. Liq.*, vol. 349, p. 118468, 2022, doi: 10.1016/j.molliq.2022.118468.

[97] L. Lu *et al.*, "An uncommon 3D 3,3,4,8-c Cd(II) metal-organic framework for highly efficient luminescent sensing and organic dye adsorption: experimental and theoretical insight," *CrystEngComm*, vol. 19, no. 46, pp. 7057–7067, 2017, doi: 10.1039/c7ce01638g.

[98] M. Sheibani, M. Ghaedi, F. Marahel, and A. Ansari, "Congo red removal using oxidized multiwalled carbon nanotubes: kinetic and isotherm study," *Desalin. Water Treat.*, vol. 53, no. 3, pp. 844–852, 2015, doi: 10.1080/19443994.2013.867540.

[99] J. Morton, N. Havens, A. Mugweru, and A. K. Wanekaya, "Detection of trace heavy metal ions using carbon nanotube modified electrodes," *Electroanalysis*, vol. 21, no. 14, pp. 1597–1603, 2009, doi: 10.1002/elan.200904588.

[100] N. Alias *et al.*, Metal oxide for heavy metal detection and removal. In Yarub Al-Douri (Ed.), *Metal oxide powder technologies* (pp. 299–332). Malaysia: Elsevier, 2020.

[101] J. A. Buledi, S. Amin, Syed, I. Haider, M. I. Bhanger, and A. R. Solangi, "Recent developments and innovative strategies in environmental sciences in Europe: a review on detection of heavy metals from aqueous media using nanomaterial-based sensors," *Environmental Science and Pollution Research*, pp. 58994–59002, 2021, doi: 10.1007/s11356-020-07865-7.

[102] T. Rasheed, M. Bilal, F. Nabeel, H. M. N. Iqbal, C. Li, and Y. Zhou, "Fluorescent sensor based models for the detection of environmentally-related toxic heavy metals," *Sci. Total Environ.*, vol. 615, pp. 476–485, 2018, doi: 10.1016/j.scitotenv.2017.09.126.

[103] S. Falina *et al.*, "Ten years progress of electrical detection of heavy metal ions (Hmis) using various field-effect transistor (fet) nanosensors: a review," *Biosensors*, vol. 11, no. 12, 2021, doi: 10.3390/bios11120478.

[104] T. Rasheed *et al.*, "Rhodamine-based multianalyte colorimetric probe with potentialities as on-site assay kit and in biological systems," *Sensors Actuators, B Chem.*, vol. 258, pp. 115–124, 2018, doi: 10.1016/j.snb.2017.11.100.

[105] L. A. Malik, A. Bashir, A. Qureashi, and A. H. Pandith, "Detection and removal of heavy metal ions: a review," *Environ. Chem. Lett.*, vol. 17, no. 4, pp. 1495–1521, 2019, doi: 10.1007/s10311-019-00891-z.

[106] Z. Barati, M. Masrournia, Z. Es'haghi, M. Jahani, and J. Ebrahimi, "Selective determination of Cr(III) by modified carbon nanotube paste electrode: a potentiometric study," *J. Chem. Technol. Biotechnol.*, vol. 97, no. 5, pp. 1234–1239, 2022, doi: 10.1002/jctb.7015.

[107] Q. Bao *et al.*, "Electrochemical performance of a three-layer electrode based on Bi nanoparticles, multi-walled carbon nanotube composites for simultaneous Hg(II) and Cu(II) detection," *Chinese Chem. Lett.*, vol. 31, no. 10, pp. 2752–2756, 2020, doi: 10.1016/j.cclet.2020.06.021.

[108] K. H. Wu, H. M. Lo, J. C. Wang, S. Y. Yu, and B. De Yan, "Electrochemical detection of heavy metal pollutant using crosslinked chitosan/carbon nanotubes thin film electrodes," *Mater. Express*, vol. 7, no. 1, pp. 15–24, 2017, doi: 10.1166/mex.2017.1351.

10 CNTs-Based Sensors for Energy Harvesting Applications

Alivia Mukherjee and Anindya Nag

10.1 INTRODUCTION

The escalating demand for energy is a direct consequence of the global population surge and the advancement of economic development. There exist two primary categories of energy sources; namely, renewable and non-renewable. The majority of energy production is derived from non-renewable sources (fossil fuels). Renewable energy sources, such as sunlight, wind, water, waves, and geothermal, are widely available. The increasing prominence of renewable green energy can be attributed to the non-renewable nature of traditional fossil fuels, which has led to escalating environmental concerns associated with their extensive utilization and over-dependence (1, 2). However, growing concerns regarding environmental safety and the detrimental consequences of the consumption of fossil fuels have led to an increased emphasis on the requirement for renewable energy generation and harnessing renewable energy sources. Harnessing waste heat from energy conversion processes is the most extensive emerging approach that can help meet the increasing demands for energy around the globe (3, 4). In this context, considerable efforts have been made to scavenge energy from ambient and renewable sources, including solar, wind, geothermal, biomass, and hydro (5, 6). It is imperative to accelerate the development of novel energy technologies capable of effectively scavenging and storing energy from surrounding sources to cater to specific applications. For instance, mechanical energy harvesters are essential in several domains, encompassing self-powered wireless sensors and human health monitoring systems (7).

The discovery of carbon nanotubes (CNTs) is attributed to the Japanese scientist, Iijima (8, 9). CNTs are composed of sp^2 carbon atoms arranged in hexagonal networks, like the structure of a graphene sheet (10). CNTs exhibit a variety of structural configurations, characterized by the presence of one or more shell structures. These configurations may be categorized into two main types: single-wall carbon nanotubes (SWCNTs) and multi-wall carbon nanotubes (MWCNTs) (11). The emergence of CNTs has played a significant role in the advancement of various frontiers in the field of nanotechnology (12), making them one of the most robust materials identified to date due to their distinctive thermal, physical, and chemical properties and mechanical and electrical characteristics (13, 14). Consequently, CNTs have become the hotspot of extensive research and exploration in the realm of nanomaterials for the purpose of energy harvesting.

DOI: 10.1201/9781003376071-10

CNTs are useful in applications such as sensors (15–18) and energy harvesters (19–21), owing to their unique features including physical properties (22), high thermal conductivity (8), soft mechanical properties (4), and excellent electronic properties (23). Owing to the inherent unique features, many CNTs-based systems have been developed to harvest ambient and renewable energies, with promising results. In recent years, there has been significant progress in the development of energy harvesting devices, mostly attributed to advancements in nanotechnology. Hence, this study highlights applications of CNTs, with a specific focus on energy harvesting.

There is a significant demand for a power source that is miniature, lightweight, and sustainable in order to enhance mobility and promote long-term energy sustainability. Self-sustaining nano-generators, such as triboelectric, piezoelectric, thermoelectric, and hydroelectric nano-generators, have the ability to harness renewable energy from the surrounding environment without the need for external energy sources. Highly efficient and ecologically benign, these power sources have emerged as a viable option for energizing future electronic wearables (23). The exceptional electrical characteristics and inherent flexibility shown by CNTs make them appropriate candidates for application as friction layers in triboelectric nano-generators (TENGs) that could eventually harness electrical energy (23). In recent times, notable breakthroughs have been made in the field of energy harvesting devices, mostly ascribed to the rapid growth in nanotechnology.

This study's primary focus is on CNTs-based nanomaterials, providing a full introduction to numerous new principles of energy harvesting. These harvesting principles include electrochemical, triboelectric (24, 25), piezoelectric (26), and thermoelectric effects. In conclusion, this chapter highlights the current challenges faced by CNTs-based nano-generators and provides an overview of potential directions for future advancements in this field. Hence, our objective is to present a comprehensive review in the field of energy harvesting using CNTs, while also pushing for further developments in this intriguing area of research.

10.2 CNTs-BASED SENSORS FOR ENERGY HARVESTING

The use of CNTs for generating and harvesting energy has been done significantly over the last decade (8, 13, 27). Flexible sensors for energy harvesting have been an uprising phenomenon in the last two decades as a result of the ever-growing population and their increased usage of electrical devices (28–30). With increased human activities in sports and recreation (31, 32), researchers have been trying to develop energy-harvesting sensors with optimized efficiency (33, 34). The use of carbon-based allotropes to develop energy-harvesting devices has been advantageous due to their excellent electrical conductivity, high mechanical flexibility and integrity, and high efficiency in terms of sensitivity, linear range, and operating range (35, 36). Among the carbon allotropes, CNTs have been used to form energy-harvesting devices due to their easy customization in accordance with the physiochemical characteristics and target application (20, 37–39). With exceptional electromechanical characteristics of CNTs, these carbon elements have been used in pure (40–42) and composite (39, 43, 44) to form the energy-harvesting devices. The small size and high aspect ratio of CNTs have helped the sensors achieve high resolution to detect very small movements (45, 46). Compared to the conventional silicon-based sensors, the other advantages of the CNTs-based energy-harvesting devices are high robustness, high output, low cost, large open-circuit output voltage, ease of fabrication, and high conversion-efficiency (47).

The sources for these sensors to harvest energy have been inspired based on the working principle of the sensors (48, 49). There have been primarily three operating principles, including thermoelectric (50, 51), photovoltaic (52, 53), and piezoelectric (54), with which the CNTs-based sensors have been employed to harvest energy. Among these types, the piezoelectric type has been the most efficient (55) due to its high linear-stability and enhanced output in terms of current and voltage. It involves the generation of voltage with respect to the intensity of applied pressure, thus converting the mechanical energy into electrical energy (56). Like the piezoelectric nature, CNTs-based sensors have also been used with the triboelectric principle (34), where the electric charges transfer between two chosen objects when they are made in contact (57). The triboelectric sensors have been further categorized based on the nature of the electrodes. Some of the basic modes include vertical contact-separation mode (58), lateral-sliding mode (59), and single-electrode mode (60). Each of these types has been developed and tested with CNTs-based sensors for harvesting energy from different sources, like body motions (61, 62), strain sensing (63, 64), environmental energy (65, 66), and vibrational energy (67, 68). One of the common areas where the CNTs-based sensors have been used to generate and harvest energy is exploiting the vibrational nature of the CNTs-based sensors. Thin-film prototypes were formed with CNTs conjugated with polymeric materials (69, 70) to obtain the resultant prototypes. Figure 10.1 (71) is the schematic diagram of some CNTs-based sensors used in the academic and industrial sectors. In addition

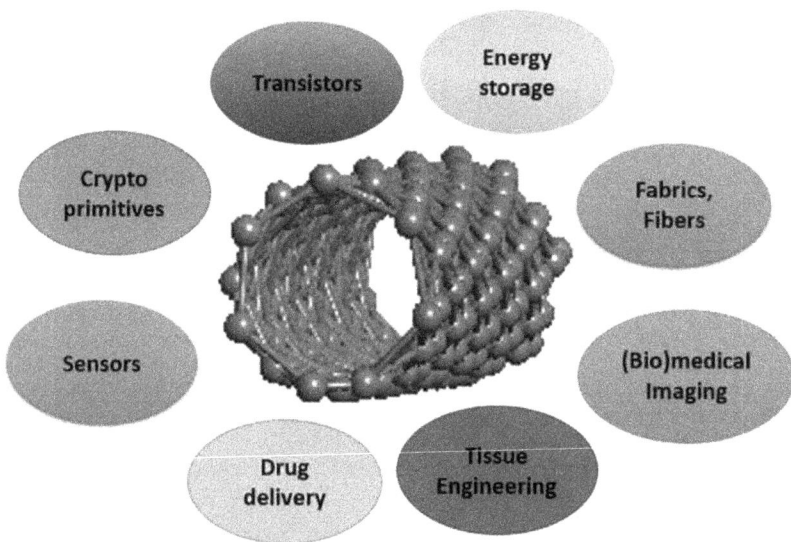

FIGURE 10.1 Application spectrum of the CNTs-based flexible sensors (71).

to CNTs, nanomaterials and polymers are particularly dedicated to forming these energy-harvesting sensors. For example, polyvinylidene fluoride (PVDF) (72, 73) and zinc oxide (ZnO) (74, 75) are two common materials that have been explicitly used to form nanogenerators due to their advantages of excellent flexibility, high toughness, chemical resistance, and biocompatible nature (76). ZnO has been developed in various forms, like nanowires and clusters, to boost the output and enhance other piezoelectric and triboelectric properties. Other than these, certain materials like natural (77, 78) rubber have also been recently considered to conjugate with CNTs due to the effectively high tensile strength, high modulus of elasticity, and stability in the responses. The generating and harvesting capability of the nanogenerators has allowed the sensors to be used as self-powered sensors (79, 80), where the embedded circuitry would assist the prototype to harvest energy for sensing and communication purposes. The self-powered nature of these CNTs-based nanogenerators has allowed the researchers to use them as wearable sensing prototypes to detect multiple nature of movements.

10.3   CNT-BASED HYDROELECTRIC ENERGY HARVESTERS

Despite the widespread utilization and the effectiveness of electromagnetic generators in deploying large-scale energy in hydro-power stations and large windmills for generations, their ability to be scaled down remains challenging in terms of gravimetric performance and providing adequate energy for wearable electronics or micro-power systems (13). Water possesses significant energy in many forms, yet the exploitation of this abundant source of renewable energy remains mostly unexplored. The utilization of water as a source for electricity generation has been recognized as a viable approach for addressing the escalating energy scarcity. In recent times, a novel hydroelectric nanotechnology has emerged, a phenomenon seen when nanostructured materials interact with water, which has the potential to enhance the technological capacity of water-energy collecting and facilitate the development of self-powered devices [28]. This innovative approach offers a fresh avenue for harnessing energy from various sources such as flowing water, water droplets, rainwater, and evaporation based on the electro-kinetic effect (23, 81, 82).

10.3.1   CNTs Film-Assisted Devices for Electricity Generation from Water Flow

Since water molecules are polar, weak electronic interactions between mobile water-dipoles and unbound charge in nanotubes generate a voltage across the opposing ends. The research findings of Kral and Shapiro (83) showcased the capacity to generate electrical current within metallic CNTs when they are immersed in a liquid medium and exposed to fluid motion. The surface of the nanotube is covered with molecular layers of the liquid, which demonstrate slide behaviour and produce a phonon wind. The phonon wind is responsible for the generation of motion in the unbound charge carriers present in the nanotube. The application of induced electric current exhibits the capacity to enable the advancement of nanoscale detectors or power cells. However, to achieve a high output voltage, the primary requirement is enhancing the coupling between water and carbon. In order to enhance the electric properties and hydrophilicity for the purpose of flow-induced energy harvesting, Kim et al. (84) employed an improved plasmonic heating technique to create networks and carbonization among customized SWCNTs. They observed improvement in voltage and current generation by 9.5 and 23.5 folds as well as a fall in the electrical sheet resistance of carbonized SWCNTs compared to the pristine SWCNTs. The improvement in output voltage and electricity generation can be attributed the dragged electron mechanism. In addition, the flow direction and sheet alignment in MWCNTs have been shown to effectively impact the output voltage. Accordingly, Liu et al. (85) reported an enhancement in voltage of 30 millivolts through vertically aligned MWNTs, even when the flow velocity was as low as 0.5 mm s^{-1}. The findings revealed that the flow-induced voltage is primarily influenced by several factors, including alignment of the nanotubes, flow velocity, and/or liquid ionic strength.

10.3.2 CNTs YARN-ASSISTED NANOGENERATORS FOR ELECTRICITY GENERATION FROM WATER FLOW

CNTs-based films and sheets provide novel approaches for the extraction of renewable energy. However, their bulky structures are sheet-shaped and lack the necessary characteristics of flexibility and durability that are essential for the integration of wearable electronics. On the other hand, CNT yarns exhibit notable flexibility and durability, rendering them well-suited for incorporation into fabrics and garments that offer both breathability and comfort. Therefore, yarn-based hydroelectric nano-generators offer a better solution for the power supply of wearable electronics and sensors. Accordingly, Zhao et al. (86) fabricated a flexible, fibre-fluidic nano-generator utilizing a composite film composed of aligned CNTs and transition metal oxides and developed a hybrid CNT yarn. When immersed in saline solution, the hybrid CNT fluidic nano-generator exhibited an OCV of 0.31 V and output power of 30 mW m^{-2}, respectively. The output power of the hydropower generators utilizing CNT surpasses that of most prior carbon nanomaterials-based hydroelectric generators. In general, it has been shown that CNT yarn fluidic nano-generators exhibit more mechanical robustness when compared to CNT films, resulting in much higher voltage outputs. Moreover, the scalability of CNT yarn fluidic nano-generators is facilitated by their compatibility with textile weaving techniques. Hence, CNTs yarn-based nano-generators provide a promising outlook for feasible applications, particularly in the domain of self-sustaining implantable devices capable of efficiently converting the kinetic energy of blood into electrical energy.

10.3.3 ELECTRICITY GENERATION BY WATER EVAPORATION

The water cycle is a fundamental and ubiquitous phenomenon in the natural world, encompassing the movement of water, its conversion into vapour through evaporation, and subsequent condensation. In this procedure, the utilization of efficient solar irradiation has the potential to significantly accelerate the evaporation or transpiration processes, hence promoting the regeneration of freshwater resources. This is crucial for meeting the fundamental requirements for the existence of many organisms. Hence, the utilization of sustainable energy through natural water evaporation has garnered significant attention in recent years due to its adaptability in different climates and geographical conditions as well as its ability to operate independently without the need for supplementary mechanical energy (81). The potential of solar-enabled water evaporation as a means of efficiently mitigating the increasingly alarming issues of water scarcity and energy depletion has been comprehensively investigated. Hydrovoltaic effect is characterized by its capacity to transform thermal energy present in the surrounding environment into electrical power by means of widespread water evaporation. Hydroelectric nano-generators enable the direct conversion of water vapours existing in the surroundings into electrical energy via the mechanism of water evaporation. However, in comparison to bulk water, water vapour has relatively weak interactions with CNT materials, owing to its lower hydrophilic properties. One viable approach to improving the interaction with water vapour involves increasing the extent of hydrophilicity of the material. Chen et al. [24] presented a novel regenerated cellulose and CNTs-based fibre produced by the wet-spinning approach. The nano-generator displayed the ability to facilitate water flow through the process of transpiration, as observed in plants. The fibre possesses conductive and hydrophilic properties, enabling the significant increase in ion flow through the macromolecular chain channel and subsequent spontaneous evaporation.

In addition, solar energy has extensive ubiquity and abundant accessibility, so enabling it to assist the phenomenon of water evaporation in a natural and continuous fashion. Solar-induced water evaporation is a promising and environmentally conscious approach that exhibits efficacy and sustainability in the utilization of energy resources. This methodology enables the processes of water purification and the generation of electrical energy. Xiao et al. (87) investigated the phenomenon of interfacial, confined water flow and evaporation. They successfully engineered a device that demonstrates the remarkable capacity to simultaneously generate electrical energy from water flow and facilitate the evaporation of interfacial water. The researchers created a bilayer CNT material that demonstrates effective solar-driven evaporation capabilities through the process of asymmetric functionalization with cellulose. The CNT paper that was initially generated was afterward adorned with hydrophobic polydimethylsiloxane (PDMS) in an asymmetric manner. This modification was employed to facilitate the collection of electricity from the unidirectional movement of water. The implementation of the monolithic design concept is expected to effectively harness the phenomenon of water evaporation to facilitate the simultaneous production of integrated energy and clean water.

The use of CNTs in the fabrication of nanomaterials has attracted considerable interest within the realm of wearable hydroelectric-energy harvesters, primarily due to their inherent flexibility. However, they have not yet achieved the expected level of competitive performance, owing to the intrinsically weak interaction of water molecules with CNTs. To enhance power production, future research endeavours should emphasize the exploration of techniques aimed at optimizing the interaction between water and CNTs-based materials. Simultaneously, efforts should be made to maximize the interaction surface area in relation to volume, while ensuring that electric conductivity is not compromised.

10.4   CNTs-BASED THERMOELECTRIC ENERGY GENERATORS

Thermal energy is an abundant and ubiquitous resource present on planet Earth, and this kind of energy can be transformed into electricity but with limited efficiency. The utilization of thermoelectric technology, which involves the conversion of residual heat into electric power, offers a viable and reversible method for transforming thermal energy into electrical energy. This approach holds significant potential as an environmentally sustainable solution for both energy conservation and power generation. In the present circumstances, wherein about two-thirds of used energy is lost as waste heat, there is a pressing need for the advancement of thermoelectric materials with superior performance characteristics. These thermoelectric materials should possess the capability to transform thermal energy efficiently and reversibly into electrical energy (88). Thermoelectric generators (TEGs) provide a viable means of converting thermal energy into electrical energy. As a result, TEGs have emerged as a compelling alternative to traditional methods of energy conversion. Extensive research has been conducted on the use of conventional inorganic thermoelectric (TE) materials and their alloys (89). Nevertheless, the inherent stiffness and restricted manipulability of these materials put limitations on their potential application in flexible devices.

CNTs exhibit considerable potential in the domain of thermal energy collection. However, it should be noted that the power output of thermal energy harvesters utilizing CNTs is still lower compared to that of conventional inorganic thermoelectric devices. This section presents an introductory overview of thermoelectric energy-harvesting systems that are based on CNTs. One specific classification of TEGs is the thermal-electrochemical energy harvester, more generally known as a thermocell. The device possesses the capacity to convert low-quality heat into electrical energy by means of a reversible, electrochemical mechanism driven by a temperature gradient. The high power-output of thermocells can be achieved by utilizing electrodes with high electrical conductivity and surface area. In this context, CNTs possess advantageous traits as electrode materials, primarily due to their high surface area and unique thermal and electrical conductivities. Successful demonstrations have been conducted on the fabrication of thermocell electrodes utilizing several CNTs-based materials. For instance, Kang and co-workers (90) employed a simple and efficient technique using rapid solvent evaporation to create three-dimensional (3D) foams composed of CNTs and polydimethylsiloxane (PDMS). These 3D foams exhibited commendable mechanical properties alongside superior thermoelectric (TE) performance. Subsequently, the authors utilized these foams to construct thermoelectric generators that were flexible and possessed remarkable durability. They presented a TE generator that exhibits commendable mechanical performance and was able to generate an output power of 3.1 µW. Im et al. (91) developed CNTs aerogel-based thermo-electrochemical cells, which exhibited the potential to be cost-effective and efficient materials for converting waste heat into electricity. They reported the areal current density of $6.6 Wm^{-2}$ was obtained for the developed CNTs. In another study, shown in (92), a range of nanocarbon-based thermocells, including CNTs, rGO, and P-SWCNTs/rGO composites, utilized an aqueous electrolyte consisting of potassium ferro/ferricyanide. These thermocells were evaluated as prospective substitutes for traditional thermoelectric devices in the context of thermal energy harvesting. The researchers noted that the pure SWCNTs-based (P-SWCNT) thermocell exhibited the highest specific power output per unit weight of electrode (6.8 W/kg), in contrast to rGO thermocell electrode. They reported ohmic resistance of the electrolyte, and the mass transport inside a porous nanocarbon electrode is the primary factor constraining the power output of nanocarbon thermocells.

While several thermocells based on CNTs have been produced, there is still a need for further improvement in energy conversion efficiency and packing convenience to make them more suitable for practical applications. In order to effectively address these difficulties, future research might prioritize the following aspects that can predominantly impact the thermocells performance: 1) one potential approach to enhance the catalytic activity of CNTs involves the introduction of functional moieties or the application of chemical treatments; 2) the integration of CNTs into engineered porous materials in order to enhance the ion transport pathways, hence promoting improved ion migration; and 3) the utilization of temperature-sensitive gel electrolytes might potentially enhance the convenience of packing

10.5 PIEZOELECTRIC, CNTs-BASED, FLEXIBLE SENSORS

The piezoelectric sensors are the first type of energy-harvesting sensors formed using CNTs. Both Single-Walled Carbon Nanotubes (SWCNTs) and Multi-Walled Carbon Nanotubes (MWCNTs) have been extensively used to form these piezoelectric sensors. As an example, the work done by Chen et al. (93) demonstrates the fabrication and implementation of piezoelectric and piezoresistive sensors formed using CNTs, polyvinyl alcohol (PVA), and ZnO nanowires. The prototypes were formed using functionalized self-assembly and hydrothermal synthesis processes. The performance of the sensors varied subsequently with the variation of the amount of PVA doped in CNT/PVA films. Figure 10.2 (93) shows the fabrication process of these sensors. Some of the advantages of the piezoelectric nature of these sensors include high sensitivity, linear stability, and high response accuracy of 4.87 mV/bf,

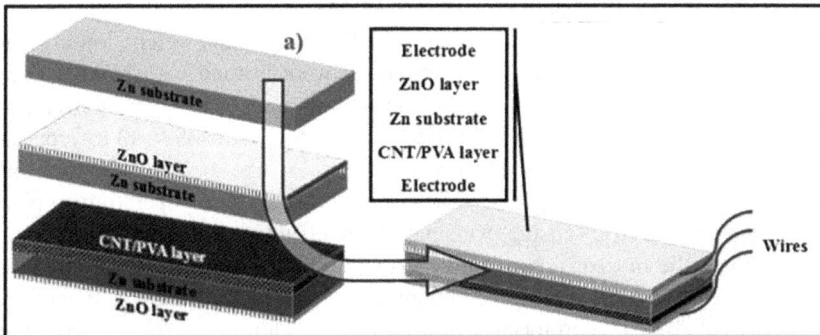

FIGURE 10.2 Schematic diagram of the CNTs/PVA/ZnO composites-based energy-harvesting device (93).

3.42% and 1.496 milliseconds, respectively. The additional advantage of operating as a piezoresistive sensor increases their application spectrum. Another example of the development of CNTs-based piezoelectric sensors can be shown in the work done by Sanati et al. (94). The nanocomposite strain sensors were formed using PVDF and CNTs, where the sensors were operated at a frequency ranging between 0 and 10 Hz. MWCNTs were used as nanofillers in the PVDF polymer matrix. The samples were treated with various processes, like magnetic stirring and heating at elevated temperatures, to obtain the final product.

Mangone et al. (95) showcased the research on designing and developing flexible energy harvesters for battery-less tyre sensors. Both carbon black and SWCNTs were used as conductive elastomer fillers in the PVDF polymer matrix to form a sandwich-like configuration. The variability in the nature of these nanogenerators lies in their capability to be fitted in objects with diverse radii of curvature. The final prototype was formed by adding 6 wt. % of SWCNTs in the PVDF matrix. The prototypes were formed by placing PVDF films between two conductive elastomers. The optimization of the sensor quality was carried out by improving the adhesion between the interfaces by treating the samples with oxygen plasma. This added two separate groups of oxygen and thiocyanate silane to the interfaces.

10.6   TRIBOELECTRIC, CNTs-BASED, FLEXIBLE SENSORS

For the triboelectric nanogenerators (TENG), the response of the sensors depends a lot on the type of materials used to form the two surfaces to be in contact. Certain polymers like polydimethylsiloxane (PDMS) (96) and polyimide (PI) (97) have been preferred due to their easy processing and biocompatible nature. Compared to the piezoelectric ones, the TENG has an additional advantage of generating temporary electrostatic charge, formed when the force is applied on the electrodes. Due to the variability in the working mechanism of TENG, the sensors can be attached to different wearables and body parts to determine the open-circuit voltage, short-circuit current, and power densities. As an example, the work done by Zhang et al. (98) highlights using single-electrode TENG to harvest energy from environmental energy and human motions. The sensors operated on a single-electrode mechanism, with the TENG layer consisting of MWCNTs and PDMS film. Some of the advantages of these sensors include high hydrophobicity, high flexibility, light weight, low cost, and human output efficiency. Homogenous mixtures were formed with MWCNTs and PDMS via stirring and curing processes. The peak-to-peak voltage, short-circuit current density, and output power density of the prototypes were 435 V, 13 μA/cm^2, and 3.7 mW/cm^2, respectively. There are interesting works, like the one done by Mousa et al. (65), where hybrid energy-harvesting devices were formed operating on piezoelectric and triboelectric sensing principles. The advantages of operating as hybrid nanogenerators lie in their non-complex structure, large output power density, high acceptability, and a high degree of freedom in terms of material selection and physical dimension (99). The prototypes were formed on aluminium foil, where the triboelectric electrode was developed using an air jet-spinning technique with α-Al$_2$O$_3$ nanoparticles and commercial Doku. The counterpart electrode was formed using an electrospinning process, where a thin-film layer of PVDF and SWCNTs nanofibres was coated to obtain the piezoelectric electrode. These hybrid prototypes were able to achieve 19.5 V and 3 V for the triboelectric and piezoelectric parts, respectively. These prototypes have very high potential as powering devices, wearable sensors, and actuators. These sensors had a high operating range and a stability of more than 3,000 seconds.

10.7 FUTURE POSSIBILITIES OF CNTs-BASED ENERGY HARVESTERS

Although significant work has been carried out to fabricate and utilize CNTs-based flexible sensors for energy-harvesting applications, some bottlenecks still need to be addressed in the current scenario. In the current era of advanced functional materials involving functionalization with various polymers and nanomaterials, the performance of pure CNTs-based sensors is getting sub-standardized. For example, after the popularization of graphene as the 'magic material', including different organic and inorganic groups in the CNTs is necessary to subsequently increase the performance of the resultant prototypes. The roll-to-roll production of CNTs-based sensors is increasingly important to incorporate with other sensing systems. The printing techniques (100) used to develop the flexible sensors need to be improvised to develop nanocomposites-based sensors that include other nanomaterials alongside CNTs. Further studies should determine the type and nature of the polymers used as matrixes to form the composites. The performance of the hybridized models should be improved, given their advantages over individualistic nanogenerators. This is achieved by customizing the materials used to form the triboelectric and piezoelectric layers in the prototypes. Apart from the regular polymers, natural and synthetic rubbers should be tried as they have higher durability, abrasion resistance, and lesser cost. The physical parameters, like the size and structure of the electrodes, should also be altered to reflect the response of the sensors.

Additionally, further research should be carried out to operate the developed energy-harvesting devices in real-time applications. Even though the sensors have performed well in controlled laboratory conditions, certain environmental interferences have curbed their output in non-ideal situations. Wireless communication with the developed energy-harvesting devices is also necessary, as they can transmit the sensed data to the monitoring unit. The wireless operation of these devices would also assist in communicating several energy-harvesting devices to store the overall energy and use it as required. The need for energy-harvesting devices in wearable applications is pivotal to processing and running the sensors used to detect anomalies in the body. This will also help the medical systems used as point-of-care devices.

ACKNOWLEDGEMENT

This work was supported by the Free State of Saxony and by the European Union (ESF Plus) by funding the research group 'MultiMOD.' This study was funded by the German Research Foundation (DFG, Deutsche Forschungsgemeinschaft) as part of Germany's Excellence Strategy – EXC 2050/1 – Project ID 390696704 – Cluster of Excellence 'Centre for Tactile Internet with Human-in-the-Loop' (CeTI) of Technische Universität Dresden.

REFERENCES

1. Zhu T, Liu Y, Fu C, Heremans JP, Snyder JG, Zhao X. Compromise and synergy in high-efficiency thermoelectric materials. Advanced Materials. 2017;29(14):1605884.
2. Koli V, Dhodamani A, More K, Acquah SF, Panda DK, Pawar S, et al. A simple strategy for the anchoring of anatase titania on multi-walled carbon nanotubes for solar energy harvesting. Solar Energy. 2017;149:188–94.
3. Romano MS, Li N, Antiohos D, Razal JM, Nattestad A, Beirne S, et al. Carbon nanotube–reduced graphene oxide composites for thermal energy harvesting applications. Advanced Materials. 2013;25(45):6602–6.
4. Hong S, Lee J, Do K, Lee M, Kim JH, Lee S, et al. Stretchable electrode based on laterally combed carbon nanotubes for wearable energy harvesting and storage devices. Advanced Functional Materials. 2017;27(48):1704353.
5. Zhang M, Cai W, Wang Z, Fang S, Zhang R, Lu H, et al. Mechanical energy harvesters with tensile efficiency of 17.4% and torsional efficiency of 22.4% based on homochirally plied carbon nanotube yarns. Nature Energy. 2023;8(2):203–13.
6. Zhang C, Tang W, Han C, Fan F, Wang ZL. Theoretical comparison, equivalent transformation, and conjunction operations of electromagnetic induction generator and triboelectric nanogenerator for harvesting mechanical energy. Advanced Materials. 2014;26(22):3580–91.
7. Kim SH, Haines CS, Li N, Kim KJ, Mun TJ, Choi C, et al. Harvesting electrical energy from carbon nanotube yarn twist. Science. 2017;357(6353):773–8.
8. Sidik NAC, Yazid MNAWM, Samion S. A review on the use of carbon nanotubes nanofluid for energy harvesting system. International Journal of Heat and Mass Transfer. 2017;111:782–94.
9. Koziel H, Williams D, Armstrong M, Richards F, Fishman J, Ezekowitz R, et al. New rapid method for the study of pneumocystis carinii interaction with alveolar macrophages. The Journal of Protozoology. 1991;38(6):173S–4S.
10. Tan CW, Tan KH, Ong YT, Mohamed AR, Zein SHS, Tan SH. Energy and environmental applications of carbon nanotubes. Environmental Chemistry Letters. 2012;10: 265–73.
11. Hoang AT, Nižetić S, Cheng CK, Luque R, Thomas S, Banh TL, et al. Heavy metal removal by biomass-derived carbon nanotubes as a greener environmental remediation: A comprehensive review. Chemosphere. 2022;287:131959.
12. Norizan MN, Moklis MH, Demon SZN, Halim NA, Samsuri A, Mohamad IS, et al. Carbon nanotubes: Functionalisation and their application in chemical sensors. RSC Advances. 2020;10(71):43704–32.
13. Hu X, Bao X, Zhang M, Fang S, Liu K, Wang J, et al. Recent advances in carbon nanotube-based energy harvesting technologies. Advanced Materials. 2023:2303035.
14. Peng J, He Y, Zhou C, Su S, Lai B. The carbon nanotubes-based materials and their applications for organic pollutant removal: A critical review. Chinese Chemical Letters. 2021;32(5):1626–36.
15. Ke K, Pötschke P, Wiegand N, Krause B, Voit B. Tuning the network structure in poly (vinylidene fluoride)/carbon nanotube nanocomposites using carbon black: Toward improvements of conductivity and piezoresistive sensitivity. ACS Applied Materials & Interfaces. 2016;8(22):14190–9.
16. Oh JH, Hong SY, Park H, Jin SW, Jeong YR, Oh SY, et al. Fabrication of high-sensitivity skin-attachable temperature sensors with bioinspired microstructured adhesive. ACS Applied Materials & Interfaces. 2018;10(8):7263–70.
17. Zhao X-H, Ma S-N, Long H, Yuan H, Tang CY, Cheng PK, et al. Multifunctional sensor based on porous carbon derived from metal–organic frameworks for real time health monitoring. ACS Applied Materials & Interfaces. 2018;10(4):3986–93.

18. Zuruzi AS, Haffiz TM, Affidah D, Amirul A, Norfatriah A, Nurmawati MH. Towards wearable pressure sensors using multiwall carbon nanotube/polydimethylsiloxane nanocomposite foams. Materials & Design. 2017;132:449–58.

19. Boudouris BW, Yee S. *Structure, properties and applications of thermoelectric polymers*. Wiley Online Library; 2017.

20. Wang H, Shi M, Zhu K, Su Z, Cheng X, Song Y, et al. High performance triboelectric nanogenerators with aligned carbon nanotubes. Nanoscale. 2016;8(43):18489–94.

21. Guo H, Li T, Cao X, Xiong J, Jie Y, Willander M, et al. Self-sterilized flexible single-electrode triboelectric nanogenerator for energy harvesting and dynamic force sensing. ACS Nano. 2017;11(1):856–64.

22. Jeong C, Joung C, Lee S, Feng MQ, Park Y-B. Carbon nanocomposite based mechanical sensing and energy harvesting. International Journal of Precision Engineering and Manufacturing-Green Technology. 2020;7:247–67.

23. Chen J, Li Y, Zhang Y, Ye D, Lei C, Wu K, et al. Knittable composite fiber allows constant and tremendous self-powering based on the transpiration-driven electrokinetic effect. Advanced Functional Materials. 2022;32(30):2203666.

24. Li Y, Zhao Z, Liu L, Zhou L, Liu D, Li S, et al. Improved output performance of triboelectric nanogenerator by fast accumulation process of surface charges. Advanced Energy Materials. 2021;11(14):2100050.

25. Salauddin M, Rana SS, Sharifuzzaman M, Rahman MT, Park C, Cho H, et al. A novel MXene/Ecoflex nanocomposite-coated fabric as a highly negative and stable friction layer for high-output triboelectric nanogenerators. Advanced Energy Materials. 2021;11(1):2002832.

26. Espinosa HD, Bernal RA, Minary-Jolandan M. A review of mechanical and electromechanical properties of piezoelectric nanowires. Advanced Materials. 2012;24(34):4656–75.

27. Blackburn JL. Semiconducting single-walled carbon nanotubes in solar energy harvesting. ACS Energy Letters. 2017;2(7):1598–613.

28. Manikkavel A, Kumar V, Kim J, Lee DJ, Park SS. Investigation of high temperature vulcanized and room temperature vulcanized silicone rubber based on flexible piezoelectric energy harvesting applications with multi-walled carbon nanotube reinforced composites. Polymer Composites. 2022;43(3):1305–18.

29. Surmenev RA, Chernozem RV, Pariy IO, Surmeneva MA. A review on piezo-and pyroelectric responses of flexible nano-and micropatterned polymer surfaces for biomedical sensing and energy harvesting applications. Nano Energy. 2021;79:105442.

30. Boopathi C, Sivaram M, Sundararajan T, Maheswar R, Yupapin P, Amiri IS. Bandenna for RF energy harvesting and flexible electronics. Microsystem Technologies. 2021;27(4):1857–61.

31. Gao S, He T, Zhang Z, Ao H, Jiang H, Lee C. A motion capturing and energy harvesting hybridized lower-limb system for rehabilitation and sports applications. Advanced Science. 2021;8(20):2101834.

32. Sahu M, Hajra S, Panda S, Rajaitha M, Panigrahi BK, Rubahn H-G, et al. Waste textiles as the versatile triboelectric energy-harvesting platform for self-powered applications in sports and athletics. Nano Energy. 2022;97:107208.

33. Bahk J-H, Fang H, Yazawa K, Shakouri A. Flexible thermoelectric materials and device optimization for wearable energy harvesting. Journal of Materials Chemistry C. 2015;3(40):10362–74.

34. Dong X, Liu Q, Liu S, Wu R, Ma L. Silk fibroin based conductive film for multifunctional sensing and energy harvesting. Advanced Fiber Materials. 2022;4(4):885–93.

35. Parvin N, Kumar V, Manikkavel A, Park S-S, Mandal TK, Joo SW. Great new generation carbon microsphere-based composites: Facile synthesis, properties and their application in piezo-electric energy harvesting. Applied Surface Science. 2023;613:156078.

36. Staaf L, Lundgren P, Enoksson P. Present and future supercapacitor carbon electrode materials for improved energy storage used in intelligent wireless sensor systems. Nano Energy. 2014;9:128–41.
37. Afsarimanesh N, Nag A, E-Alahi ME, Sarkar S, Mukhopadhyay S, Sabet GS, et al. A critical review of the recent progress on carbon nanotubes-based nanogenerators. Sensors and Actuators A: Physical. 2022:113743.
38. Liu Z, Muhammad M, Cheng L, Xie E, Han W. Improved output performance of triboelectric nanogenerators based on polydimethylsiloxane composites by the capacitive effect of embedded carbon nanotubes. Applied Physics Letters. 2020;117(14).
39. Han JK, Jeon DH, Cho SY, Kang SW, Yang SA, Bu SD, et al. Nanogenerators consisting of direct-grown piezoelectrics on multi-walled carbon nanotubes using flexoelectric effects. Scientific Reports. 2016;6(1):29562.
40. Sun H, Tian H, Yang Y, Xie D, Zhang Y-C, Liu X, et al. A novel flexible nanogenerator made of ZnO nanoparticles and multiwall carbon nanotube. Nanoscale. 2013;5(13):6117–23.
41. Choi D, Choi M-Y, Shin H-J, Yoon S-M, Seo J-S, Choi J-Y, et al. Nanoscale networked single-walled carbon-nanotube electrodes for transparent flexible nanogenerators. The Journal of Physical Chemistry C. 2010;114(2):1379–84.
42. Ani MH, Zulkeflee MZ, Kaderi A, Sutjipto AGE, Salim N, editors. *Multiwall carbon nanotubes based triboelectric nanogenerators*. Materials Science Forum, Trans Tech Publ; 2022.
43. Zhang R, Örtegren J, Hummelgård M, Olsen M, Andersson H, Olin H. A review of the advances in composites/nanocomposites for triboelectric nanogenerators. Nanotechnology. 2022;33(21):212003.
44. Hou R, Jin Z, Sun D, Shi B, Wang L, Shan X. Carbon nanotubes doped cellulose nanocomposite film for high current flexible piezoelectric nanogenerators. Journal of Alloys and Compounds. 2023;965:171422.
45. Nag A, Alahi M, Eshrat E, Mukhopadhyay SC, Liu Z. Multi-walled carbon nanotubes-based sensors for strain sensing applications. Sensors. 2021;21(4):1261.
46. Nag A, Mukhopadhyay SC, Kosel J. Flexible carbon nanotube nanocomposite sensor for multiple physiological parameter monitoring. Sensors and Actuators A: Physical. 2016;251:148–55.
47. Pan M, Yuan C, Liang X, Zou J, Zhang Y, Bowen C. Triboelectric and piezoelectric nanogenerators for future soft robots and machines. iScience. 2020;23(11).
48. Fan FR, Tang W, Wang ZL. Flexible nanogenerators for energy harvesting and self-powered electronics. Advanced Materials. 2016;28(22):4283–305.
49. Zhu J, Cho M, Li Y, He T, Ahn J, Park J, et al. Machine learning-enabled textile-based graphene gas sensing with energy harvesting-assisted IoT application. Nano Energy. 2021;86:106035.
50. Karthikeyan V, Surjadi JU, Wong JC, Kannan V, Lam K-H, Chen X, et al. Wearable and flexible thin film thermoelectric module for multi-scale energy harvesting. Journal of Power Sources. 2020;455:227983.
51. Cheng H, Du Y, Wang B, Mao Z, Xu H, Zhang L, et al. Flexible cellulose-based thermoelectric sponge towards wearable pressure sensor and energy harvesting. Chemical Engineering Journal. 2018;338:1–7.
52. Park J, Joshi H, Lee HG, Kiaei S, Ogras UY. Flexible PV-cell modeling for energy harvesting in wearable IoT applications. ACM Transactions on Embedded Computing Systems (TECS). 2017;16(5s):1–20.
53. Zhao J, Ghannam R, Htet KO, Liu Y, Law Mk, Roy VA, et al. Self-Powered implantable medical devices: Photovoltaic energy harvesting review. Advanced Healthcare Materials. 2020;9(17):2000779.

54. Dagdeviren C, Joe P, Tuzman OL, Park K-I, Lee KJ, Shi Y, et al. Recent progress in flexible and stretchable piezoelectric devices for mechanical energy harvesting, sensing and actuation. Extreme Mechanics Letters. 2016;9:269–81.
55. Sezer N, Koç M. A comprehensive review on the state-of-the-art of piezoelectric energy harvesting. Nano Energy. 2021;80:105567.
56. Hwang GT, Byun M, Jeong CK, Lee KJ. Flexible piezoelectric thin-film energy harvesters and nanosensors for biomedical applications. Advanced Healthcare Materials. 2015;4(5):646–58.
57. Su M, Brugger J, Kim B. Simply structured wearable triboelectric nanogenerator based on a hybrid composition of carbon nanotubes and polymer layer. International Journal of Precision Engineering and Manufacturing-Green Technology. 2020;7(3): 683–98.
58. Kang B-C, Choi H-J, Park S-J, Ha T-J. Wearable triboelectric nanogenerators with the reduced loss of triboelectric charges by using a hole transport layer of bar-printed single-wall carbon nanotube random networks. Energy. 2021;233:121196.
59. Yang HJ, Lee J-W, Seo SH, Jeong B, Lee B, Do WJ, et al. Fully stretchable self-charging power unit with micro-supercapacitor and triboelectric nanogenerator based on oxidized single-walled carbon nanotube/polymer electrodes. Nano Energy. 2021;86:106083.
60. Kim HS, Kim DY, Kwak JH, Kim JH, Choi M, Kim DH, et al. Microwave-welded single-walled carbon nanotubes as suitable electrodes for triboelectric energy harvesting from biomaterials and bioproducts. Nano Energy. 2019;56:338–46.
61. Liu R, Kuang X, Deng J, Wang YC, Wang AC, Ding W, et al. Shape memory polymers for body motion energy harvesting and self-powered mechanosensing. Advanced Materials. 2018;30(8):1705195.
62. Proto A, Penhaker M, Bibbo D, Vala D, Conforto S, Schmid M. Measurements of generated energy/electrical quantities from locomotion activities using piezoelectric wearable sensors for body motion energy harvesting. Sensors. 2016;16(4):524.
63. Maurya D, Kumar P, Khaleghian S, Sriramdas R, Kang MG, Kishore RA, et al. Energy harvesting and strain sensing in smart tire for next generation autonomous vehicles. Applied Energy. 2018;232:312–22.
64. Harnden R, Carlstedt D, Zenkert D, Lindbergh G. Multifunctional carbon fiber composites: A structural, energy harvesting, strain-sensing material. ACS Applied Materials & Interfaces. 2022;14(29):33871–80.
65. Jayababu N, Kim D. Co/Zn bimetal organic framework elliptical nanosheets on flexible conductive fabric for energy harvesting and environmental monitoring via triboelectricity. Nano Energy. 2021;89:106355.
66. Shi Q, He T, Lee C. More than energy harvesting–combining triboelectric nanogenerator and flexible electronics technology for enabling novel micro-/nano-systems. Nano Energy. 2019;57:851–71.
67. Neri I, Travasso F, Mincigrucci R, Vocca H, Orfei F, Gammaitoni L. A real vibration database for kinetic energy harvesting application. Journal of Intelligent Material Systems and Structures. 2012;23(18):2095–101.
68. Chalasani S, Conrad JM, editors. *A survey of energy harvesting sources for embedded systems*. IEEE SoutheastCon. IEEE; 2008.
69. Kumar V, Kumar P, Deka R, Abbas Z, Mobin SM. Recent development of morphology-controlled hybrid nanomaterials for triboelectric nanogenerator: A review. The Chemical Record. 2022;22(9):e202200067.
70. Gabris MA, Ping J. Carbon nanomaterial-based nanogenerators for harvesting energy from environment. Nano Energy. 2021;90:106494.
71. Venkataraman A, Amadi EV, Chen Y, Papadopoulos C. Carbon nanotube assembly and integration for applications. Nanoscale Research Letters. 2019;14(1):1–47.

72. Liu X, Shang Y, Zhang J, Zhang C. Ionic liquid-assisted 3D printing of self-polarized β-PVDF for flexible piezoelectric energy harvesting. ACS Applied Materials & Interfaces. 2021;13(12):14334–41.

73. Chamankar N, Khajavi R, Yousefi AA, Rashidi A, Golestanifard F. A flexible piezoelectric pressure sensor based on PVDF nanocomposite fibers doped with PZT particles for energy harvesting applications. Ceramics International. 2020;46(12):19669–81.

74. Saravanakumar B, Kim S-J. Growth of 2D ZnO nanowall for energy harvesting application. The Journal of Physical Chemistry C. 2014;118(17):8831–6.

75. Malakooti MH, Patterson BA, Hwang H-S, Sodano HA. ZnO nanowire interfaces for high strength multifunctional composites with embedded energy harvesting. Energy & Environmental Science. 2016;9(2):634–43.

76. Lu L, Ding W, Liu J, Yang B. Flexible PVDF based piezoelectric nanogenerators. Nano Energy. 2020;78:105251.

77. Alam MN, Kumar V, Jung H-S, Park S-S. Fabrication of high-performance natural rubber composites with enhanced filler–rubber interactions by stearic acid-modified diatomaceous earth and carbon nanotubes for mechanical and energy harvesting applications. Polymers. 2023;15(17):3612.

78. Esmaeili A, Hossain M, Masters I. Comparison of two compounding techniques for carbon nanotubes filled natural rubbers through microscopic and dynamic mechanical characterizations. Materials Letters. 2023;335:133786.

79. Wang ZL. Nanogenerators, self-powered systems, blue energy, piezotronics and piezo-phototronics–a recall on the original thoughts for coining these fields. Nano Energy. 2018;54:477–83.

80. Wang S, Lin L, Wang ZL. Triboelectric nanogenerators as self-powered active sensors. Nano Energy. 2015;11:436–62.

81. Zhang Z, Li X, Yin J, Xu Y, Fei W, Xue M, et al. Emerging hydrovoltaic technology. Nature Nanotechnology. 2018;13(12):1109–19.

82. Xue G, Xu Y, Ding T, Li J, Yin J, Fei W, et al. Water-evaporation-induced electricity with nanostructured carbon materials. Nature Nanotechnology. 2017;12(4):317–21.

83. Král P, Shapiro M. Nanotube electron drag in flowing liquids. Physical Review Letters. 2001;86(1):131.

84. Kim J, Lee J, Kim S, Jung W. Highly increased flow-induced power generation on plasmonically carbonized single-walled carbon nanotube. ACS Applied Materials & Interfaces. 2016;8(44):29877–82.

85. Liu J, Dai L, Baur JW. Multiwalled carbon nanotubes for flow-induced voltage generation. Journal of Applied Physics. 2007;101(6).

86. Zhao T, Hu Y, Zhuang W, Xu Y, Feng J, Chen P, et al. A fiber fluidic nanogenerator made from aligned carbon nanotubes composited with transition metal oxide. ACS Materials Letters. 2021;3(10):1448–52.

87. Xiao P, He J, Ni F, Zhang C, Liang Y, Zhou W, et al. Exploring interface confined water flow and evaporation enables solar-thermal-electro integration towards clean water and electricity harvest via asymmetric functionalization strategy. Nano Energy. 2020;68:104385.

88. Biswas K, He J, Blum ID, Wu C-I, Hogan TP, Seidman DN, et al. High-performance bulk thermoelectrics with all-scale hierarchical architectures. Nature. 2012;489(7416):414–18.

89. Zhao L-D, Lo S-H, Zhang Y, Sun H, Tan G, Uher C, et al. Ultralow thermal conductivity and high thermoelectric figure of merit in SnSe crystals. Nature. 2014;508(7496):373–7.

90. Kang YH, Bae EJ, Lee MH, Han M, Kim BJ, Cho SY. Highly flexible and durable thermoelectric power generator using CNT/PDMS foam by rapid solvent evaporation. Small. 2022;18(5):2106108.

91. Im H, Kim T, Song H, Choi J, Park JS, Ovalle-Robles R, et al. High-efficiency electrochemical thermal energy harvester using carbon nanotube aerogel sheet electrodes. Nature Communications. 2016;7(1):10600.

92. Kang TJ, Fang S, Kozlov ME, Haines CS, Li N, Kim YH, et al. Electrical power from nanotube and graphene electrochemical thermal energy harvesters. Advanced Functional Materials. 2012;22(3):477–89.
93. Chen S, Luo J, Wang X, Li Q, Zhou L, Liu C, et al. Fabrication and piezoresistive/piezo-electric sensing characteristics of carbon nanotube/PVA/nano-ZnO flexible composite. Scientific Reports. 2020;10(1):1–12.
94. Sanati M, Sandwell A, Mostaghimi H, Park SS. Development of nanocomposite-based strain sensor with piezoelectric and piezoresistive properties. Sensors. 2018;18(11):3789.
95. Mangone C, Kaewsakul W, Gunnewiek MK, Reuvekamp LA, Noordermeer JW, Blume A. Design and performance of flexible polymeric piezoelectric energy harvesters for battery-less tyre sensors. Smart Materials and Structures. 2022;31(9):095034.
96. Yun BK, Kim JW, Kim HS, Jung KW, Yi Y, Jeong M-S, et al. Base-treated poly-dimethylsiloxane surfaces as enhanced triboelectric nanogenerators. Nano Energy. 2015;15:523–9.
97. Kim Y, Wu X, Oh JH. Fabrication of triboelectric nanogenerators based on electrospun polyimide nanofibers membrane. Scientific Reports. 2020;10(1):2742.
98. Zhang H, Zhang D-Z, Wang D-Y, Xu Z-Y, Yang Y, Zhang B. Flexible single-electrode triboelectric nanogenerator with MWCNT/PDMS composite film for environmental energy harvesting and human motion monitoring. Rare Metals. 2022;41(9):3117–28.
99. Xie L, Zhai N, Liu Y, Wen Z, Sun X. Hybrid triboelectric nanogenerators: From energy complementation to integration. Research. 2021;2021, Article ID: 9143762.
100. Nag A, Mukhopadhyay SC, Kosel J. *Printed flexible sensors*. Springer; 2019.

11 A Comparative Study on the Characteristics and Applications of Carbon Nanotubes, Carbon Nanofibres, and Carbon Nanoparticles

Kavithanjali M, Mahalakshmi, Merlin R. Charlotte, Alisha Mary Manoj, Suresh Nuthalapati, Anindya Nag, and Leema Rose Viannie

11.1 INTRODUCTION

A nanoscale material is defined as a substance with at least one dimension, either in three dimensions or as one whose composition is similar to that of a nanoscale (1–100 nm). Nanostructured materials and nanostructured elements are the two categories into which nanomaterials are most divided. In nanostructured materials, the structures are on the nanoscale dimensions and consists of a minimum of one primary structural nanoscale component having an outward diameter in the nanometer range (1). Nanomaterials offer unique characteristics when compared to other materials (2). Nanomaterial has drawn a lot of research attraction because of their small size, quantum tunnelling effect, possible use in conventional materials, electronic devices, coatings, and other sectors, as well as their potential for use in these and other fields. The effects of grain size, production method, and interfacial structures influence the mechanical characteristics of nanomaterials. Owing to their substantial optical, electrical, thermal, mechanical, and chemical properties, carbon-based nanomaterials have received a great deal of attention over the last three decades. Carbon nanomaterials with zero, one, two, and three dimensions, such as fullerenes, carbon nanotubes, graphene, carbon quantum dots, carbon nanohorns, nanodiamonds, and carbon nanofibres, have inherent properties that allow them to be used in a wide range of applications, including medicine, nano-, and microelectronics (fuel cells, supercapacitors, electrolysers, and lithium-ion

DOI: 10.1201/9781003376071-11

batteries), ecological treatments, and technology development. Zero-dimensional CNPs are made of pure carbon, which gives them exceptional mechanical properties (extreme stiffness, strength, and toughness), high stability, significant electronic and optical properties, outstanding electrical and heat conductivity, and low toxicity. They are also highly biocompatible and hydrophobic because of their sp^2 hybridisation. The CDs were created utilising the laser ablation method, pyrolysis, hydrothermal synthesis, electrochemical procedures, and microwave synthesis, among other techniques. Carbon quantum dots (CQDs) and graphene quantum dots are two imperative, different types of 17CDs (GQDs). Researchers preferred porous, structured carbon nanomaterials because of their superior characteristics, including their customisable pore structure, elevated specific surface area, high porosity, good stability, and surface modifiability. Considering the outstanding qualities, such as great electrical and thermal conductivity, high surface area, good mechanical stability, and exceptional corrosion resistance against acid, one-dimensional carbon nanotubes and carbon fibres are of special interest for catalytic support or current collectors (3). In general, carbon nanotubes (CNTs) are made up of concentric shells of carbon atoms joined by intra-shell sp^2 hybridised covalent bonds, with van der Waals forces predominating in their inter-shell interactions (4). The dimensions, chirality vectors, and stack of CNT layers can all be customised (the symmetry of the nulled graphite sheet). CNTs can be split into two fundamental classes on structural basis: single-walled carbon nanotubes (SWCNTs) and multi-walled carbon nanotubes (MWCNTs). In comparison with other fibrous materials, CNTs exhibit superior physical qualities, including stiffness, strength, and flexibility. They possess higher aspect-ratios than other materials. Another class of one-dimensional carbon nanomaterials, carbon nanofibres (CNFs), has fascinated the research community with their excellent thermal, mechanical, and electrical properties. The electronic capabilities and reasonable mechanical strength of CNFs have enabled their use in areas such as nanodevices, energy, environmental sciences, and biosensors. CNF has developed various biosensors for different purposes. Recently, there has been an active research effort to the development of electrical and chemical performance of CNFs by dopant inclusion into carbon and changing the structure of CNFs using the electrospinning process. In instance, incorporation of apertures and multi-channel architectures to CNFs forms the nanostructured carbon material with a maximum surface area. Additionally, since metal and metal oxides can be easily incorporated into CNFs and micropores can be created through chemical and thermomechanical treatment methods like chemical vapour deposition, carbonisation (pyrolysis) of polymeric precursors, and electrospinning techniques, CNFs are extensively used as energy storage and sensor electrode materials (3, 5).

In this chapter, the implications of several properties (optical, thermal, electrical, mechanical, etc.) and the behaviour of carbon nanoparticles, carbon nanotubes, and carbon nanofibres and their applications in energy storage and sensing systems will be explained in detail.

11.2 CHARACTERISTICS OF CARBON NANOTUBES, CARBON NANOFIBRES, AND CARBON NANOPARTICLES

11.2.1 OPTICAL PROPERTIES

An object's optical characteristics are determined by how it reacts to electromagnetic radiation in the visible spectrum. Some of the most important optical properties of nanomaterials are

- Absorption
- Dispersion
- Scattering
- Refractive index
- Transmission coefficient

Doudou et al. investigated the refractive indices (n) and dispersion coefficients of polyvinyl alcohol (PVA)/MWCNT composite films, and it shows the insertion of 0.1 weight percent nanotubes does not help improve the refractive features of PVA; nevertheless, when the quantity of nanotubes is between 0.1 and 1 weight percent, there is a rise in the refractive index. The refractive index depends on temperature and wavelength. In this case, PVA's refractive index falls as temperature rises and wavelength increases with decreasing refractive indices, which indicates the normal dispersion of all materials. Figure 11.1 explains the relationship between the refractive index and temperature (6). Rehammar R et al. fabricated vertically aligned carbon nanofibre using ellipsometry, where finite difference time domain simulations were

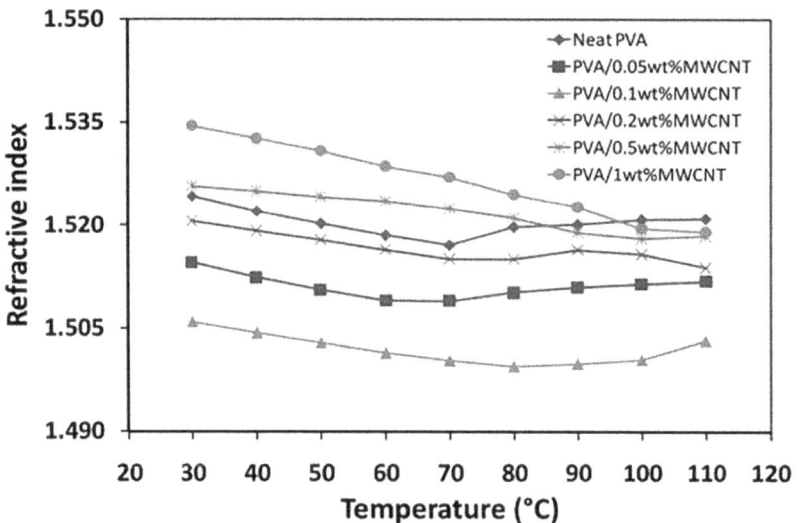

FIGURE 11.1 Demonstrates how the in-plane index refraction values for PVA/MWCNT nanomaterials at $\lambda = 633$ nm is temperature dependent (6).

used to determine a 2D, square, photonic crystal's band structure. It is composed of dielectric pillars with the refractive index of 4.1 for a single nanofibre (7). The behaviour of the light absorption spectrum was studied in carbon nanoparticles, and the results gave information regarding the optical band gap. These results were found by analysing the spectrum in high and low-energy areas.

11.2.2 Thermal Properties

Nanomaterials' thermal characteristics rely on a variety of parameters that are often negligible in bulk materials. In nano-systems, large interfaces are present. For example, in comparison to bulk silicon, silicon nanowires have substantially lower thermal conductivities. The carbon nanofibre has a heat conductivity of 2,000 W/m-K based on direct measurements of carbon nanofibres' parent object or massive vapour-grown carbon nanofibres. The thermal treatment of nanofibre significantly improves the polymer composite's thermal conductivity. Leong KY et al. concluded that highly thermal conductive nanoparticles increase the phase transition material's heat conductivity. Paraffin wax/0.08 wt% MWCNT/Gum Arabic (GA) has poor thermal conductivity, as compared to the results of paraffin wax/0.08 wt% MWCNT with different types of surfactants. Figure 11.2 results show that paraffin wax with 0.08 weight percent MWCNT/GA has greatest latent heat of melting of 100.5 J/g, which is the maximum value achievable (8). By using the one-step electrode location approach, Chen W et al. fabricated the nickel-cobalt-layered double hydroxides nanosheet arrays on carbon cloth by the rational introduction of carbon quantum dots (CQD/NiCoLDH@CC). The CQD/NiCoLDH@CC electrodes were made using 0.50 g of CQD and labelled as CQD/NiCoLDH-3@CC. It retained 60.1% of its capacitance after 10,000 cycles at 10 A g^{-1} and exhibited remarkable consistency in cycling. It also has a capacitance with a high specific value of 1,587.1 F g^{-1} at 1 A g^{-1} and retains 1,281.2 F g1 at 20 A g^{-1} (9).

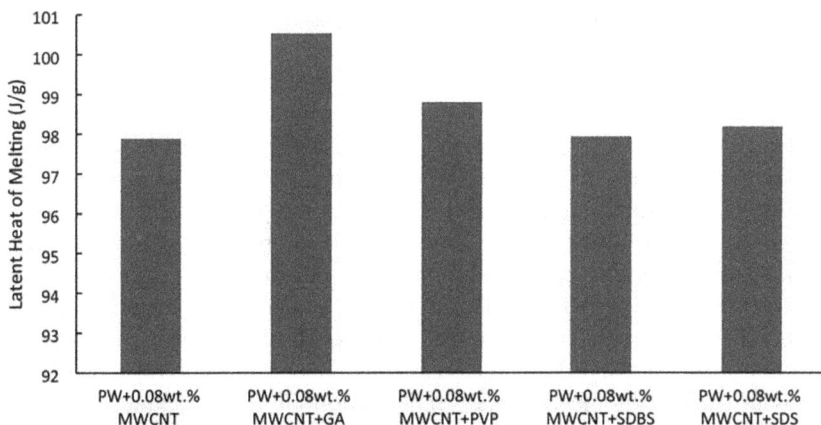

FIGURE 11.2 Depicts the effect of different surfactants on paraffin wax's latent heat of melting/0.08 weight percent MWCNT (8).

11.2.3 PHYSICAL PROPERTIES

Graphite is a well-known allotrope of carbon. The sp2 carbons in the graphitic structure form groups of planar sheets, each of which is called graphene. Carbon nanotubes (CNTs) are defined as sheets of graphene that are rolled over the ends to form cylindrical structures whose length varies from nanometres to micrometres with a diameter that falls within 100 nm. These cylindrical structures are considered derivatives of both fullerenes and carbon fibres. CNTs are divided into two primary categories based on their physical properties: single-walled carbon nanotubes (SWCNTs) and multi-walled carbon nanotubes (MWCNTs) (10). SWCNTs comprise a single graphene layer and have diameters ranging from 0.4 to 2 nm, occurring in the form of hexagonally packed bundles. These structures usually contain only around 10 atoms in circumference and a single atom in thickness. In the case of MWCNTs, two or more co-axial cylinders comprising graphene sheets occur, with a diameter ranging from 1 to 3 nm. In all cases, the graphitic layers are separated by 3.4 Å. The outer diameter of MWCNTs is restricted to less than 15 nm. Structures with higher diameters are called carbon nanofibres (CNFs). The MWCNTs are not as defined as the SWCNTs due to their structural complexities. However, the advantages include low product costs and the ease of bulk production techniques. Another class of CNTs that falls between SWCNTs and MWCNTs is double-walled carbon nanotubes (DWNTs), a combination of both. These are a combination of two nanotubes, separated by 0.34 nm to 0.40 nm. This sufficient band-gap enables their applications in field effect transistors (FETs). The inner and outer walls exhibit different optical and Raman characteristics as well as electrical behaviour. This limits their applications in thin-film electronics. The major difference between CNTs and CNFs is that CNTs are not single molecules but strands of graphene layers. The three other possible structures of CNTs include the armchair, zigzag, and chiral CNTs, as shown in Figure 11.3(a). The basic difference between these structures depends on the way the graphene sheets are rolled up during their formation.

Another structure of carbon that has gained a lot of attention recently is carbon nanofibres. These are basic structures whose length ranges from 0.1 m to 1000 m, with diameters ranging from 3 m to 100 nm. However, the surface structure of these carbon fibres is controlled, based on the type of synthesis used (13). CNFs are highly graphitic in nature, long, and discontinuous. The main feature that makes them different from CNTs is their length, and they are highly compatible with polymer-processing techniques, making them highly dispersible in polymer matrices (14). These structures exhibit high mechanical strength and good electrical and thermal conductivity, which makes them extensively used in applications like thermosets, ceramics, thermoplastics, etc. Another important feature is that it enables surface functionalisation and is, hence, suitable for various chemical and engineering applications. They are generally free-flowing powders in fibrous form (15). The manufacturing processes of CNFs now include several techniques, ranging from the vaporisation of graphite to chemical vapour deposition (CVD). However, based on the synthesis conditions and catalyst used in the process, different morphologies of fibres are obtained. The major types include tubular, fishbone, platelet, ribbon, etc., as shown in Figure 11.3(b). Tubular-type fibres with hollow cores are considered

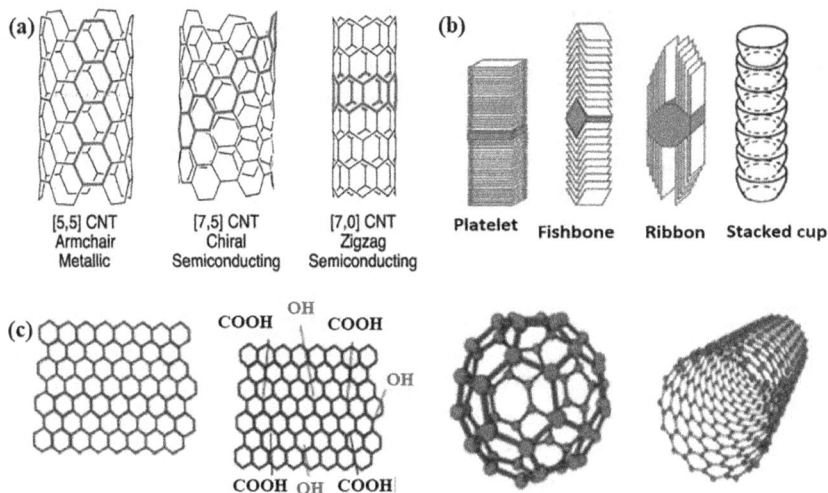

FIGURE 11.3 **(a)** Classification of CNTs as armchair, chiral, and zigzag based on their physical structure. The difference in their structure leads to a difference in electronic properties (11), **(b)** rolling up in different CNT structures (12), and **(c)** different types of carbon nanostructures.

CNTs. Some of the early structural studies showed graphitic basal planes when grown from iron. The fishbone-type fibres were commonly found with nickel and nickel-iron alloys with methane as the carbon source. Details of these structures are discussed in Chapter 1 and in the reviews (16).

The carbon nanoparticles include a whole class of nanostructures from the carbon family, including graphene and its close derivatives like fullerenes, CNTs, CNFs, and carbon quantum dots (CQDs), including fullerene, carbon nanotubes (CNTs), graphene (G) and its derivatives like graphene oxide (GO) and reduced GO, and nanodiamonds (NDs) (Figure 11.3 (c)). The hexagonal, single layer of graphite constitutes graphene. It consists of carbon atoms with a bond distance of 0.142 nm. The chemical oxidation of graphene or graphite leads to the graphene-oxide structure (GO). Here, graphene layers are stacked one over the other and has oxygen-containing functional groups attached to the surface, making them highly hydrophilic. The chemical reduction of GO results in (rGO). To classify graphene as metal, non-metal, or semiconducting is still under debate. Fullerenes are one of the widely studied carbon nanostructures (C_{60}) and contain 60 carbon atoms, in a football shape, a truncated icosahedron with 20 hexagons and 12 pentagons. A single C_{60} has a diameter of 1 nm. Nanodiamonds (ND) are carbon nanoparticles of 2–8 nm and exhibit a truncated, octahedral structure, mainly consisting of a core with sp3 hybridisation and defect-rich sp^2 hybridisation at the surface. These are further classified into two types of reported structures: NDs (DNDs) and high-pressure, high-temperature (HPHT), based on the synthesis method used and that find applications in the medical field (17). Another highly attractive structure is that of graphene nanoplatelets (GNPs). These are usually found in a range of 1–15 mm with a thickness ranging from 100 μm.

11.2.4 ELECTRICAL PROPERTIES

The electrical behaviour of these nanomaterials is highly structure-dependent. Maintaining the electrical properties of CNTs on a macroscopic scale is a difficult task due to the non-uniformity in their diameter, length, chirality, and number of walls. The electrical properties are also defect-dependent, which affects electron transport. Electrical conductivity is found to increase with the presence of amorphous carbon structures and surface functionalities. According to Mintmire, Hamada, and Saito, the tight binding coefficients (n1 and n2) are related to the translation vector ch as ch = n1a1 + n2a2, connecting two sites that are crystallographically equivalent, and hence, determining the conducting behaviour. When 2n1 + n2 is an integer multiple of 3, metallic behaviour is exhibited by CNTs, and non-metallic or semiconducting behaviour is exhibited in all other cases. In other words, mod(n1-n2,3) = 0,1,2. Here, mode1 and mode2 SWCNTs are semiconducting in nature, and mode0 (n1) behave like metals at room temperature, and the energy gap depends on the chiral nature, exhibiting a quasi-metallic behaviour. In low temperature regions, n1 = n2 signifies that the armchair nanotubes are truly metallic. The zig-zag nanotubes can be metallic or semiconducting.

The electronic properties also change for the SWCNTs and MWCNTs. SWCNTs usually behave as p-type semiconducting materials. This property is usually induced by the higher work function of the contacts, which results in electron transfer from the NTs to the contact-generating holes in the NTs. In the case of field effect transistor-based devices (FETs), with a top gate configuration, CNT-based devices show ambipolar characteristics with higher drive current properties [18]. The determination of Schottky-barrier heights for electron holes at the metal-CNT interfaces helps determine the device characteristics accurately. Reports also show that vacuum annealing, oxygen removal, and doping with alkali metals help convert p-type conductivity to n-type, and hence, provide easy tunability for various device applications [19].

The spun fibres produced from CNTs exhibit a range of conductivities from around 10 S cm 1 to 67,000 S cm^{-1}. DWCNTs of smaller diameter as spun from CVD reactors exhibit a higher conductivity of around 20,000 S cm^{-1} [20]. CNT structures are highly affected by the synthesis conditions. These are also found to be sensitive to the amount of oxygen present in the atmosphere. The oxygen-related defects can lead to increases in conductivity by inducing p-type doping in CNTs [21]. The intrinsic electrical and mechanical properties of SWCNTs change on functionalisation due to changes in C=C bond structures, unlike those of MWCNTs, where chemical changes happen only at the outer walls.

In the case of CNFs, all the electrical conductance properties are influenced by the highly connected nanofibre network. A better dispersion in the matrix medium and sufficient length of the fibres help in attaining high electrical conductivity suitable for various applications. Since the main applications of CNFs are in composite making for various applications, they act as an excellent candidate due to their very high aspect ratio. They exhibit excellent conductivity, even with lower fibre loadings. For the very first time, the room-temperature electrical conductivity of CNFs was described by Endo et al. as 5×10^{-5} cm [22] For different applications, different resistivity ranges of CNFs are used, such as EMI shielding preferentially with a resistance of 103 to 101 cm, electrostatic dissipation (ESD) (106–108 cm), protection from lightning strikes (10 cm), etc. (14).

The two pi-electrons that are found in the hexagonal structure of graphene sheets give rise to superior conductivity characteristics for graphene. However, the incomparable electrical conductivity of any non-metallic element makes them highly promising for next-generation technologies. Zero-band-gap characteristics and little overlap between valence and conduction bands make them attractive. The charge carriers in graphene are called Dirac fermions, since they behave as semi-metals and exhibit the quantum hall effect (QHE) (23). In GO structures, the existence of electron-rich oxygen species and a highly conducting graphene backbone makes them excellent candidates for electrical applications ranging from materials chemistry to quantum physics (23).

11.2.5 CHEMICAL PROPERTIES

Generally, nanomaterials possess predominant chemical parameters like defined molecular structure, high reactive sites, pH range, zeta potential, and ionic stability. In particular carbon nanomaterials are differentiated into carbon nanotubes, carbon nanoparticles and carbon nanofibres whose chemical behaviour has been compared in the following sections.

11.2.5.1 Carbon Nanotubes

Electrochemical sensors utilise a variety of nanomaterials. Carbon nanotubes are regarded as remarkable nanostructures due to their distinctive mechanical, electrical, and chemical properties. CNTs having sp^2 carbon groups were several nanometres wide and few micrometres long. Multi-wall (MW) and single-wall CNTs can be produced using chemical vapour deposition, arc discharge, and laser evaporation methods. CNTs act like metals or semiconductors, relying on their dimension and degree of helicity. It is suitable for exploring multiple electrodes in both aqueous and non-aqueous solutions due to its exceptional electrochemical and chemical stability as well great electronic properties for electron transfer reactions (24).

Wu et al. prepared multi-walled carbon nanotubes by a microwave irradiation method with carboxylation, varying the operating time to change the degree of functionalisation. Critical coagulation concentration (CCC) values have been increased gradually with the presence of NaCl (CCC ~ 142.14 to 268.69 mM) and $MgCl_2$ (CCC ~ 0.97 to 5.32 mM). The significant chemical factors such as dispersibility, reagent strength, operating time, and temperature governs the functionalisation degree via generation of carboxylate groups in CNT structure. Dispersibility tests have been conducted to evaluate the surface charge and crystalline size in the presence of salts. The results show that f-CNTs produced tiny particles with a negative zeta potential and higher critical coagulation concentration values, indicating strong colloidal stability (25).

Gosh D et al. investigate the alteration in carbon nanotube transport mechanism through wet absorbent media due to the variation of chemical parameters such as pH range (5–9) ionic strength (10–100mM), natural organics (0.1–10 mg/L), and particle concentration (10–50 mg/L). As ionic strength and particle concentration increased, charge transmission became more difficult. Ghosh et al. predicted the interaction between molecules by DLVO theory (26).

11.2.5.2   Carbon Nanofibres

Having extensive tensile behaviour, carbon nanofibres are formed with high chemical and thermal stability irrespective of oxidising agents which renders for superior electrical and thermal conductivity. Zhang Li (27) and team has explored the structural control, functionalisation, and use of CNF in electrochemical sensors. The geometry and porosity of CNF are easily altered by adjusting parameters, employing sintering process, or using the co-axial spinning technologies. One of the effective techniques for modifying CNF is heteroatomic doping, which could improve the hydrophilicity and permeability of CNF-related electrodes and enhance their ionic conductivity. N, B, O, P, and S are the heteroatoms that have been investigated the most. NiCo-modified carbon nanofibres with layered double hydroxides (LDH) were investigated for their structure and electrochemical characteristics (28). Successively synthesising -Fe_2O_3/-CD-CNFs, Sebastian et al. (29) discovered that these materials had a higher defect density than pure CNFs, implying fewer defects and disarrays. Therefore, it is widely accepted that CNFs can link efficiently to heteroatoms, CNTs, graphene, and various metals, enhancing their structural and chemical characteristics for use in particular cutting-edge applications.

11.2.5.3   Carbon Nanoparticles

Carbon dots represent the small (< 10 nm) carbon compounds with sp^2/sp^3 carbon chains and reactive hydroxyl groups on their surfaces. Contrary to inorganic nanoparticles, carbon dots are extremely water soluble and are not carcinogenic. Hua et al. prepared carbon dots via microwave-assisted method to study the influence of sonication and hydrothermal treatment on the material's behaviour. It is noted that the ultrasonic irradiation degrades the radiating element like chromophore by accelerating the oxidation of surface functional groups, according to FT-IR and XPS studies [Figure 11.4(a) and (b)].

However, the Hg^{2+} ion reactivity and photochemical reduction of C-dots MW considerably decline as a result of the oxidation of surface functional groups. The findings further validated the theory that carbon dots synthesized by microwave assisted method remain either in the early stages of decomposition or the combustion phase, and their characteristics may be enhanced by thermal treatment at a low temperature, making them useful for ion-sensing experiments. Fluorescence quantum yield (ϕ_f) of C-dots MW was found to increase despite reducing their reactivity to Hg^{2+} ions by autoclaving them at 100°C (30).

11.2.6   MECHANICAL PROPERTIES

Mechanical characteristics describe how a material behaves mechanically under various external loads and environmental conditions. The mechanical properties of metals, which are considered to be classic materials, are typically comprised of hardness, flexibility, plasticity, tensile point, durability, tenacity, coarseness, etc. Nanoparticle volume, surface, and quantum effects are responsible for the exceptional mechanical behaviour of nanomaterials (31).

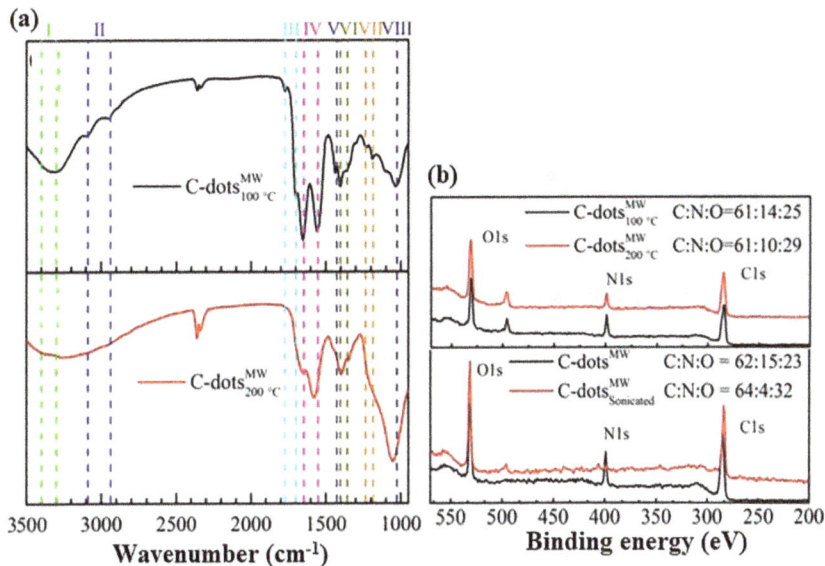

FIGURE 11.4   Spectral analysis from **(a)** FTIR and **(b)** XPS of microwave assisted C-dots, treated at 100°C and 200°C, are shown (30)

11.2.6.1   Carbon Nanotubes

To improve compressive and flexural strengths, the ideal ranges of CNT length, diameter, and concentration were found. CNTs with short and tiny diameters might help to improve tensile strength. Besides, flexural strength was considerably improved by relatively stretching and widening the diameter of CNTs. In general, CNTs having lengths between 10 and 20 m and a diameter between 20 and 32 nm were found to have the best long-term mechanical performance. In order to obtain compressive and flexural strengths, the optimal concentrations were found to be 0.15 and 0.20 c-wt% (32).

Antonio R et al. (33) prepared double-walled and triple-walled carbon nanotubes (DTWCNTG) via chemical vapour deposition to tune the mechanical property of the conservative, methacrylate-based dental polymer. The team made a study on physicochemical and morphology varying the concentrations. The mechanical characteristics of modified DTWCNTG at various concentrations, including tensile strength, elastic modulus fracture toughness, and modulus, were investigated. Figure 11.5 depicts that DTWCNTG with 0.001% has greater elastic modulus (~ 165%) and compressive strength (~ 250%).

11.2.6.2   Carbon Nanofibres

Zhou H et al. experimented the curing process for epoxy resin using chlorinated fluorene amine and compared it with other curing agents. The study reveals that inclusion of Al_2O_3 nano powder into epoxy-fluorene amine and epoxy-DDS (Diaminodipephenylsulfone) to reduce temperatures also has a slight impact on the adhesive matrix's viscosity. The compression strength and modulus range of the

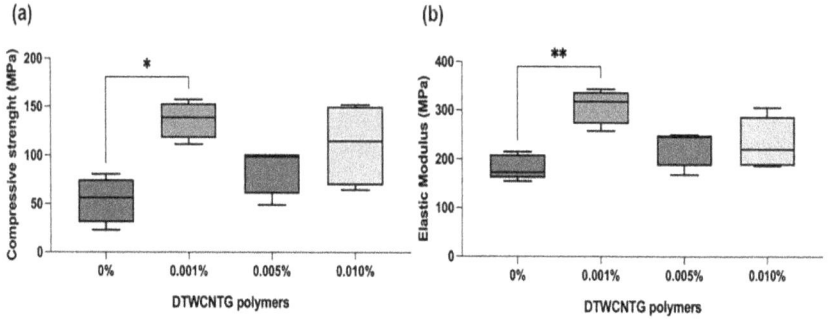

FIGURE 11.5 (a) Compressive strength and (b) elastic modulus of DTWCNTG at different concentration (33).

system containing epoxy-fluorine amine is uplifted compared to epoxy-DDS system; as well, Al_2O_3 nano powder inclusion can improve the compression modulus of both systems while having differing impacts on the compression strength. When compared to DDS as hardening compound, the application of reinforced skeleton structure of fluorene amine can greatly increase the tensile strength of epoxy resin, while the shear resistance increases and the impact strength somewhat decreases. With the gradual increment in the concentration of Al_2O_3 nanoparticles leads to the rise in compression modulus of epoxy resin.

11.2.6.3 Carbon Nanoparticles

Ataol et al. investigated the mechanical properties of fluorescent carbon nanoparticles (FCN) with the incorporation of an acrylic-based tissue conditioner and a silicon-based, soft denture liner. By performing tests, it is found that significant factors like toughness, durability, and shear bond modulus are decreased. Ataol et al. predicted that inclusion of licorice root extract-based FCN to acrylic-based tissue conditioner had no hinderance on the stress, strain, and elasticity of the material that is recognised for its antioxidant and antibacterial characteristics (34).

11.3 APPLICATION OF CARBON NANOTUBE, NANOFIBRE, AND NANOPARTICLES

11.3.1 ENERGY STORAGE DEVICES

Due to the quick growth of the global economy, energy has taken centre stage in both scientific and industrial circles. Due to worries about contamination of the environment, rising mining costs, and the depletion of fossil fuels, there is an urgent need for continual material for energy storage. Natural high-carbon materials have attracted a lot of scientific attention due to their many advantages, which include their light weight, large surface area, strong electrical conductivity, thermally stable, semiconducting nature, and inherent structural flexibility, all of which help to increase the effectiveness of energy conversion and storage (35, 36). Because of these benefits, carbon-derived materials are preferred for a broad range of applications. As depicted in Figure 11.6, these carbon materials include

1) Carbon nanotube
2) Carbon nanofibre
3) Carbon nanowire
4) Graphene quantum dots
5) Graphene nano-particles

Beyond these, several carbon derivatives have been investigated as useful components for enhancing the energy-conversion and storage performances in devices: supercapacitors, battery, and fuel cells (37).

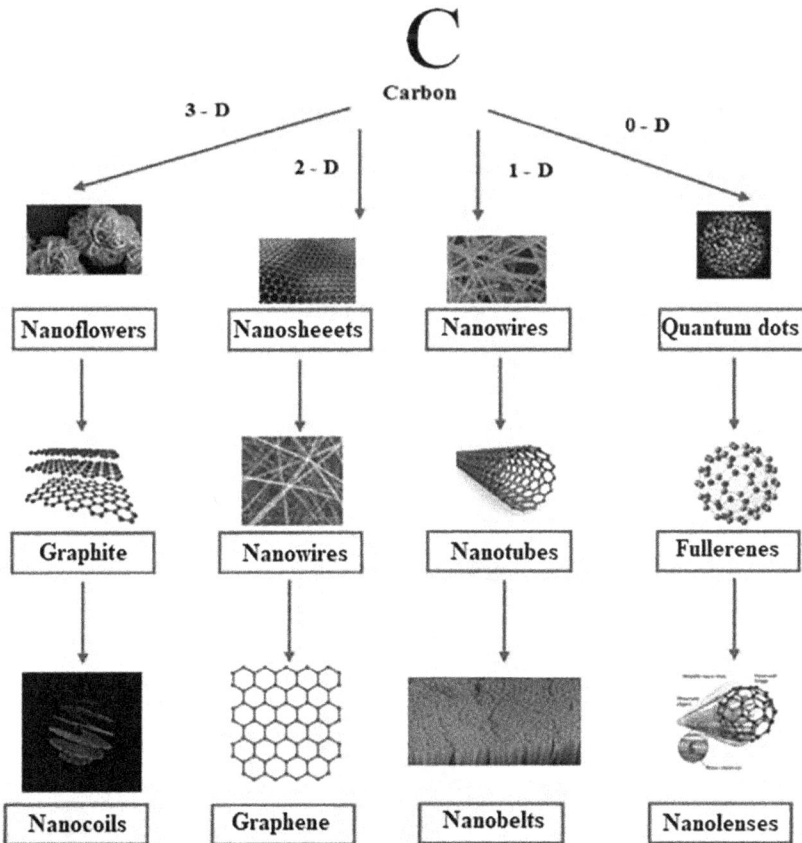

FIGURE 11.6   Various forms of carbon and derived structures (38).

11.3.1.1 Supercapacitor

Supercapacitors have higher energy densities and capacitance values than ordinary capacitors, but they also have lower voltage thresholds. Generally, the greatest single-cell voltage rating is 3.0 Volts DC. We can store electrical energy in it, and it should be a sustainable energy source, since it won't produce any dangerous gases for humans. Supercapacitors have received a lot of attention in comparison to batteries because of their quick charge-discharge rate, prolonged life cycle, high power density, and absence of short-circuit problems (39).

11.3.1.1.1 Carbon Nanotube

CNTs are an interesting electrode material for creating high-performance supercapacitors because of their inexpensive cost, excellent charge transfer capacity, stable thermal strength, large porous surface area, and good conductive nature. To increase the strength, stability, and longevity of supercapacitors, research has been done to create several kinds of CNT electrode materials (37). The deposition achieved by the chemical bathing process was employed by Pandit B et al. to encapsulate V_2O_5 across a network of linked MWCNTs. Figure 11.7(a) demonstrates the built-in device

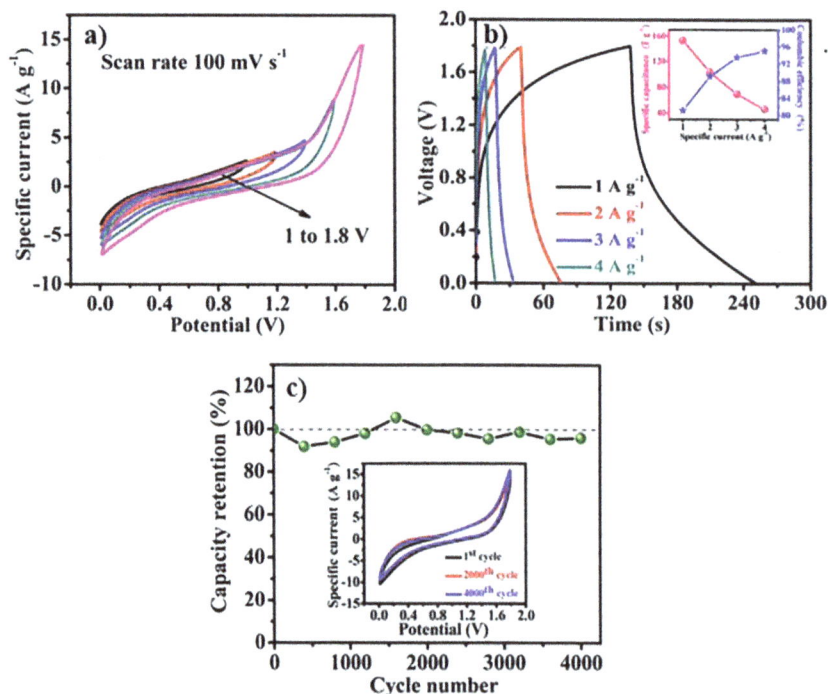

FIGURE 11.7 (a) Shows cyclic voltammetry (CV) curves for various voltages between 1 and 1.8V at a 100mV/s scan rate; (b) charge-discharge (CD) curve is shown at mulitple current densities, ranging from current density of 1 to 4 A/g; (c) displays the consistency in cycling for 4,000 cycles at 100mV/s scan; the picture illustrates CV curves for various cycle counts at this same scan rate (40).

exhibits perfect supercapacitive action throughout an enlarged potential range from 0 to 1.8 V. Figure 11.7(b) depicts a flexible V_2O_5/MWCNTs symmetric supercapacitor with a maximum energy density of 72.07 Wh/kg and a specific power density of 2.3 kW/kg at current density of 1 A/g. Solid-state devices had 81% Coulombic efficiency (CE) at 1 A/g and raised to 95% at 4 A/g. Figure 11.7(c) illustrates the cycling stability, which is greater than that of the similar liquid electrolyte design and displays 96% charge retention after 4,000 cycles (40).

Chen Y et al. reported the electrochemical performance of CNTs made from low-density polyethylene (LDPE)-as electrodes and 6M KOH solution as electrolyte. The initial pyrolysis temperature was 450 °C, while the second, with carbonisation temperatures (850 °C, 800 °C, 750 °C, 700°C, 650° C) for 6, 4, 2, and 0 hours, were examined. The CNTs produced at 750 °C have a bamboo-like structure, the greatest yield (41.9%), and the maximum carbon conversion (61.2%), as assessed by FETEM and FESEM techniques. Figure 11.8(a) and (b) displays galvanic charge-discharge (GCD) and CV curves for 4H-CNT electrodes under various circumstances. A test of 10,000 CD cycles at a current density of 5 A/g was performed on an electrode to determine its durability. Figure 11.8(c) shows the electrode maintained at 92.85% Coulomb efficiency and it has 93.16% capacitance with 10,000 CD cycles (41).

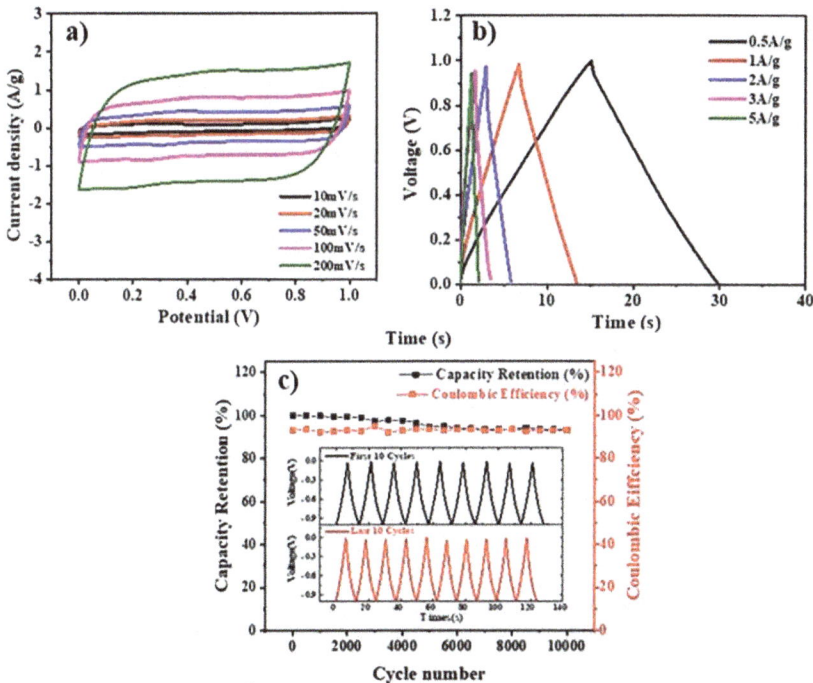

FIGURE 11.8   (a) Represents the 4H-CNTs capacitor's CV curves at a range of voltage sweep rates, (b) illustrates the 4H-CNTs capacitor's GCD curves at various current densities, (c) displays the capacitance retention and Coulomb efficiency curves of the 4H-CNTs electrodes following charge–discharge (41).

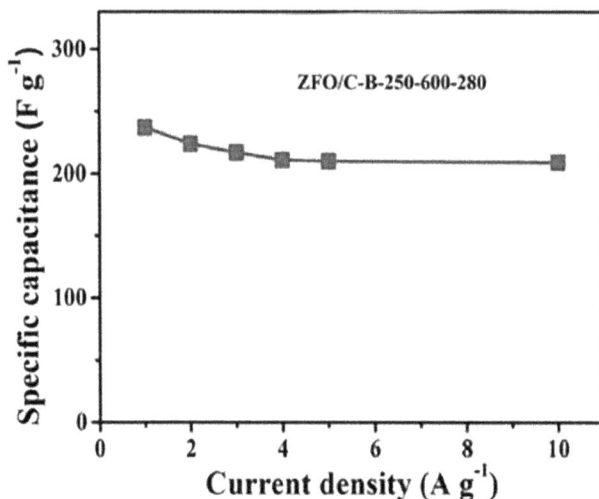

FIGURE 11.9 Displays the ZFO/C–B-250–600–280 capacitance curve shown at varied current densities (43).

11.3.1.1.2 Carbon Nanofibres (CNF)

Carbon nanofibre are cylinder-shaped nanostructures made up of stacked cones, cups, or plates made of graphene. CNF is a 1D carbon nanomaterial with a diameter of 50–200 nanometres and lengths up to few micrometres (42).

Using the electrospinning technique, Yang S et al. created stable $ZnFe_2O_4$/carbon (ZFO/C) nanofibres. The ZFO/C-B-250–600–280 electrode is made using the electrospinning technique, which was subjected to 250 °C stabilisation treatment, 600 °C thermal carbonisation, and 280 °C further-induced annealing. The ZFO/C-B-250–600–280 nanofibre electrode yields a specific capacitance of 237 F/g at 1A/g by using 2M KOH electrolyte solution. Figure 11.9 depicts the specific capacitance as a function of current density for the ZFO/C-B-250–600–280 electrode, illustrating the electrode's high-rate capability (43).

11.3.1.1.3 Carbon Nanoparticles

Carbon quantum dots (CQDs, or C-dots) were unexpectedly found by Xu et al. in 2004 while purifying the single-walled carbon nanotubes. Their diameter is less than 10 nm (44). Sun et al. used a hydrothermal technique to synthesise the arrays of nickel hydroxide nanosheet adorned with carbon quantum dots (CQDs/-Ni $(OH)_2$). At 3 A/g, the composite's specific capacitance of 1724.0 F/g is 2.3 times higher than the 750.7 F/g of bare -Ni $(OH)_2$. Increasing the current density from 3 to 50 A/g yields a capacitance retention rate of 45.8%. The NiOH-C-0.2 GCD curves at a various current densities are shown in Figure 11.10(a). The built-in asymmetric supercapacitor NiOH-C-0.2/AC displays maximum energy densities of 44.0 and 29.03 Wh/kg at 3000 and 18,015 W/kg, respectively. After 5,000 cycles, 99.91% of the capacitance

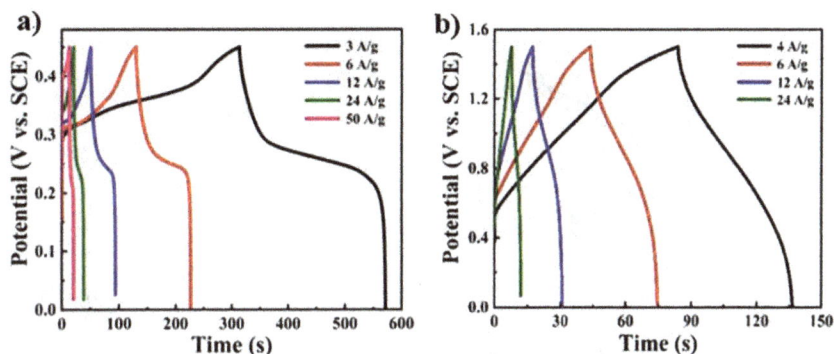

FIGURE 11.10   Displays the GCD curves of NiOH-C-0.2 (**a**) and NiOH-C-0.2/AC (**b**) at various current densities (45).

retention rate and 99.46% of CE are still present at 10 A/g R-15. Figure 11.10(b) shows the GCD curves for different current densities in asymmetric NiOH-C-0.2// AC cells (45).

11.3.1.2   Battery

Batteries store a lot of energy compared to supercapacitors. A positive cathode, a negative anode, and an electrolyte make up every battery. There are two types of batteries:

- Primary and
- Secondary battery

Single-use batteries are the primary battery. Secondary batteries are recharged up to 1,000 times. Lithium-ion batteries are among the most significant types due to their low discharge rate, strong performance across a broad temperature range, lightweight, and high specific energy density (46).

11.3.1.2.1   Carbon Nanotube

CNT has a more porous surface area, stable electrochemical characteristics, and is more affordable. CNT have been investigated as electrode materials for lithium-ion battery anode applications. By using the chemical vapour deposition approach, Kaneko K et al. produced the graphite-CNT/(lithium cobalt oxide) LCO-CNT electrodes. In order to make negative and positive electrodes, active materials (graphite and LCO) with 97-mass-percent concentration were trapped within a self-supporting CNT matrix. While selecting the characteristics of CNT-based electrodes, the specific surface area and length of a raw CNT material are very important. Area capacities of 1 mA h cm^{-2} were used for both the negative and positive electrodes (137 mA h g^{-1} and 7.3 mg cm^{-2} for LCO; 372 mA h g^{-1} and 2.7 mg cm^{-2} for graphite) (47). As seen in Figure 11.11(a), the entire cells exhibit a normalised charge capacity between 0.2 and 0.4 at voltages ranging from 3.1 to 3.7 volts. CNT with a large surface area causes

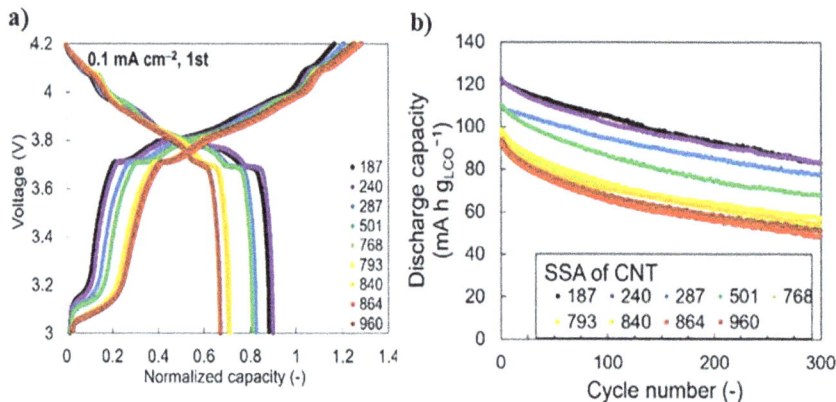

FIGURE 11.11 (a) Displays charge-discharge curves at the first cycle of the graphite-CNT/LCO-CNT complete cells, which have electrodes based on CNTs with different specific surface areas. (b) Depicts the discharge capacity of a complete graphite/LCO cell throughout a cycle (47).

too much electrolyte to break down at the negative electrode, which leads to a permanent capacity loss and a deficiency in active Li ions. Figure 11.11(b) illustrates the performance where CNTs with relatively small surface areas of 300 m^{-2} g^{-1} are preferable for attaining a higher rate efficiency, while those with tiny surface areas of 187–240 m^{-2} g^{-1} are better for maximum reversible capacity at lower charge-discharge rates.

11.3.1.2.2   Carbon Nanofibres

In this study, Li Y et al. explored carbon-coated Si/N-doped porous-carbon nanofibre composites (Si/pCNF@C) with a double-buffered structure that serves as an active anode material. With a specific capacity of 740 mAh g^{-1} at 1 A/g and a 66% capacity retention rate after 400 cycles for Si/pCNF@C electrode, Figure 11.12(a) shows that it has strong discharging specific capacitance at a current density of 0.1 A g^{-1} and Figure 11.12(b) compares the performance of Si/CNF@C and Si/pCNF@C. This reveals that the inclusion of the extremely porous and the improved N-doping composition facilitates both transmission of charge at the electrode interfaces and Li-ion diffusion inside the electrode (48).

11.3.1.2.3   Carbon Nanoparticle

This section discusses the use of quantum dots in batteries for electrochemical energy storage systems. Electrodes treated with carbon dots offer a significant increase in capacitance, performance, and high stability. According to research by Javed M et al., synthesised carbon quantum dots (CQDs) from D-(+)-glucose after 500 CD cycles at a current rate of 0.5 C, the CQD-anode material still had 864.9 mA h g^{-1} of Li-storage capacity compared to 153.7 mA h g^{-1} at 20 C. When CQDs were examined as an anode material for sodium-ion batteries, they showed exceptional capacity retention of 72.4% with a good cyclic stability (after 500 cycles) and displaying the

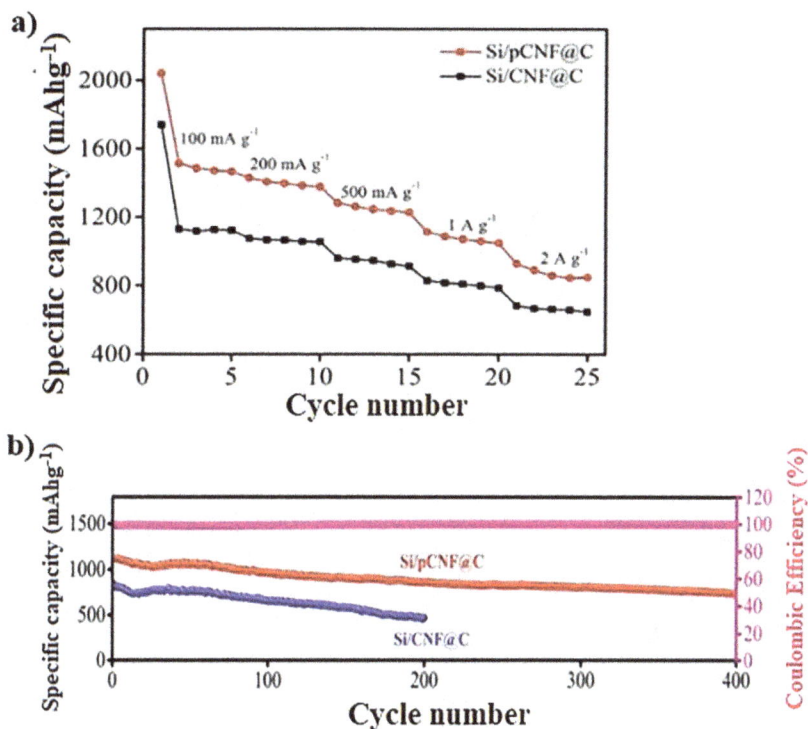

FIGURE 11.12 Shows **(a)** rate and **(b)** cycle performance of Si/CNF@C and Si/pCNF@C (48).

specific-capacity values of 323.9 mAh g^{-1} at 0.5 C (Figure 11.13(b)) and 55.7 mAh g^{-1} at 20 C. It nevertheless recovers a substantially bigger capacity (853.3 mAh g^{-1}) and has an exceptional efficiency (99.5%), as shown in Figure 11.13(a), even after cycling at high rates and setting the current density to 0.5 C (49).

11.3.1.3 Fuel Cell

More people are paying attention to fuel cells because of the numerous benefits they may provide, such as great energy efficiency and minimal environmental load. In a fuel cell, chemical energy is transformed into electrical energy. Carbon nanomaterials show potential for fuel cell applications as well.

11.3.1.3.1 Carbon Nanotube

CNT are broadly used materials due to their thermal stability, distinctive electrical structure, and steady charge transfer capabilities. Wang W et al. hydrothermally synthesised Pt-CoFe@NCNT/CFC and 20 wt% Pt/C, as cathodes were used. Figure 11.14 displays the power density and I-V polarization curves at 20, 30, 40, and 80°C, and it shows a peak power density of 73.41 mW cm^{-2} 80°C. Due to the direct drop-coating of commercial 20 wt% Pt/C on carbon fibre fabric, the performance of

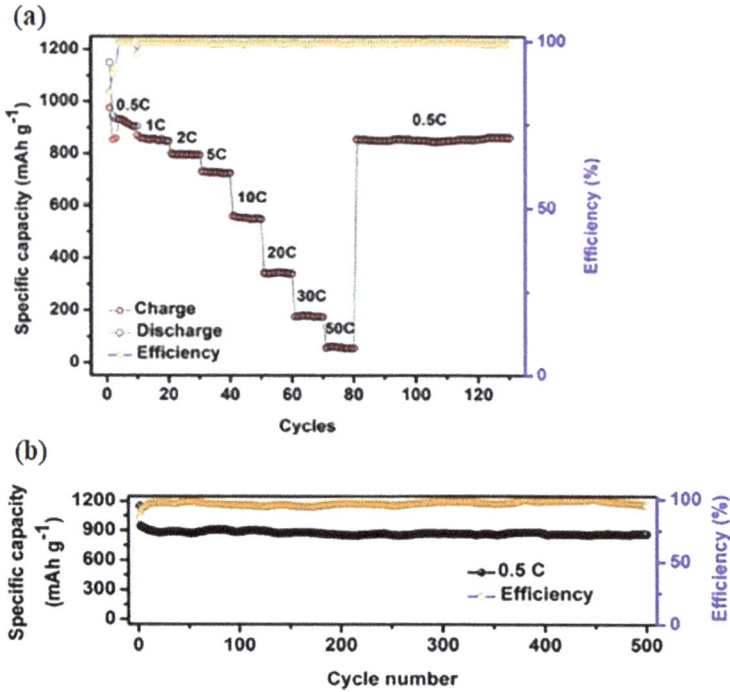

FIGURE 11.13 (a) Displays the CQD rate performance and related Coulombic efficiency (CE) of the electrode at various current densities at 0.5 C, whereas (b) displays the cyclic stability with the associated CEs of the CQD electrode at 0.5 C (49).

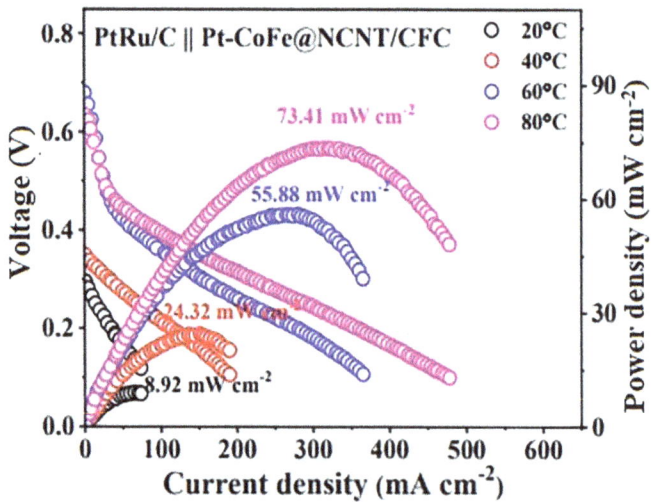

FIGURE 11.14 Shows the power density and I-V curves of DMFCs using Pt-CoFe@NCNT/CFC (50).

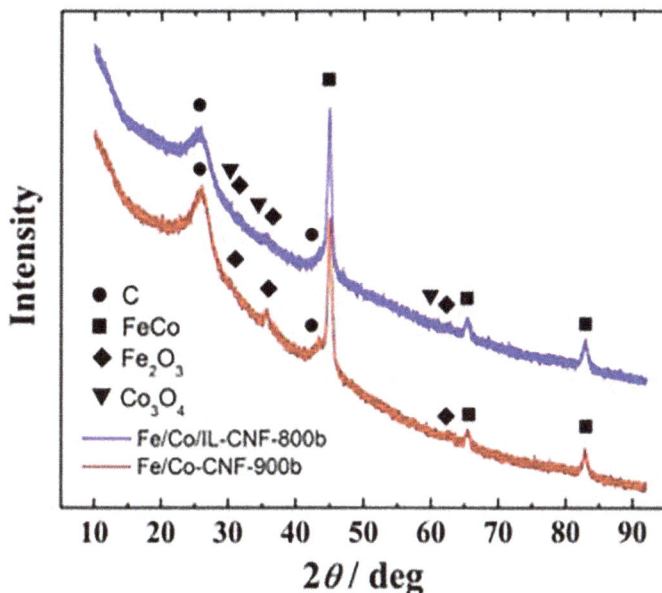

FIGURE 11.15 Displays the XRD patterns for the catalysts Fe/Co/IL-CNF-900b and Fe/Co-CNF-800b (51).

direct methanol fuel cell (DMFC) was improved and has the large power density of 137.46 mW cm^{-2} (50).

11.3.1.3.2 Carbon Nanofibres

CNFs got more attractive than CNTs because of their flexibility, unique textural/morphology, and hierarchical pores, and these offer decorative active sites. Owing to the attention-seeking behaviour of nanofibres, Sokka A and his colleagues decided to investigate the assembly of membrane-electrodes containing a cathode with a nanofibre-based catalyst manufactured using (Fe/Co/IL-CNF-900b), (Fe/Co-CNF-800b), commercial Pt/C, and ionic liquid (IL). All of them operate as cathode catalysts, exhibiting Pmax values of 195, 168, and 250 mW cm^{-2}. A commercial Pt/C catalyst produced a Pmax value for the Fe/Co/ILeCNFe800b that was 78% higher. XRD analysis was used to assess the produced non-precious metal catalyst's crystalline structure (Figure 11.15) (51).

11.3.1.3.3 Carbon Nanoparticle

Hyperbranched macromolecule (HBM) membranes filled by graphene oxide (GO-HBM) and HBM filled by graphene quantum dot (GQD-HBM) composites were created by Xu G et al. utilising the drop-casting technique.

The GQD-HBM composite tested here had a proton conductivity of 0.12 S/cm at 30 °C (Figure 11.16) and with 100% absolute humidity, it had a max power density of 20.5 mW/cm². This all indicates that GQD-HBM is a trustworthy material for the proton exchange membrane in DMFC (52).

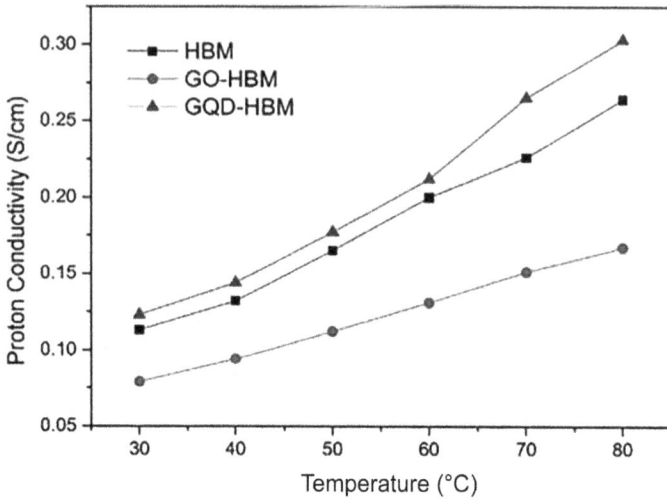

FIGURE 11.16 Illustrates the relationship between proton conductivity and temperature for the purest HBM, GO-HBM, and GQD-HBM membranes (52).

11.3.2 SENSOR APPLICATION

Sensors aren't usually designed for fashionable reason and are software acquainted. In order for various types of sensing to function, sensors must be compatible and comparable as speed sensors for coinciding with the velocity of multiple motors, temperature detectors used for controlling the temperature, and ultrasonic sensors for measuring the distance. Being an integral part of modern living, a sensor is a module known for the detection of changes that occurred in the environment. A sensing device is mainly used to collect the data as the physical input from the atmosphere, analyse ('sense') the event, and provide the information as signals through electronics like a computer processor. Primarily, sensors are made up of chief components, as shown in Figure 11.17.

1. Sensing element
2. Processor
3. Transmission unit
4. Power supply

This electronic subsystem measures temperature, gauge distance, smoke, pressure, pH range, stress/strain, speed, torque, and the presence of chemical elements. Sensors are used in day-to-day life, especially in smartphone displays, smart doors, and lamps in buildings and transports with soft-touch technology development. Some other examples of sensing devices used are aerospace and airplanes, medicines, cars, robotics, manufacturing, and machinery. Different types of sensors were developed recently to detect the physicochemical properties of the materials (such as

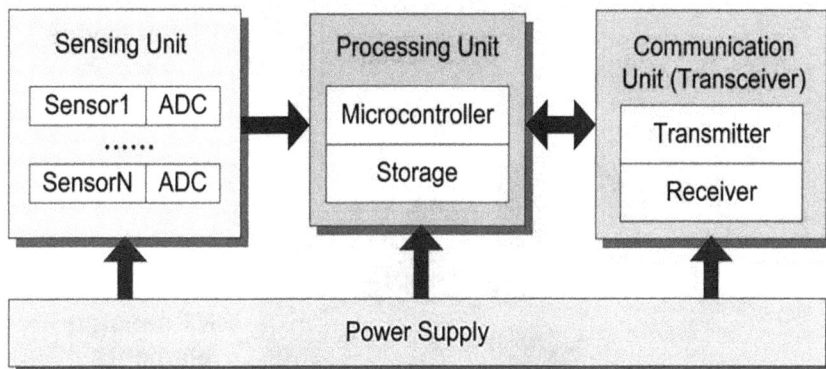

FIGURE 11.17 Pictorial representation of sensor components (53).

temperature, pressure, resistance, capacitance, conduction, heat transfer, etc.). High sensitivity, quick reaction, cheap cost, high volume production, and high reliability are the primary requirements for an excellent sensor (54).

Carbon-based nanomaterial's detection signal for analytical discovery resolutions is of the utmost curiosity among nano-structured materials (55). It is designed to render an extensive number of applications from the perspective of eco-friendly and biomedical monitoring to develop research with current trends in this desired sector. These applications have been made possible by the distinctive features of CNTs (like microstructure, functional, electronic, and optical). The predominant carbon nanotubes, like single-walled carbon nanotubes (SWCNTs) and multiwall carbon nanotubes (MWCNTs), are the potential candidates for trailblazing the sustainable and biomedical applications (56, 57).

11.3.2.1 Types of Sensors

Sensors are found to be used in electrical and electronic appliances to measure the physical properties like pressure, temperature, light, resistance, capacitance, conductance in our living environments like home, cars, trains, buses, offices, laboratory, power plants, food factories, industries, etc. There is an enormous number of types of sensors developed, based on the parameters such as position, pressures, temperature, strain, electrochemical reactions, humidity, etc.

11.3.2.2 Temperature Sensor

Scientists have been fascinated by the evolution of sensors recently, while the use of nanostructures has resulted in the creation of innovative types of nanoscale sensors. The miniaturisation of sensors is a result of the demand for small, rapid devices with high responsiveness, reduced usage of power, and quick reaction times. Without disrupting the environment of the neighbours, it can also be installed in precise placements in difficult-to-reach places or places with rapid temperature variations (such as in cryogenic systems). It is still difficult to accurately measure temperature with minimal power consumption, the maximum sensitivity, and the smallest element.

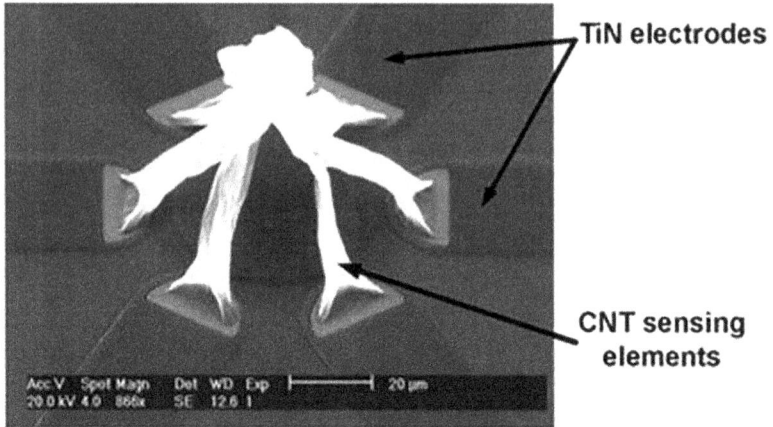

FIGURE 11.18   SEM image of a carbon nanotube-based temperature sensor (58).

The thermal, mechanical, and electrical behaviour of nanomaterials, such as carbon nanotubes (CNTs), can be used to make a variety of sensors that perform significantly better than other devices of a similar size.

11.3.2.2.1   Carbon Nanotubes

CNTs are, indeed, an essential candidate as a novel material that provides solutions for the creation of temperature sensors in the future due to their good electrical responsiveness to temperature changes. Temperature sensors that exhibit metallic or semiconducting behaviour have been described for both single-walled and multi-walled CNTs (58).

Zhao et al. prepared the carbon nanotube graphene oxide (CNT-GO) to fabricate the temperature sensors with PET tapes as temperature sensitive traces. In this study, Figure 11.18 represents the SEM image of temperature sensor developed by self-assembly process in a 3D model, which shows MWCNT networks consisting of six TiN electrodes over the substrate that possess faster response time than thin films. Figure 11.19 shows the results obtained from temperature-resistivity characteristics of the carbon nanotubes whose temperature coefficient is said to be better than -001%/K after numbers of cycles under the operating temperature between 20°C -130°C.

11.3.2.2.2   Carbon Nanoparticles

In a work published by Nguyen et al., carbon nanodots were created by introducing graphite powder in ethylenediamine via fractional photoablation. Sensitivity to temperature has been plotted on the graph. The thermal characteristic peaks of C-dots at 320 nm excitation are depicted in Figure 11.20(a). As temperature rises, non-radiative decay mechanisms become active, which causes the emission strengths of the 400 and 465 nm peak to steadily drop. The peak intensities vary exponentially as the temperature increases from 5 to 85 degrees, as shown in Figure 11.20(b). For the 400 and 465 nm peaks, respectively, the fluorescence intensity of temperature-sensitive

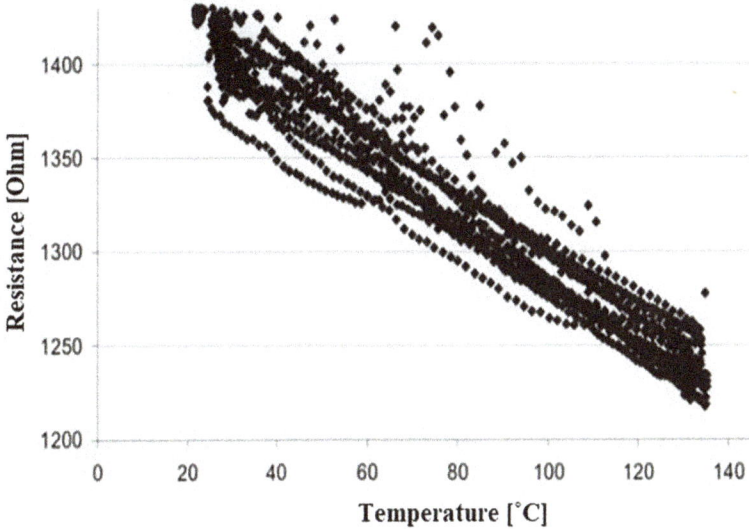

FIGURE 11.19   Temperature-resistance curve of CNT-based temperature sensor (58).

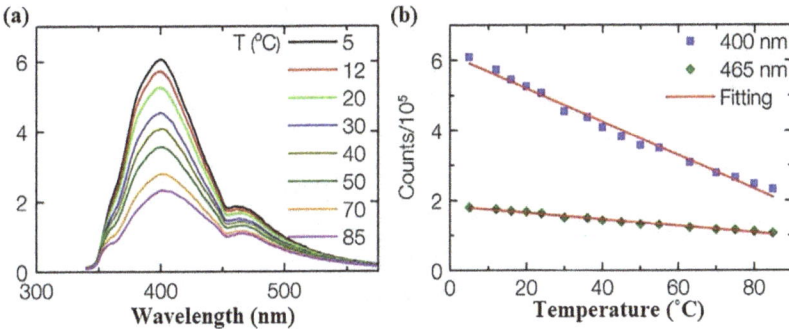

FIGURE 11.20   (a) Temperature-dependent emission spectra. (b) Peak intensities with respect to temperature (59).

carbon dots dropped from 3.3 to 2.1% per °C. Therefore, the obtained results recommend that a temperature sensor might be created using carbon dots whose sensitivity and intensity are greater (59)

11.3.2.3 Electrochemical Sensors

A chemical sensor that uses an electrode as a transducer when an analyte is present is the electrochemical sensor. Modern electrochemical sensors can identify a range of physical, chemical, or biological features. Modern sensing systems have profited from recent advances in microelectronics and microengineering, particularly by the development of smaller-scale sensors exhibiting better sensitivity, selectivity, and cheaper production and maintenance costs.

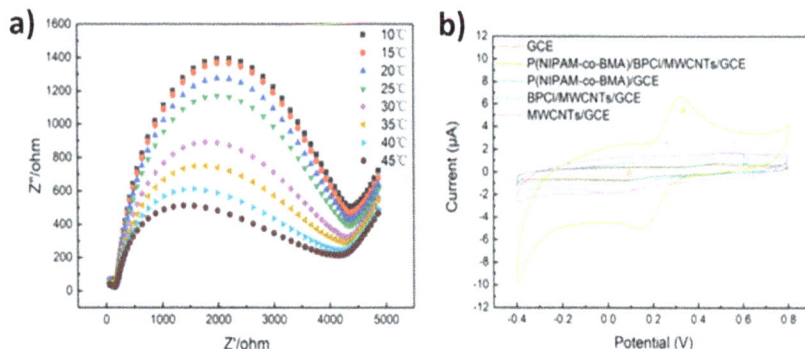

FIGURE 11.21 (a) EIS Studies of P(NIPAM-co-BMA)/BIPO$_4$/BiOCl/MWCNTs for different temperature. (b) Comparative study on I-V curve of GCE, P(NIPAM-co-BMA)/GCE, MWCNTs/BPCl/GCE, MWCNTs/GCE, and P(NIPAM-co-BMA)/BPCl/MWCNTs/GCE (60).

11.3.2.3.1 Carbon Nanotubes

To detect the harmful pollutant catechol, Wang Y et al. developed an electrochemical sensor using synthesised P(NIPAM-co-BMA)/BIPO4/BiOCl/MWCNTs nanocomposite. The semicircle's diameter has shrunk, improving the electric conductance of the nanocomposite electrode with rise in temperature, according to the EIS investigations (Figure 11.21(a)) (10°C-45°C). The cyclic voltammetry (CV) curve in Figure 11.21 was employed for studying the electrocatalytic impact of P(NIPAM-co-BMA)/BIPO4/BiOCl/MWCNTs. The comparative study has made for GCE, P(NIPAM-co-BMA)/GCE, MWCNTs/BPCl/GCE, MWCNTs/GCE, and P(NIPAM-co-BMA)/BPCl/MWCNTs/GCE, which depicts that nanocomposites obtained better electrical conductivity than individual samples, and P(NIPAM-co-BMA)/BPCl/MWCNTs/GCE composites show better electrocatalytic activity.

11.3.2.3.2 Carbon Nanofibres

Owing to the substantial surface area, CNFs are frequently utilised in electrochemical sensor electrodes. In the studies of Jahromi et al., electro spun CNF was used to enhance the superficial layer of a screen-printed electrode. Square wave voltammetry (SWV) was utilised to determine the tramadol concentration using a modified electrode. It was shown that electro-spun CNFs/SPE electrodes perform improved electrical conductivity and boosted electron transfer rate, offering a dependable and uncomplicated method for tramadol measurement (61).

For the highly sensitive and accurate identification of metronidazole, Zhang et al. proposed an electrochemical sensor technology, utilising a modified carbon nanofibre with gold nanoparticles (CNF@AuNPs) as the working electrode. Metronidazole (MNZ) has a rectilinear result ranging between 0.1M and 2000 M and a reduced amount of sensing range of about 0.024 M. The sensor displays good performance in MNZ analysis of tap water and is straightforward to prepare and is reproducible. In order to assess the electrocatalytic effectiveness of various electrodes, CV

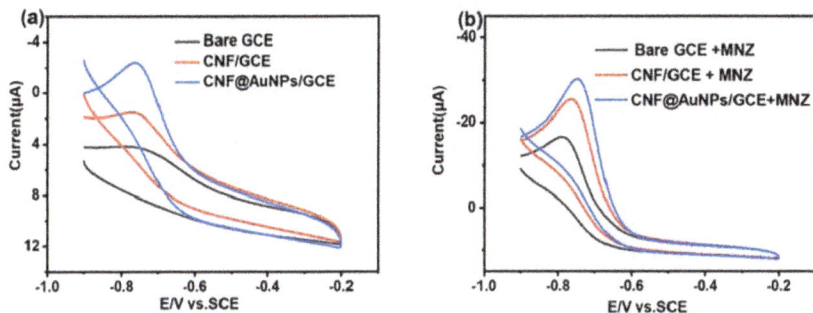

FIGURE 11.22 Electrochemical studies of **(a)** bare GCE, CNF/GCE, and CNF@AuNPs/GCE without MNZ, **(b)** bare GCE, CNF/GCE, and CNF@AuNPs/GCE with 200 μM MNZ in 0.2 M PBS of pH = 7.4. Scan rate:100 mV s $^{-1}$ (62).

examined their electrochemical response. The peak is invisible for the bare GCE electrode implying the absence of metronidazole (Figure 11.22(a)) in the region of 200 to 900 mV, but weak peaks like 749 mV and 753 mV are visible at CNF/GCE and CNF@AuNPs/GCE, respectively. The cathodic levels are broader, implying a sluggish migration of electrons. Inclusion of 200 M MNZ demonstrates the significant increment in the minimal current obtained by CNF@AuNPs/GCE. And there was a large movement to the right in the reduction peak voltage. This can be as a result of the expansive specific surface.

11.3.2.3.3 Carbon Nanoparticles

Thara et al. reported a unidirectional synthesis of carbon nitride via hydrothermal method. Utilising multiple characterisation techniques, the material's structure and composition were investigated. Cyclic voltammetry (CV) experiments were used to determine the effectiveness of the CND-enhanced electrode towards the specific diagnosis of Hg^{2+} and TC. The improved sensing capability of the proposed device in the CV trials with the voltage range within -1.0 to1.0 V is clearly shown in Figure 11.23. The tiny cathodic peak at about 0.26 V that the unmodified GCE and CND electrodes displayed may be the consequence of oxygen elimination at the electrode surface. A redox peak in the voltage ranges of about 0.56 V and 0.031 V is produced by the improved electrode when Hg^{2+} is present. Peaks that develop and shift when Hg^{2+} is present are connected to the GCE's CND and its favourable interaction with Hg^{2+}. The CV measure with tetracycline show a similar rise in trend. A chronoamperometric technique was utilised within 200 sec to investigate the behaviour of the improved electrode to potential intrusions during the metal ion and drug detection. For the Hg^{2+} selectivity experiment, metal ions like Mn^{2+}, Cu^{2+}, Pb^{2+}, Cd^{2+}, Cr^{6+}, Co^{2+}, Cu^{2+}, Fe^{2+}, and Ni^{2+} were selected as probable organic pollutants, and investigations were carried out with these ions present at concentrations three times greater than [Hg^{2+}]. The precise detection of TC was made possible with the existence of interferents (e.g., vancomycin, 5-fluorouracil, amoxicillin trihydrate, doxycycline, and cloxacillin) (63).

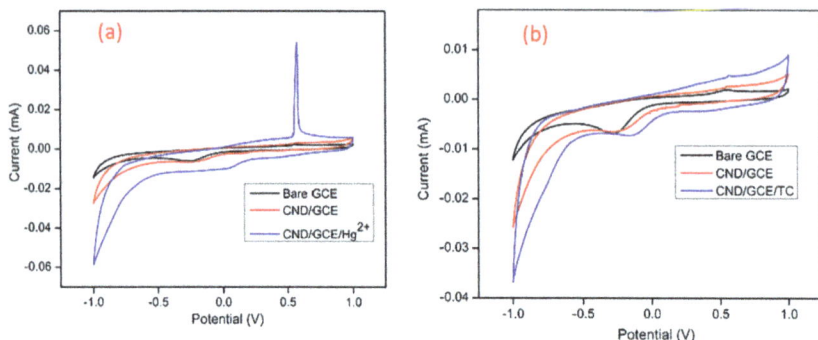

FIGURE 11.23 CV curve of (a) bare GCE, CND/GCE, and CND/GCE/Hg^{2+}; (b) bare GCE, CND/GCE, and CND/GCE/TC (63).

11.3.2.4 Gas Sensor

It is crucial to detect nitrogen dioxide (NO_2), since it's a conventionally toxic gas that is bad for both individuals and the environment. Recently, numerous of nanomaterials, which includes CNTs, graphene, nanowires, and nanoparticles, have, indeed, been widely employed to construct the platform of NO_2 gas sensors. On the account of remarkable electrical and physical properties, which are necessary for sensing of NO_2, carbon nanostructures have drawn more and more interest among these materials (64).

As per Li et al. findings, a nanomesh of SWCNTs on IDEs offers enough places for the molecules of NO_2 gas to bind. Figure 11.24(a) depicts the network that SWCNTs create on IDEs. The IDE promotes efficient electrical connection across significant areas amongst SWCNTs and electrodes, resulting in enhanced gas molecule availability to all SWCNTs. From Figure 11.24(b), the gas sensor picks up NO_2 gas at concentrations of 6–100 ppm with a LOD of 44 ppb. Despite having outstanding sensitivity, this sensor's recovery period is extraordinarily long (10 h) because of the strong affinity of NO_2 gas molecules with SWCNTs. Along with the sensitivity of the sensor, conductance recovery is a pivotal component that needs to be considered. A very hot working environment or exposure to ultraviolet is needed to shorten the recuperation period (Figure 11.24(c)) (65).

Im et al. created a gas sensor by electrospinning a polyacrylonitrile and carbon black compound. In order to create carbon fibres, the electrospun fibres underwent thermal treatment first. The adsorption sites for gas were subsequently increased by chemically energising the carbon fibres. The fluorination process altered the surface of the activated sample. The activation technique increased the particular area of the permeability by approximately 100 times. The addition of carbon black compounds increased the electrical conductivity.

In Figure 11.25(a), the resistive response was plotted versus the duration of CO gas exposure. Due to its non-porous nature and inactive CO gas adsorption sites, the FCCF sample only displayed a 1% resistive response. As demonstrated in FCACF-6 and FCACF-8, the formation of a void-filled structure improved carbon-monoxide-detection

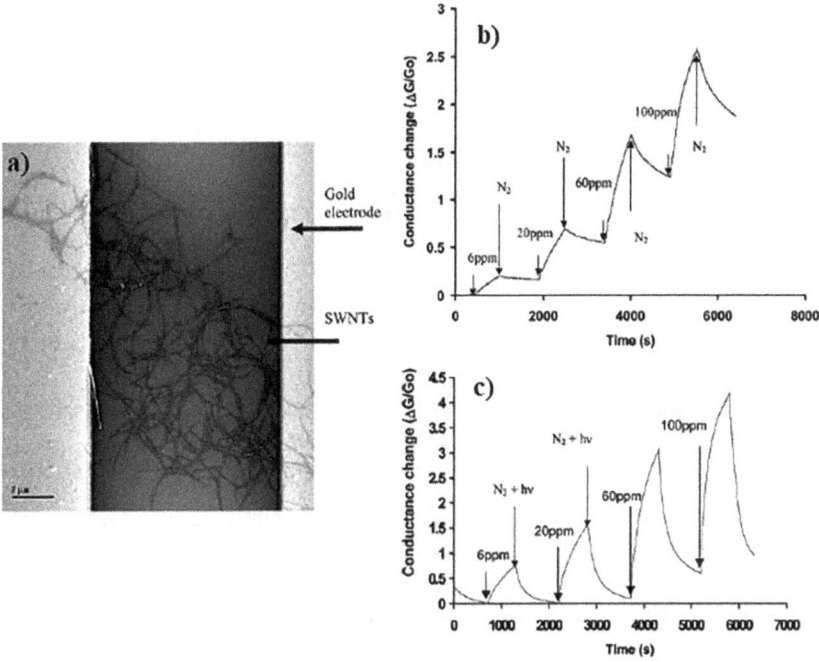

FIGURE 11.24 (a) SEM micrograph of SWNTs over two gold electrodes; sensor response for various concentration of NO_2 (400 cm³/min nitrogen flow at room temperature). (b) Without using UV light during recuperation. (c) Using UV light to retrieve sensors (65).

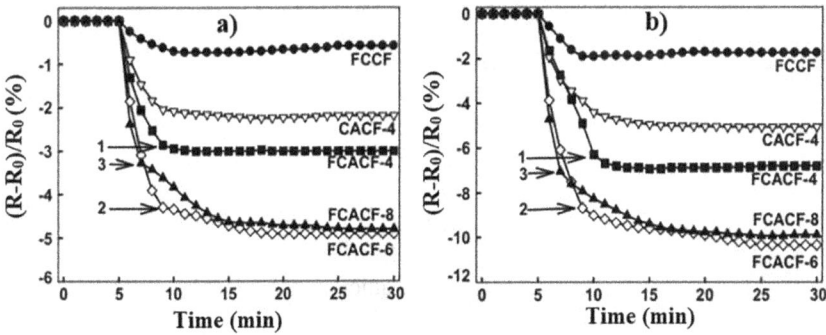

FIGURE 11.25 Changes in resistance over a period of time for detecting (a) carbon monoxide gas and (b) nitrogen oxide gas (66).

capabilities by around 5%. This data suggests that the gas adsorption sites were significantly enhanced by the porosity produced by chemical energising. The electrical resistance in the FCACF-4 sample significantly fell to about 3% from time zero within 3 minutes, and afterwards, there was no discernible change. To be more precise, within a span of 5 minutes, the targeted gas rapidly adheres to the micropores,

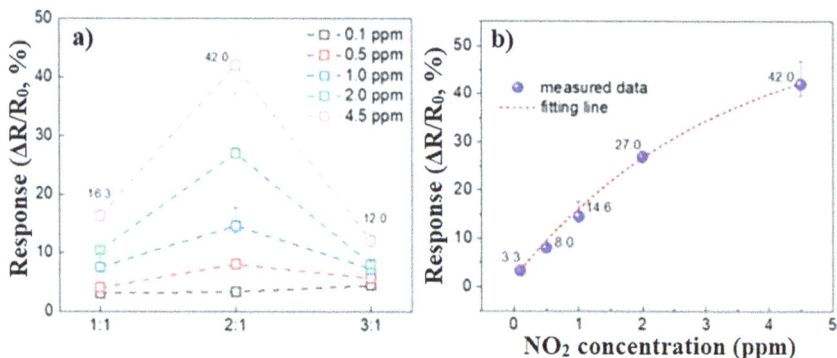

FIGURE 11.26 (a) Differences in the response that occur due to the ratio of CDs to SWCNTs. (b) Variation in 2:1 device's response in relation to NO_2 concentration (67).

causing a significant decrease in resistance, and subsequently, accumulates slowly into the mesopores, leading to a gradual reduction in resistance. The resistive response versus exposure time to NO gas is shown in Figure 11.25(b). Similar to the CO gas, the resistive response followed this pattern. The responsiveness to NO was, however, more than twice as sensitive. For NO, the impacts of the permeable structure and fluorination processing were proven. This finding suggests that the NO gas has a larger electron hopping effect than the CO gas (66).

Lim et al. developed a chemical resistive nitrogen dioxide gas sensor that is highly selective. They used SWCNTs coated with CDs to create the sensor. Unlike prior studies on NO_2 gas sensors using SWCNTs, the resistance of the sensor increased when exposed to the gas. The increase in resistance was due to the electron-rich CDs that significantly altered the polarity of the SWCNTs from p-type to n-type. The researchers tested three different suspensions of CD-decorated SWCNTs (1:1, 2:1, and 3:1) to determine their reactions to the NO_2 gas using a specially designed gas detecting device. The results highlighted the importance of the CDs to SWCNTs ratio in influencing the response values.

The overview of how sensors respond based on their response curves is presented in Figure 11.26(a). The sensor shows the highest response value of 42.0% to NO_2 gas at 4.5 ppm when CDs to SWCNTs ratio is 2:1. Figure 11.26(b) shows the response as a fraction of NO_2 concentration, with a quick rise from 3.3% to 27.0% as NO_2 concentration increases from 0.1 to 2.0 ppm. However, as NO_2 concentration exceeds 2.0 ppm, the sensor becomes gradually saturated, causing a reduction in available adsorption sites. The sensor's LOD was determined as 18 ppb from the curve (67).

11.3.3 BIOMEDICAL APPLICATIONS

Due to their outstanding mechanical and electrical properties, carbon nanostructures are frequently used in biomedical applications. They are used for the preparation of nanocomposites (68), nanosensors, and in the medical field (10). Among the biocompatibility assessments of carbon nanostructures, the effects of them on the immune system and immune-related pathways have been studied (69). A major reason for their toxicity is their highly hydrophobic character, which makes them penetrate membranes during cellular uptake. This limitation has been overcome by the chemical modifications of the surfaces with hydroxyl, carboxyl, amino groups, etc., leading to improved biocompatibility (70). The high surface area, chemical functionalisation, and tunability of carbon structures make them compatible with many polymeric structures, enabling a wide range of applications. Some of the most common applications of carbon nanostructures in the medical field are listed as follows:

1. Therapeutic applications
   1.1 Chemotherapy drugs
   1.2 Applications in proteins and peptides
   1.3 Gene delivery applications
   1.4 Photothermal therapy
   1.5 Photodynamic therapy
2. Bioimaging applications
   2.1 Optical and non-optical bioimaging
3. Biosensing, diagnosis, and theragnostic
4. Tissue engineering and repair

Cancer is a heterogeneous disease affecting a wide range of populations. The EPR effect, as observed in carbon nanostructures, shows high efficacy at lower dosages. ND-anthracycline-based drugs have been found to reduce toxicity, as compared to conventional structures. CNTs are also extensively used for drug delivery systems, owing to their unique structural properties. Further in-vivo cancer treatments using CNT-modified drugs show higher inhibition of tumour growth, with fewer side effects (76). A famous chemotherapy drug, doxorubicin (DOX), loaded with CNTs and used recently, was loaded onto CNTs. This showed stable behaviour at neutral pH, which was desirable for in-vivo drug delivery applications when coated with PEG (77). Protein-based drugs have a great ability to be used in therapeutic applications; however, they suffer many limits in physiological conditions. For this purpose, carbon nanostructures have recently been used. CNTs are effective choices because of their distinctive optical characteristics. SWCNTs exhibit intrinsic fluorescence emission in the long wavelength range (1000–1700 nm), which helps reduce photon scattering. CNTs were used for monitoring the signal from tumours and other organs based on the principle of principal component analysis (PCA). The results showed that signals were observed within 20 s of administration and were persistent up to 72 h (78). Biosensors are devices used for the determination of biological analytes like biomolecules, microorganisms, and other biological structures in various environments. The excellent biocompatibility and chemical stability make them excellent candidates for biosensing with a longer lifespan. A label-free field-effect transistor (FET) biosensing platform, based on a semiconducting carbon nanotube (CNT) thin film, as shown in Figure 11.27, was developed by Li et al. for the detection of exosomal miRNA 21 (71). An optical ultrasound transmitter was designed by electrospinning a MWCNT-polymer composite. An absorbent polymer composite was electrospun directly onto the end face of optical fibres to create the device. The resulting structure, dip-coated

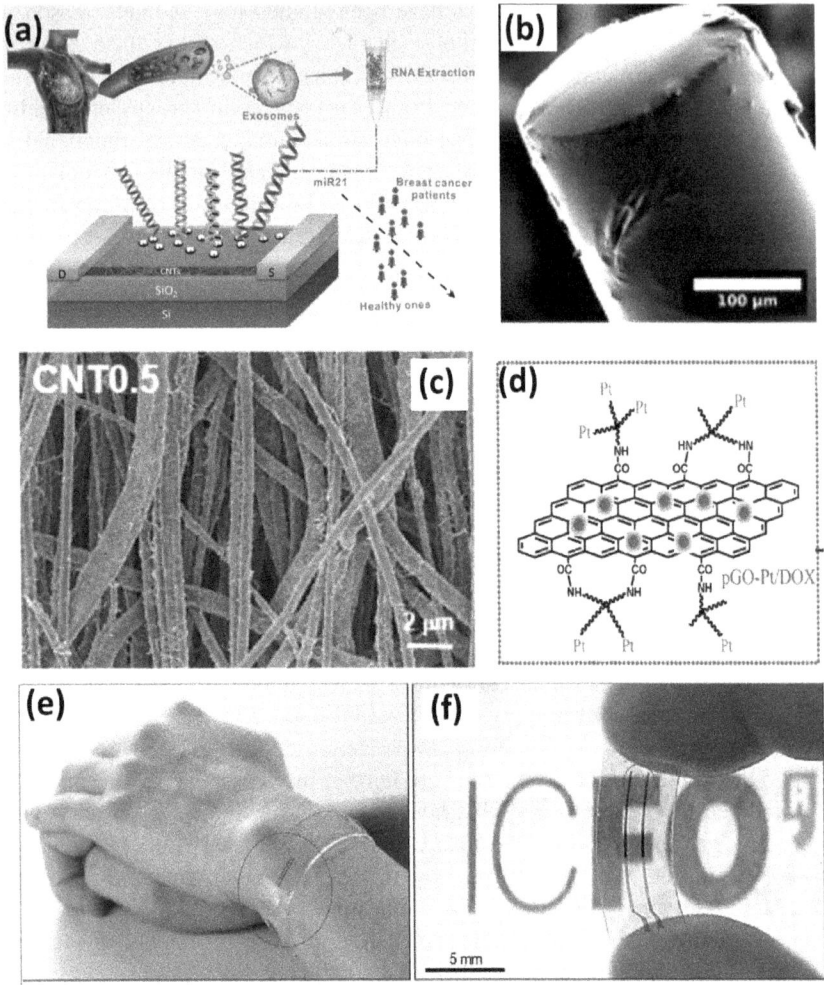

FIGURE 11.27 (a) An ultrasensitive platform for sensing exosomal miR21 with the help of a CNT field effect transistor (71); (b) a MWCNT-PVA composite dip-coated with PDMS for fibre optic ultrasound transmission (72); (c) PCL nanofibres coated with CNTs used for bone tissue regeneration (73); (d) a pGO-Pt/DOX dual-drug delivery system consisting of PEG covalently functionalised on GO for anti-cancer drug delivery applications (74); (e) photograph of a flexible, transparent bracelet enhanced with GQD for heart rate measurements; (f) a 1 mm2 graphene tunnel on a PbS QD-coated PEN substrate (30 nm in thickness) (75).

with PDMS, showed high optical absorption characteristics (72). Due to their easy tunability by chemical processes, fullerenes are highly used in the medical diagnostic field (79). Applications for nanodiamonds include tissue scaffolds, surgical implants, medication delivery, biosensing, imaging, and protein mimics (80). The fluorescent behaviour exhibited by nanodiamonds over semi-conducting quantum dots with a non-toxic nature makes them a potential candidate to revolutionise in-vivo bioimaging applications. They are also used to address indicators of liver toxicity by addressing them intravenously (69). In tissue engineering applications, NDs prove to be an excellent platform for neuronal growth, like protein-coated materials. In addition, a new generation of transparent, flexible wearables includes graphene that has been sensitised by semiconducting quantum dots (GQD). A prototype of a gadget for non-invasive vital-sign monitoring was created. The device successfully monitors vital health signs like heart rate, oxygen saturation (SpO_2), etc. The device offers low power consumption. The device also has an attached, flexible UV photodetector, and wireless operation in ambient light conditions showed lower power requirements. Additionally, the use of a flexible, ultraviolet (UV)-sensitive photodetector in conjunction with a near-field communication circuit board allows for wireless communication, enables power transfer, and allows for battery-free operation between the photodetectors, in addition to a smartphone.

CONCLUSION

The study aims to provide a comprehensive understanding of the size-dependent properties of carbon nanomaterials. The results demonstrate that the properties of carbon-based materials are inclined by their dimensions and morphology. The study highlights the potential applications of carbon nanomaterials in various fields, including optoelectronics, thermal management, mechanical engineering, and energy storage. The findings suggest that carbon nanotubes, carbon nanofibres, and nanoparticles can provide unique solutions to some of the most critical challenges facing modern science and technology. Thus, the study serves as a valuable resource for researchers and engineers interested in designing and developing novel nanomaterials with enhanced properties and performance.

REFERENCES

1. Koutsawa Y, Tiem S, Yu W, Addiego F, Giunta G. A micromechanics approach for effective elastic properties of nano-composites with energetic surfaces/interfaces. Compos Struct. 2017 Jan 1;159:278–87.
2. Wu Q, Miao WS, du Zhang Y, Gao HJ, Hui D. Mechanical properties of nanomaterials: A review. Nanotechnol Rev. 2020 Mar 24;9(1):259–73.
3. Long A, Liu H, Xu S, Feng S, Shuai Q, Hu S. Polyacrylic acid functionalized biomass-derived carbon skeleton with highly porous hierarchical structures for efficient solid-phase microextraction of volatile halogenated hydrocarbons. Nanomaterials. 2022 Dec 1;12(24).
4. Nguyen TD, Lee JS. *Electrospinning-based carbon nanofibers for energy and sensor applications. Vol. 12, applied sciences (Switzerland)*. MDPI; 2022.
5. Holmannova D, Borsky P, Svadlakova T, Borska L, Fiala Z. *Carbon nanoparticles and their biomedical applications. Vol. 12, applied sciences (Switzerland)*. MDPI; 2022.
6. ben Doudou B, Chiba I, Daoues HS. Optical and thermo-optical properties of polyvinyl alcohol/carbon nanotubes composites investigated by prism coupling technique. Opt Mater (Amst). 2022 Sept 1;131.
7. Rehammar R, Magnusson R, Fernandez-Dominguez AI, Arwin H, Kinaret JM, Maier SA, et al. Optical properties of carbon nanofiber photonic crystals. Nanotechnology. 2010 Nov 19;21(46).

8. Leong KY, Hasbi S, Ku Ahmad KZ, Mat Jali N, Ong HC, Md Din MF. Thermal properties evaluation of paraffin wax enhanced with carbon nanotubes as latent heat thermal energy storage. J Energy Storage. 2022 Aug 25;52.

9. Chen W, Quan H, Chen D. Carbon quantum dots boosted structure stability of nickel cobalt layered double hydroxides nanosheets electrodeposited on carbon cloth for energy storage. Surf. Interfaces. 2023 Feb 1;36.

10. Anzar N, Hasan R, Tyagi M, Yadav N, Narang J. *Carbon nanotube – a review on synthesis, properties and plethora of applications in the field of biomedical science. Vol. 1, sensors international.* KeAi Communications Co.; 2020.

11. Sisto TJ, Zakharov LN, White BM, Jasti R. Towards pi-extended cycloparaphenylenes as seeds for CNT growth: Investigating strain relieving ring-openings and rearrangements. Chem Sci. 2016;7(6):3681–8.

12. Restivo J, Soares OSGP, Pereira MFR. *Processing methods used in the fabrication of macrostructures containing 1d carbon nanomaterials for catalysis. Vol. 8, processes.* MDPI AG; 2020. pp. 1–36.

13. Guadagno L, Raimondo M, Vittoria V, Vertuccio L, Lafdi K, De Vivo B, et al. The role of carbon nanofiber defects on the electrical and mechanical properties of CNF-based resins. Nanotechnology. 2013 Aug 2;24(30).

14. Feng L, Xie N, Zhong J. *Carbon nanofibers and their composites: A review of synthesizing, properties and applications. Vol. 7, materials.* MDPI AG; 2014. pp. 3919–45.

15. Grandgirard J, Poinsot D, Krespi L, Nénon JP, Cortesero AM. *Costs of secondary parasitism in the facultative hyperparasitoid Pachycrepoideus dubius: Does host size matter?* Vol. 103. Entomologia Experimentalis et Applicata; 2002.

16. Zhang M, Li J. Carbon nanotube in different shapes. Mater Today. 12;2009:12–18.

17. Mochalin VN, Shenderova O, Ho D, Gogotsi Y. *The properties and applications of nanodiamonds. Vol. 7, nature nanotechnology.* Nature Publishing Group; 2012. pp. 11–23.

18. Singh DV., Jenkins KA, Appenzeller J, Neumayer D, Grill A, Wong HSP. Frequency response of top-gated carbon nanotube field-effect transistors. IEEE Trans Nanotechnol. 2004 Sept;3(3):383–7.

19. Derycke V, Martel R, Appenzeller J, Avouris P. Controlling doping and carrier injection in carbon nanotube transistors. Appl Phys Lett. 2002 Apr 15;80(15):2773–5.

20. Zhao Y, Wei J, Vajtai R, Ajayan PM, Barrera E V. Iodine doped carbon nanotube cables exceeding specific electrical conductivity of metals. Sci Rep. 2011;1.

21. Lekawa-Raus A, Patmore J, Kurzepa L, Bulmer J, Koziol K. Electrical properties of carbon nanotube based fibers and their future use in electrical wiring. Adv Funct Mater. 2014 Jun 25;24(24):3661–82.

22. Structural improvement of carbon fibers prepared from benzene. Available from: https://iopscience.iop.org/article/10.1143/JJAP.15.2073

23. Tiwari SK, Kumar V, Huczko A, Oraon R, Adhikari A De, Nayak GC. *Magical allotropes of carbon: Prospects and applications. Vol. 41, critical reviews in solid state and materials sciences.* Taylor and Francis Inc.; 2016. pp. 257–317.

24. Saleh Ahammad AJ, Lee JJ, Rahman MA. Electrochemical sensors based on carbon nanotubes. Sensors. 2009;9:2289–319.

25. Wu Z, Wang Z, Yu F, Thakkar M, Mitra S. Variation in chemical, colloidal and electrochemical properties of carbon nanotubes with the degree of carboxylation. J Nanopart Res. 2017 Jan 1;19(1).

26. Ghosh D, Chakraborty K, Bharti, Pulimi M, Anand S, Chandrasekaran N, et al. The effects of pH, ionic strength, and natural organics on the transport properties of carbon nanotubes in saturated porous medium. Colloids Surf A Physicochem Eng Asp. 2022 Aug 20;647.

27. Zhang L, Yin M, Wei X, Sun J, Xu D. *Recent advances in morphology, aperture control, functional control and electrochemical sensors applications of carbon nanofibers. Vol. 656, analytical biochemistry.* Academic Press Inc.; 2022.

28. Xie L, Su F, Xie L, Guo X, Wang Z, Kong Q, et al. *Effect of pore structure and doping species on charge storage mechanisms in porous carbon-based supercapacitors. Vol. 4, materials chemistry frontiers.* Royal Society of Chemistry; 2020. pp. 2610–34.

29. Ruiz-Cornejo JC, Sebastián D, Lázaro MJ. *Synthesis and applications of carbon nanofibers: A review. Vol. 36, reviews in chemical engineering.* De Gruyter; 2020. pp. 493–511.

30. Tsai IH, Li JT, Chang CW. Effects of sonication and hydrothermal treatments on the optical and chemical properties of carbon dots. ACS Omega. 2021 Jun 8;6(22):14174–81.

31. Wu Q, Miao WS, Zhang Y du, Gao HJ, Hui D. Mechanical properties of nanomaterials: A review. Nanotechnol Rev. 2020 Mar 24;9(1):259–73.

32. Ramezani M, Kim YH, Sun Z. Mechanical properties of carbon-nanotube-reinforced cementitious materials: Database and statistical analysis. Mag Concr Res. 2020 Oct 1;72(20):1047–71.

33. Rodrigues RAA, Silva RMF da C e., Ferreira L de AQ, Branco NTT, Ávila É de S, Peres AM, et al. Enhanced mechanical properties, anti-biofilm activity, and cytocompatibility of a methacrylate-based polymer loaded with native multiwalled carbon nanotubes. J Mech Behav Biomed Mater. 2022 Dec 1;136.

34. Ataol AS, Ergun G, Cekic-Nagas I, Alas MO, Genc R. The effects of adding fluorescent carbon nanoparticles on various mechanical properties of denture liners. Dent Mater J. 2021;40(3):573–83.

35. Ren X, Chen C, Nagatsu M, Wang X. Carbon nanotubes as adsorbents in environmental pollution management: A review. J Chem Eng. 2011;170:395–410.

36. Ni J, Li Y. Carbon nanomaterials in different dimensions for electrochemical energy storage. Adv Energy Mater. 2016 Sept 7;6(17).

37. Dai L, Chang DW, Baek JB, Lu W. Carbon nanomaterials for advanced energy conversion and storage. Small. 2012; 8:1130–66.

38. al Sheheri SZ, Al-Amshany ZM, al Sulami QA, Tashkandi NY, Hussein MA, El-Shishtawy RM. The preparation of carbon nanofillers and their role on the performance of variable polymer nanocomposites. Des Monomers Polym. 2019 Jan 1;22(1):8–53.

39. Jyoti J, Gupta TK, Singh BP, Sandhu M, Tripathi SK. *Recent advancement in three dimensional graphene-carbon nanotubes hybrid materials for energy storage and conversion applications. Vol. 50, journal of energy storage.* Elsevier Ltd; 2022.

40. Pandit B, Dubal DP, Gómez-Romero P, Kale BB, Sankapal BR. V2O5 encapsulated MWCNTs in 2D surface architecture: Complete solid-state bendable highly stabilized energy efficient supercapacitor device. Sci Rep. 2017 Mar 3;7.

41. Chen Y, Wang X, Lin H, Vogel F, Li W, Cao L, et al. Low-density polyethylene-derived carbon nanotubes from express packaging bags waste as electrode material for supercapacitors. J Ind Eng Chem. 2022;119:633–46.

42. Merino C, Soto P, Vilaplana-Ortego E, de Gomez Salazar JM, Pico F, Rojo JM. Carbon nanofibres and activated carbon nanofibres as electrodes in supercapacitors. Carbon NY. 2005;43(3):551–7.

43. Yang S, Ai J, Han Z, Zhang L, Zhao D, Wang J, et al. Electrospun ZnFe2O4/carbon nanofibers as high-rate supercapacitor electrodes. J Power Sources. 2020 Sept 1;469.

44. Baslak C, Demirel S, Kocyigit A, Alatli H, Yildirim M. Supercapacitor behaviors of carbon quantum dots by green synthesis method from tea fermented with kombucha. Mater Sci Semicond Process. 2022 Aug 15;147.

45. Sun W, Lu Q. Self-supported α-Ni(OH)2 nanosheet arrays modified with carbon quantum dots for high-performance supercapacitors. Scr Mater. 2023 Feb 1;224.

46. Choi JH, Lee C, Park S, Hwang M, Embleton TJ, Ko K, et al. Improved electrochemical performance using well-dispersed carbon nanotubes as conductive additive in the Ni-rich positive electrode of lithium-ion batteries. Electrochem Commun. 2023 Jan 1;146.

47. Kaneko K, Li M, Noda S. Appropriate properties of carbon nanotubes for the three-dimensional current collector in lithium-ion batteries. Carbon Trends. 2023 Mar 1;10.

48. Li Y, Liu X, Zhang J, Yu H, Zhang J. Carbon-coated Si/N-doped porous carbon nanofibre derived from metal–organic frameworks for Li-ion battery anodes. J Alloys Compd. 2022 May 5;902.

49. Javed M, Saqib ANS, Ata-Ur-Rehman, Ali B, Faizan M, Anang DA, et al. Carbon quantum dots from glucose oxidation as a highly competent anode material for lithium and sodium-ion batteries. Electrochim Acta. 2019 Feb 20;297:250–7.

50. Wang W, Jiang Z, Tian X, Maiyalagan T, Jiang ZJ. Self-standing CoFe embedded nitrogen-doped carbon nanotubes with Pt deposition through direct current plasma magnetron sputtering for direct methanol fuel cells applications. Carbon NY. 2023 Jan 5;201:1068–80.

51. Sokka A, Mooste M, Käärik M, Gudkova V, Kozlova J, Kikas A, et al. Iron and cobalt containing electrospun carbon nanofibre-based cathode catalysts for anion exchange membrane fuel cell. Int J Hydrogen Energy. 2021 Sept 3;46(61):31275–87.

52. Xu G, Wu Z, Xie Z, Wei Z, Li J, Qu K, et al. Graphene quantum dot reinforced hyperbranched polyamide proton exchange membrane for direct methanol fuel cell. Int J Hydrogen Energy. 2021 Feb 26;46(15):9782–9.

53. Institute of Electrical and Electronics Engineers. *2020 IEEE international conference on communications: Proceedings*. Institute of Electrical and Electronics Engineers. 2020 Jun 7–11.

54. Sinha N, Ma J, Yeow JTW. Carbon nanotube-based sensors. J Nanosci Nanotechnol. 2006 Mar;6(3):573–90.

55. Sánchez Arribas A, Moreno M, González L, Blázquez N, Bermejo E, Zapardiel A, et al. A comparative study of carbon nanotube dispersions assisted by cationic reagents as electrode modifiers: Preparation, characterization and electrochemical performance for gallic acid detection. J Electroanal Chem. 2020 Jan 15;857.

56. Bilal M, Rasheed T, Mehmood S, Tang H, Ferreira LFR, Bharagava RN, et al. *Mitigation of environmentally-related hazardous pollutants from water matrices using nanostructured materials – a review. Vol. 253, chemosphere*. Elsevier Ltd; 2020.

57. Raphey VR, Henna TK, Nivitha KP, Mufeedha P, Sabu C, Pramod K. *Advanced biomedical applications of carbon nanotube. Vol. 100, materials science and engineering C*. Elsevier Ltd; 2019. pp. 616–30.

58. Monea BF, Ionete EI, Spiridon SI, Ion-Ebrasu D, Petre E. *Carbon nanotubes and carbon nanotube structures used for temperature measurement. Vol. 19, sensors (Switzerland)*. MDPI AG; 2019.

59. Nguyen V, Yan L, Xu H, Yue M. One-step synthesis of multi-emission carbon nanodots for ratiometric temperature sensing. Appl Surf Sci. 2018 Jan 1;427:1118–23.

60. Wang Y, Fu Q, Chen J, Lin Y, Yang Y, Wang C, et al. Temperature-controlled electrochemical sensor based on environmentally responsive polymer/BiPO4/BiOCl/multiwalled carbon nanotube composite for the detection of catechol in water. Colloids Surf A Physicochem Eng Asp. 2023 Jan 20;657.

61. Zhang L, Yin M, Wei X, Sun J, Xu D. *Recent advances in morphology, aperture control, functional control and electrochemical sensors applications of carbon nanofibers. Vol. 656, analytical biochemistry*. Academic Press Inc.; 2022.

62. Zhang L, Yin M, Qiu J, Qiu T, Chen Y, Qi S, et al. An electrochemical sensor based on CNF@AuNPs for metronidazole hypersensitivity detection. Biosens Bioelectron X. 2022 May 1;10.

63. Thara CR, Mathew S, Rose Chacko A, Mathew B. Dual functional carbon nitride dots as electrochemical sensor and anticancer agent with chemotherapic and photodynamic effect. Microchem J. 2023 Apr 1;187.

64. Lee SW, Lee W, Hong Y, Lee G, Yoon DS. *Recent advances in carbon material-based NO2 gas sensors. Vol. 255, sensors and actuators, B: Chemical*. Elsevier B.V.; 2018. pp. 1788–804.

65. Li J, Lu Y, Ye Q, Cinke M, Han J, Meyyappan M. Carbon nanotube sensors for gas and organic vapor detection. Nano Lett. 2003 Jul 1;3(7):929–33.

66. Im JS, Kang SC, Lee SH, Lee YS. Improved gas sensing of electrospun carbon fibers based on pore structure, conductivity and surface modification. Carbon NY. 2010 Aug;48(9):2573–81.

67. Lim N, Lee JS, Byun YT. Negatively-doped single-walled carbon nanotubes decorated with carbon dots for highly selective NO2 detection. Nanomaterials. 2020 Dec 1;10(12):1–11.

68. Liu Z, Fan AC, Rakhra K, Sherlock S, Goodwin A, Chen X, et al. Supramolecular stacking of doxorubicin on carbon nanotubes for in vivo cancer therapy. Angew Chem Int Ed Engl. 2009;48.

69. Chow EK, Zhang XQ, Chen M, Lam R, Robinson E, Huang H, et al. Nanodiamond therapeutic delivery agents mediate enhanced chemoresistant tumor treatment. Sci Transl Med. 2011 Mar 9;3(73).

70. Mura S, Nicolas J, Couvreur P. Stimuli-responsive nanocarriers for drug delivery. Nat Mater. 2013;12:991–1003.

71. Li T, Liang Y, Li J, Yu Y, Xiao MM, Ni W, et al. Carbon nanotube field-effect transistor biosensor for ultrasensitive and label-free detection of breast cancer exosomal miRNA21. Anal Chem. 2021 Nov 23;93(46):15501–7.

72. Poduval RK, Noimark S, Colchester RJ, Macdonald TJ, Parkin IP, Desjardins AE, et al. Optical fiber ultrasound transmitter with electrospun carbon nanotube-polymer composite. Appl Phys Lett. 2017 May 29;110(22).

73. Patel KD, Kim TH, Mandakhbayar N, Singh RK, Jang JH, Lee JH, et al. Coating biopolymer nanofibers with carbon nanotubes accelerates tissue healing and bone regeneration through orchestrated cell- and tissue-regulatory responses. Acta Biomater. 2020 May 1;108:97–110.

74. Pei X, Zhu Z, Gan Z, Chen J, Zhang X, Cheng X, et al. PEGylated nano-graphene oxide as a nanocarrier for delivering mixed anticancer drugs to improve anticancer activity. Sci Rep. 2020 Dec 1;10(1).

75. Polat EO, Mercier G, Nikitskiy I, Puma E, Galan T, Gupta S, et al. *Flexible graphene photodetectors for wearable fitness monitoring* [Internet]. 2019. Available from: www. science.org

76. Liu Z, Chen K, Davis C, Sherlock S, Cao Q, Chen X, et al. Drug delivery with carbon nanotubes for in vivo cancer treatment. Cancer Res. 2008 Aug 15;68(16):6652–60.

77. Liu Z, Jiao L, Yao Y, Xian X, Zhang J. Aligned, ultralong single-walled carbon nanotubes: From synthesis, sorting, to electronic devices. Adv Mater. 2010; 22:2285–310.

78. Murjani BO, Kadu PS, Bansod M, Vaidya SS, Yadav MD. *Carbon nanotubes in biomedical applications: Current status, promises, and challenges. Vol. 32, carbon letters.* Springer; 2022. pp. 1207–26.

79. Kai-Hua Chow E, Gu M, Xu J. Carbon nanomaterials: Fundamental concepts, biological interactions, and clinical applications. In: *Nanoparticles for biomedical applications: Fundamental concepts, biological interactions and clinical applications.* Elsevier; 2019. pp. 223–42.

80. Mochalin VN, Shenderova O, Ho D, Gogotsi Y. *The properties and applications of nanodiamonds. Vol. 7, nature nanotechnology.* Nature Publishing Group; 2012. pp. 11–23.

12 Conclusion and Future Work

Anindya Nag, Alivia Mukherjee, and
Mehmet Ercan Altinsoy

12.1 INTRODUCTION

Followed by the invention of CNTs by Iijima in 1991 (1), the last three decades have seen exponential growth in the integration of CNTs with daily life. CNTs have been synthesized and optimized in various manners to develop their credibility for form-ing efficient, flexible sensors. The impact of CNTs has been primarily realized in the form of sensors, where the analytical performance of the prototypes has been a direct reflection of the electromechanical characteristics of the carbon allotropes (2, 3). The researchers have tried to formulate fabrication techniques to generate high-quality CNTs. CNTs have been chosen to form flexible sensors because of high electrical conductivity, small dimension, high aspect ratio, low parasitic capaci-tance, thermal stability and excellent mechanical flexibility (4–6). Some of the com-mon fabrication techniques used to develop CNTs are chemical vapour deposition (CVD) (7, 8), laser arc discharge (9, 10), laser ablation (11, 12) and liquid electrolysis (13, 14). Each of these methodologies has varied based on catalysts, temperature, atmosphere, pressure and current (15). Different physicochemical forms of CNTs, including single-walled carbon nanotubes (SWCNTs) (16, 17) and multi-walled carbon nanotubes (MWCNTs) (18, 19), have been extensively developed for various electro-chemical (20, 21), strain (22, 23) and electrical (24, 25) applications. Since the popu-larization of CNTs, other types of carbon allotropes like graphene (26–28), graphene oxide (GO) (29–31), reduced graphene oxide (rGO) (32–34) and graphite (35–37) have been devised academically and industrially to form the sensors. Each carbon-based element has been quite efficient and competitive to CNTs while being deployed to form flexible sensors. Due to this technicality, researchers have been constantly try-ing to optimize the quality of CNTs to improve the analytical characteristics of the CNTs-based flexible sensors. This has been primarily achieved by functionalizing the SWCNTs and MWCNTs with various organic and inorganic groups to strengthen their covalent bonding with other nanomaterials and polymers. While SWCNTs and MWCNTs have been significantly used for forming electrodes, further work on other forms of CNTs like double-walled carbon nanotubes (DWCNTs) (38, 39) and few-walled carbon nanotubes (FWCNTs) (40, 41) should also be done to optimize their properties and subsequently use them for forming flexible sensors. Similar to the need for CNTs with improved characteristics, it is a state-of-the-art to develop CNTs-based flexible sensors with high functionality, longevity and robustness. This is also dic-tated by the fabrication techniques and embedded signal conditioning circuits that are

DOI: 10.1201/9781003376071-12

considered to form the CNTs-based flexible sensors. In the era of printing techniques (42–46), certain processes like 3D printing (47–49) and laser ablation (50–52) have gained high popularity due to certain advantages, like high customization, minimal waste, cost effectiveness, ease of access, rapid prototyping, rapid sample analysis turnaround time and high spatial resolution (53–55). Each of these techniques have been chosen based on the processed materials, target application and required physical dimension of the sensors. The association of these printing techniques to CNTs needs to be supervised to minimize the alteration of the chemical properties of the nanomaterial and simultaneously form the thin-film sensors. For the communication of the sensed data, the wireless communication protocols used to transfer the data are a significant parameter in real-time applications. With the rise of machine learning and artificial intelligence (56, 57), these flexible sensors should be integrated efficiently with the devised neural networks for signal validation, high signal-to-noise ratio, excellent dynamic response, fast response and high sensitivity (58).

12.2 CURRENT CHALLENGES AND FUTURE PERSPECTIVE

Although both CNTs and CNTs-based flexible sensors have been developed and deployed on a large scale, some bottlenecks still need to be addressed in the current scenario. One of the limitations of CNTs has been their hydrophobic nature, which leads to their agglomeration in the aqueous solutions. This creates heterogeneous mixtures, thus deterring the sensitivity of the resultant prototypes. When surfactants like sodium dodecyl sulfonate (SDS) (59) are mixed with CNTs to improve their hydrophilicity, it interferes with the mechanical integrity of the CNTs and overall composites. A certain way to address this would be to functionalize the CNTs with organic groups that are hydrophilic in nature. The covalent bonding formed by the CNTs and other nanomaterials in the nanocomposites-based flexible sensors is another issue that needs to be addressed. When the covalent bond is formed, there is a decrease in the mobility of charges and transfer of charge carriers, which subsequently decreases the overall electrical conductivity. The type of covalent bond formed between the CNTs and other nanomaterials is another area that should be researched. While covalent bonds are stronger than ionic bonds, the changes in the response of the sensors with respect to the corresponding morphological changes during the formation of CNTs-based composites should also be considered. This is particularly necessary for multifunctional sensors, where the CNTs are responsible for functioning in both an electrochemical and strain-sensing manner. The fabrication techniques that form the thin-film flexible sensors should also be optimized to address certain analytical characteristics like hysteresis, surface tension and viscosity. The physical dimension of the CNTs should be optimized in accordance with the target application. While the literature has allowed to development of a database for SWCNTs and MWCNTs, the variation in the physicochemical nature of CNTs can increase the dynamic nature of the CNTs in terms of sensor development and their application spectrum. While there has been a gradual shift from the microelectronics industry to the nanoelectronics industry, the miniaturization in the size of CNTs compared to the currently developed ones could improve their demand and market value. The toxicity of CNTs is another issue that must be addressed at the fabrication and application levels. When CNTs are developed using certain techniques like CVD, there is a drawback due to the generation of toxic by-products. This causes problems when the industries produce CNTs on a large scale. Alternative synthesis techniques should be devised, optimized and implemented to develop the CNTs to avoid producing toxic substances. The biocompatibility of the CNTs-based sensors should also be researched, as the inclusion of other processed materials intervenes with the degree of biocompatibility of the resultant sensors. Since certain nanomaterials like metallic nanowires have an issue with their biocompatibility, there should be detailed research on the type of nanomaterial that can be integrated with CNTs to make the biocompatibility of the prototypes. The nanomaterials, including CNTs, when mixed with polymers forming nanocomposites, have a limitation on the toughness and their performances. The choice of polymers used as matrixes for the nanocomposites should be further investigated to analyze the percolation threshold and mechanical attributes of the nanocomposites-based prototypes. From an application point of view, further encouragement should be done for testing the CNTs-based sensors in real-time applications. After successfully implementing the sensors in controlled temperature and humidity conditions, they should be integrated with the Internet of Things (IoT)-embedded signal-conditioning circuits and tried in real-time conditions. This would compare their performances with the existing sensors designed for the same application and validate their capability as commercial sensors. The comparison of the performances of the wireless fidelity (Wi-Fi)/CNTs-based sensors should be made in terms of their analytical characteristics and the applications, covered by each prototype. With the increase in the use of sensors associated with domestic and industrial sectors like defence, medicine and agriculture, the quality of service of the commercial CNTs-based sensors should be checked thoroughly before deploying them. The robustness and longevity of the sensors should be checked to minimize the generated electronic waste. With the ever-growing population and shortage of required energy, the energy harvesting application of the CNTs-based sensors should be further researched. When these sensors are used as wearable prototypes (60, 61), the change in their dimension based on the applied strain should be able to harvest and store energy. This can be done cumulatively, and the stored energy can be subsequently used as the input energy for other sensing prototypes and sensing systems. The wearable nature of the CNTs-based sensors should also be increased to determine the acute and chronic problems in the human body. Instead of attaching them to the body, they can be integrated into the fabric to reduce the discomfort of the patient. These wearable sensors should target to and monitor the issues periodically and ubiquitously with a high operating range, high sensitivity, low response and recovery times and a high linear range. The application spectrum of the wearable sensors also includes the diagnosis of the issues happening inside the human body. For example, the prototypes can be attached to catheters to only monitor the anatomical changes but also delivery drugs to improve the conditions. These sensors should be used as drug delivery system to address certain human diseases like cancer and acquired immunodeficiency syndrome.

12.3 CURRENT CHALLENGES OF CNTS-ASSISTED SENSORS

The chapter presents a comprehensive analysis of the notable research conducted on CNTs-based sensors in the context of environmental remediation applications. CNTs exhibit significant potential in the domain of environmental remediation for industrial and commercial applications. In brief, this study provides an overview of sensors based on CNTs and their significant environmental prospectives for monitoring various emerging toxic contaminants (such as gaseous components, vapour phase aromatic compounds, dyes and heavy metals from aqueous solution) emitted by the industrial sectors. The subsequent findings derived based on the review conducted, it is evident that there is a pressing need to develop a robust platform for the accurate and efficient assessment and measurement of risks associated with emerging environmental contaminants. Despite significant advancements in the production of CNTs, the widespread implementation of these materials still poses a barrier due to the limited ability to regulate the production of nanostructures on a large scale. It was evident that CNTs generated by the chemical vapour deposition (CVD) technique have mostly been employed for the purpose of removal of contaminants. MWCNTs have been widely employed, in comparison to SWCNTs, due to their cost-effectiveness, with MWCNTs being about 100 times more economical than SWCNTs. The contaminant adsorption capabilities of MWCNTs are seen to be lower, in comparison to the higher surface areas exhibited by SWCNTs. This finding highlights the reliance of adsorption capacity on the surface area and porous structure of the nanomaterials. At ambient temperature, pristine CNTs exhibit similar behaviour to graphite and demonstrate resistance to interaction with chemical constituents. However, CNTs have demonstrated enhanced sensing capabilities when subjected to functionalization with metal and/or metal-oxide particles, polymer layers or compounds that possess selective binding properties towards certain toxic compounds. Functionalization of the surface and side chains/side walls of CNTs is a promising strategy to achieve the target commercial applications with CNTs-based nanostructured sensors. The use of functionalized CNTs and CNT-based composites has been shown to have superior capabilities in adsorbing contaminants compared to pristine CNTs, although the major challenge associated with chemical functionalization is that the toxicity potential of CNTs can be considerably influenced and enhanced by the functionalization process. One of the constraints associated with the utilization of magnetic CNTs as one of the forms of functionalized CNTs in contaminant removal is to their reliance on an external magnetic field in order to facilitate the separation of organic and inorganic contaminants from a liquid phase. Also, the functionalization of CNTs with oxygenated, functional groups influences their sensitivity to environmental effects. The efficacy of CNTs-based sensors in detecting volatile organic compounds (VOCs) may be adversely impacted by several factors when exposed to humidity. The cross-sensitivity between water vapour and any given volatile chemical compound of interest can be complex. Several research studies have been conducted to elucidate the adsorption process of toxicants on the surface of CNTs using kinetics, thermodynamics and adsorption isotherm. However, there is a pressing necessity to report the regeneration studies of the spent CNTs to indicate its reusability potential.

12.4 FUTURE RECOMMENDATIONS FOR CNTS-ASSISTED SENSORS

Despite the aforementioned obstacles, the literature clearly demonstrates the significant potential of CNTs in the field of chemical sensing. Potential applications of these portable and energy-efficient sensors lie in the realm of detecting and identifying hazardous substances such as toxic gases, aromatic compounds, dyes and heavy transition metal ions, with detection limits extending to low parts per billion (ppb) levels. Regardless of the notable progress in this specific field, there remain some critical issues that require resolution and more work to get a deeper appreciation. Carbon nanostructured materials, which are derived from nanotechnology, have the potential to serve as a viable alternative to traditional technologies in the removal of pollutants from both gaseous and aqueous environments. However, extensive research should be conducted to thoroughly investigate the availability of feedstocks and energy requirements for large-scale productions and applications and to compare the adsorption capacities of CNTs generated using several techniques, including CVD, arc discharge method and laser ablation techniques. The preparation technique can have an impact on the physiochemical properties of CNTs. Therefore, in the context of synthesizing CNTs, functionalizing them and employing them for environmental remediation applications, it is imperative to practice care and to manufacture economically feasible quantities of pristine, functionalized and CNTs-composite at an economically feasible cost while minimizing environmental consequences. Developing effective methods for detecting various toxic molecules within a mixture is a necessity to study the performance of CNTs in terms of target selectivity and sensitivity in the presence of multitoxicants. To de-coat the surface of CNTs with metal nanoparticles – namely, iron – is well-recognized as an economically beneficial and highly attractive functionalization technique for the purpose of environmental remediation. This can be attributed to their inherent ferromagnetic properties. The metal nanoparticles have a role in increasing the surface area, active sites and functional groups, thus resulting in enhanced efficacy in the removal of contaminants from aqueous solutions. To fabricate CNTs in combination with other conductive materials as nanocomposite materials as the amalgamation enhances their functionality and promotes sustainability in removing contaminants. In order to enhance the outcomes obtained via recent advancements in multiplex functionalizations of CNTs-based sensor arrays, it is imperative to have a thorough comprehension of the underlying factors contributing to the delayed recovery time of the sensors to their initial state as well as the instances when the recovery process is accelerated. To study the CNTs-based sensor performance in the presence of moisture in the background, as it has the potential to affect the sensor performance to detect aromatic compounds. To improve the utilization of CNTs-based chemical sensors in monitoring toxic constituents in industrial and commercial sectors, extensive research is crucial to get a comprehensive understanding of the underlying interaction mechanism between CNTs and the specific molecules of interest, particularly under normal environmental circumstances. To forecast the efficacy of contaminant removal and comprehend the adsorption mechanism in actual industrial effluents under different operating conditions, using both batch and column processes, more research work should be focused on the sustainable regeneration of CNTs adsorbents. Referring to various research works, it has been concluded that CNTs have many potential applications in the field of chemical sensing for environmental remediation, but additional exploration is still needed. The rapid pace of research and innovation in developing newer techniques make CNTs a cutting-edge agent applicable across various fields. In general, a comprehensive analysis of the literature presented in this discussion reveals that carbon nanostructures – specifically, carbon nanotubes – have demonstrated their efficacy in the laboratory-scale mitigation of various environmental pollutants. However, further investigation and development are necessary to facilitate the transition to large-scale industrial applications in the field of environmental remediation. This advancement would significantly enhance the utilization of CNTs in both industrial and commercial domains. It is imperative for researchers to investigate more towards the development of highly pure CNTs with reduced toxicity and defects. In the near future, novel methodologies will arise; hence, enhancing the affordability and viability of CNTs across many fields of application is of profound interest.

ACKNOWLEDGEMENT

This work was supported by the Free State of Saxony and by the European Union (ESF Plus) by funding the research group "MultiMOD." This study was funded by the German Research Foundation (DFG, Deutsche Forschungsgemeinschaft) as part of Germany's Excellence Strategy – EXC 2050/1 – Project ID 390696704 – Cluster of Excellence "Centre for Tactile Internet with Human-in-the-Loop" (CeTI) of Technische Universität Dresden.

REFERENCES

1. Iijima S. Synthesis of carbon nanotubes. Nature. 1991;354(6348):56–8.
2. Gao J, He S, Nag A, Wong JWC. A review of the use of carbon nanotubes and graphene-based sensors for the detection of aflatoxin M1 compounds in milk. Sensors. 2021;21(11):3602.
3. Nag A, Alahi M, Eshrat E, Mukhopadhyay SC, Liu Z. Multi-walled carbon nanotubes-based sensors for strain sensing applications. Sensors. 2021;21(4):1261.
4. Nag A, Afsarimanesh N, Nuthalapati S, Altinsoy ME. Novel surfactant-induced MWCNTs/PDMS-based nanocomposites for tactile sensing applications. Materials. 2022;15(13):4504.
5. Magar HS, Hassan RY, Abbas MN. Non-enzymatic disposable electrochemical sensors based on CuO/Co3O4@ MWCNTs nanocomposite modified screen-printed electrode for the direct determination of urea. Scientific Reports. 2023;13(1):2034.
6. Nurazzi N, Sabaruddin F, Harussani M, Kamarudin S, Rayung M, Asyraf M, et al. Mechanical performance and applications of CNTs reinforced polymer composites – a review. Nanomaterials. 2021;11(9):2186.
7. Shah KA, Najar FA, Sharda T, Sreenivas K. Synthesis of multi-walled carbon nanotubes by thermal CVD technique on Pt–W–MgO catalyst. Journal of Taibah University for Science. 2018;12(2):230–4.
8. Yao Y, Zhang S, Yan Y, editors. *CVD synthesis and purification of multi-walled carbon nanotubes.* 2008 2nd IEEE International Nanoelectronics Conference. IEEE; 2008.
9. Ribeiro H, Schnitzler MC, da Silva WM, Santos AP. Purification of carbon nanotubes produced by the electric arc-discharge method. Surfaces and Interfaces. 2021;26:101389.

10. Shoukat R, Khan MI. Carbon nanotubes/nanofibers (CNTs/CNFs): A review on state of the art synthesis methods. Microsystem Technologies. 2022;28(4):885–901.

11. Alamro FS, Mostafa AM, Al-Ola KAA, Ahmed HA, Toghan A. Synthesis of Ag nanoparticles-decorated CNTs via laser ablation method for the enhancement the photocatalytic removal of naphthalene from water. Nanomaterials. 2021;11(8):2142.

12. Alheshibri M, Elsayed K, Haladu SA, Magami SM, Al Baroot A, Ercan İ, et al. Synthesis of Ag nanoparticles-decorated on CNTs/TiO2 nanocomposite as efficient photocatalysts via nanosecond pulsed laser ablation. Optics & Laser Technology. 2022;155:108443.

13. Schwandt C, Dimitrov AT, Fray DJ. High-yield synthesis of multi-walled carbon nanotubes from graphite by molten salt electrolysis. Carbon. 2012;50(3):1311–15.

14. Abbasloo S, Ojaghi-Ilkhchi M, Mozammel M. *Synthesis of carbon nanotubes by molten salt electrolysis: A review*. Proceedings of the [0]5th International Biennial Conference on Ultrafine Grained and Nanostructured Materials, 2015.

15. Han T, Nag A, Mukhopadhyay SC, Xu Y. Carbon nanotubes and its gas-sensing applications: A review. Sensors and Actuators A: Physical. 2019;291:107–43.

16. Jena SK, Chakraverty S, Malikan M, Tornabene F. Effects of surface energy and surface residual stresses on vibro-thermal analysis of chiral, zigzag, and armchair types of SWCNTs using refined beam theory. Mechanics Based Design of Structures and Machines. 2020:1–15.

17. Guo X, Huang Y, Zhao Y, Mao L, Gao L, Pan W, et al. Highly stretchable strain sensor based on SWCNTs/CB synergistic conductive network for wearable human-activity monitoring and recognition. Smart Materials and Structures. 2017;26(9):095017.

18. Akhter F, Siddiquei H, Alahi MEE, Mukhopadhyay SC. An IoT-enabled portable sensing system with MWCNTs/PDMS sensor for nitrate detection in water. Measurement. 2021;178:109424.

19. Nag A, Mukhopadhyay SC, Kosel J. Flexible carbon nanotube nanocomposite sensor for multiple physiological parameter monitoring. Sensors and Actuators A: Physical. 2016;251:148–55.

20. Nag A, Mukhopadhyay S, Kosel J, editors. *Influence of temperature and humidity on carbon based printed flexible sensors*. 2017 Eleventh International Conference on Sensing Technology (ICST). IEEE; 2017.

21. Naik SS, Lee SJ, Theerthagiri J, Yu Y, Choi MY. Rapid and highly selective electrochemical sensor based on ZnS/Au-decorated f-multi-walled carbon nanotube nanocomposites produced via pulsed laser technique for detection of toxic nitro compounds. Journal of Hazardous Materials. 2021;418:126269.

22. Yan T, Wu Y, Yi W, Pan Z. Recent progress on fabrication of carbon nanotube-based flexible conductive networks for resistive-type strain sensors. Sensors and Actuators A: Physical. 2021;327:112755.

23. Chen C-B, Kao H-L, Chang L-C, Cho C-L, Lin Y-C, Huang C-C, et al. Fabrication of inkjet-printed carbon nanotube for enhanced mechanical and strain-sensing performance. ECS Journal of Solid State Science and Technology. 2021;10(12):121001.

24. Wang Q, He S, Bowen CR, Xiao X, Oh JAS, Sun J, et al. Porous pyroelectric ceramic with carbon nanotubes for high-performance thermal to electrical energy conversion. Nano Energy. 2022;102:107703.

25. Li C, Chen H, Zhang L, Zhong J. Electrical properties of carbon nanotube/liquid metal/rubber nanocomposites. AIP Advances. 2020;10(10):105106.

26. He S, Zhang Y, Gao J, Nag A, Rahaman A. Integration of different graphene nanostructures with PDMS to form wearable sensors. Nanomaterials. 2022;12(6):950.

27. Nag A, Simorangkir RB, Gawade DR, Nuthalapati S, Buckley JL, O'Flynn B, et al. Graphene-based wearable temperature sensors: A review. Materials & Design. 2022:110971.

28. Nag A, Alahi MEE, Mukhopadhyay SC. Recent progress in the fabrication of graphene fibers and their composites for applications of monitoring human activities. Applied Materials Today. 2021;22:100953.

29. Nag A, Mitra A, Mukhopadhyay SC. Graphene and its sensor-based applications: A review. Sensors and Actuators A: Physical. 2018;270:177–94.

30. Garkani Nejad F, Beitollahi H, Sheikhshoaie I. Graphene oxide–PAMAM nanocomposite and ionic liquid modified carbon paste electrode: An efficient electrochemical sensor for simultaneous determination of catechol and resorcinol. Diagnostics. 2023;13(4):632.

31. Karami-Kolmoti P, Beitollahi H, Modiri S. Voltammetric detection of catechol in real samples using MnO2 nanorods-graphene oxide nanocomposite modified electrode. Journal of Food Measurement and Characterization. 2023;17(2):1974–84.

32. Nag A, Simorangkir RB, Sapra S, Buckley JL, O'Flynn B, Liu Z, et al. Reduced graphene oxide for the development of wearable mechanical energy-harvesters: A review. IEEE Sensors Journal. 2021; 21.

33. Olivieri F, Rollo G, De Falco F, Avolio R, Bonadies I, Castaldo R, et al. Reduced graphene oxide/polyurethane coatings for wash-durable wearable piezoresistive sensors. Cellulose. 2023:1–20.

34. Gao S, Li H, Zheng L, Huang W, Chen B, Lai X, et al. Superhydrophobic and conductive polydimethylsiloxane/titanium dioxide@ reduced graphene oxide coated cotton fabric for human motion detection. Cellulose. 2021;28(11):7373–88.

35. Nag A, Alahi MEE, Feng S, Mukhopadhyay SC. IoT-based sensing system for phosphate detection using graphite/PDMS sensors. Sensors and Actuators A: Physical. 2019;286:43–50.

36. Nag A, Feng S, Mukhopadhyay S, Kosel J, Inglis D. 3D printed mould-based graphite/PDMS sensor for low-force applications. Sensors and Actuators A: Physical. 2018;280:525–34.

37. Nag A, Afasrimanesh N, Feng S, Mukhopadhyay SC. Strain induced graphite/PDMS sensors for biomedical applications. Sensors and Actuators A: Physical. 2018;271:257–69.

38. Thanh CT, Binh NH, Duoc PND, Thu VT, Van Trinh P, Anh NN, et al. Electrochemical sensor based on reduced graphene oxide/double-walled carbon nanotubes/octahedral Fe 3 O 4/chitosan composite for glyphosate detection. Bulletin of Environmental Contamination and Toxicology. 2021;106:1017–23.

39. Duoc PND, Binh NH, Van Hau T, Thanh CT, Van Trinh P, Tuyen NV, et al. A novel electrochemical sensor based on double-walled carbon nanotubes and graphene hybrid thin film for arsenic (V) detection. Journal of Hazardous Materials. 2020;400:123185.

40. Ma Y, Li P, Sedloff JW, Zhang X, Zhang H, Liu J. Conductive graphene fibers for wire-shaped supercapacitors strengthened by unfunctionalized few-walled carbon nanotubes. ACS Nano. 2015;9(2):1352–9.

41. Zhang Y, Gregoire JM, Van Dover R, Hart AJ. Ethanol-promoted high-yield growth of few-walled carbon nanotubes. The Journal of Physical Chemistry C. 2010;114(14):6389–95.

42. Sapra S, Chakraborthy A, Nuthalapati S, Nag A, Inglis DW, Mukhopadhyay SC, et al. Printed, wearable e-skin force sensor array. Measurement. 2023;206:112348.

43. Nag A, Mukhopadhyay SC. *Printed flexible sensors for academic research. Printed and flexible sensor technology: Fabrication and applications.* IOP Publishing; 2021. pp. 2-1–2-16.

44. He S, Feng S, Nag A, Afsarimanesh N, Han T, Mukhopadhyay SC. Recent progress in 3D printed mold-based sensors. Sensors. 2020;20(3):703.

45. Han T, Kundu S, Nag A, Xu Y. 3D printed sensors for biomedical applications: A review. Sensors. 2019;19(7):1706.

46. Han T, Nag A, Afsarimanesh N, Mukhopadhyay SC, Kundu S, Xu Y. Laser-assisted printed flexible sensors: A review. Sensors. 2019;19(6):1462.

47. Chen H, Liu J, Cao W, He H, Li X, Zhang C. 3D printing CO2-activated carbon nanotubes host to promote sulfur loading for high areal capacity lithium-sulfur batteries. Nano Research. 2023;16(6):8281–9.

48. Cao K, Wu M, Bai J, Wen Z, Zhang J, Wang T, et al. Beyond skin pressure sensing: 3D printed laminated graphene pressure sensing material combines extremely low detection limits with wide detection range. Advanced Functional Materials. 2022;32(28):2202360.

49. Kalkal A, Kumar S, Kumar P, Pradhan R, Willander M, Packirisamy G, et al. Recent advances in 3D printing technologies for wearable (bio) sensors. Additive Manufacturing. 2021;46:102088.

50. Han T, Nag A, Simorangkir RB, Afsarimanesh N, Liu H, Mukhopadhyay SC, et al. Multifunctional flexible sensor based on laser-induced graphene. Sensors. 2019;19(16):3477.

51. Nag A, Mukhopadhyay SC, Kosel J. Sensing system for salinity testing using laser-induced graphene sensors. Sensors and Actuators A: Physical. 2017;264:107–16.

52. Nag A, Mukhopadhyay SC, Kosel J. Tactile sensing from laser-ablated metallized PET films. IEEE Sensors Journal. 2016;17(1):7–13.

53. He S, Yuan Y, Nag A, Feng S, Afsarimanesh N, Han T, et al. A review on the use of impedimetric sensors for the inspection of food quality. International Journal of Environmental Research and Public Health. 2020;17(14):5220.

54. Afsarimanesh N, Nag A, Sarkar S, Sabet GS, Han T, Mukhopadhyay SC. A review on fabrication, characterization and implementation of wearable strain sensors. Sensors and Actuators A: Physical. 2020;315:112355.

55. Nag A, Mukhopadhyay SC, Kosel J. Wearable flexible sensors: A review. IEEE Sensors Journal. 2017;17(13):3949–60.

56. Chen X, Zhang D, Luan H, Yang C, Yan W, Liu W. Flexible pressure sensors based on molybdenum disulfide/hydroxyethyl cellulose/polyurethane sponge for motion detection and speech recognition using machine learning. ACS Applied Materials & Interfaces. 2022;15(1):2043–53.

57. Zhu J, Cho M, Li Y, He T, Ahn J, Park J, et al. Machine learning-enabled textile-based graphene gas sensing with energy harvesting-assisted IoT application. Nano Energy. 2021;86:106035.

58. Nor ASM, Faramarzi M, Yunus MAM, Ibrahim S. Nitrate and sulfate estimations in water sources using a planar electromagnetic sensor array and artificial neural network method. IEEE Sensors Journal. 2014;15(1):497–504.

59. Harnchana V, Ngoc HV, He W, Rasheed A, Park H, Amornkitbamrung V, et al. Enhanced power output of a triboelectric nanogenerator using poly (dimethylsiloxane) modified with graphene oxide and sodium dodecyl sulfate. ACS Applied Materials & Interfaces. 2018;10(30):25263–72.

60. Mukhopadhyay SC, Suryadevara NK, Nag A. *Wearable sensors and systems in the IoT*. MDPI; 2021. p. 7880.

61. Nag A, Simorangkir RB, Valentin E, Björninen T, Ukkonen L, Hashmi RM, et al. A transparent strain sensor based on PDMS-embedded conductive fabric for wearable sensing applications. IEEE Access. 2018;6:71020–7.

Index

For Product Safety Concerns and Information please contact our EU
representative GPSR@taylorandfrancis.com
Taylor & Francis Verlag GmbH, Kaufingerstraße 24, 80331 München, Germany

9 7 8 1 0 3 2 4 5 2 3 2 6